Operator Theory
Advances and Applications
Vol. 102

Editor:
I. Gohberg

Differential and Integral Operators

International Workshop on Operator Theory and Applications, IWOTA 95, in Regensburg, July 31–August 4, 1995

I. Gohberg
R. Mennicken
C. Tretter
Editors

Springer Basel AG

Authors:

I. Gohberg
School of Mathematical Sciences
Raymond and Beverly Sackler
Faculty of Exact Sciences
Tel Aviv University
Ramat Aviv 69978
Israel
e-mail: gohberg@math.tau.ac.il

R. Mennicken
NWFI-Mathematik
Universität Regensburg
D-93040 Regensburg
Germany
e-mail: reinhard.mennicken@mathematik.uni-regensburg.de

C. Tretter
NWFI-Mathematik
Universität Regensburg
D-93040 Regensburg
Germany
e-mail: christiane.tretter@mathematik.uni-regensburg.de

1991 Mathematics Subject Classification 47-06, 93-06

A CIP catalogue record for this book is available from the
Library of Congress, Washington D.C., USA

Deutsche Bibliothek Cataloging-in-Publication Data
Differential and integral operators / International Workshop on Operator Theory
and Applications, IWOTA 95, in Regensburg, July 31 – August 4, 1995.
I. Gohberg ... ed.. – Basel ; Boston ; Berlin : Birkhäuser, 1998
(Operator theory ; Vol. 102)
ISBN 978-3-0348-9774-7 ISBN 978-3-0348-8789-2 (eBook)
DOI 10.1007/978-3-0348-8789-2

Originally published by Birkhäuser Verlag in 1998
Softcover reprint of the hardcover 1st edition 1998

Printed on acid-free paper produced from chlorine-free pulp. TCF ∞
Cover design: Heinz Hiltbrunner, Basel

ISBN 978-3-0348-9774-7

9 8 7 6 5 4 3 2 1

Table of contents

Editorial introduction

This and the next volume of the OT series contain the proceedings of the Workshop on Operator Theory and its Applications, IWOTA 95, which was held at the University of Regensburg, Germany, July 31 to August 4, 1995. It was the eigth workshop of this kind. Following is a list of the seven previous workshops with reference to their proceedings:

1981	Operator Theory (Santa Monica, California, USA)
1983	Applications of Linear Operator Theory to Systems and Networks (Rehovot, Israel), OT 12
1985	Operator Theory and its Applications (Amsterdam, The Netherlands), OT 19
1987	Operator Theory and Functional Analysis (Mesa, Arizona, USA), OT 35
1989	Matrix and Operator Theory (Rotterdam, The Netherlands), OT 50
1991	Operator Theory and Complex Analysis (Sapporo, Japan), OT 59
1993	Operator Theory and Boundary Eigenvalue Problems (Vienna, Austria), OT 80

IWOTA 95 offered a rich programme on a wide range of latest developments in operator theory and its applications. The programme consisted of 6 invited plenary lectures, 54 invited special topic lectures and more than 100 invited session talks. About 180 participants from 25 countries attended the workshop, more than a third came from Eastern Europe.

The conference covered different aspects of linear and nonlinear spectral problems, starting with problems for abstract operators up to spectral theory of ordinary and partial differential operators, pseudodifferential operators, and integral operators. The workshop was also focussed on operator theory in spaces with indefinite metric, operator functions, interpolation and extension problems. The applications concerned problems in mathematical physics, hydrodynamics, magnetohydrodynamics, quantum mechanics, astrophysics as well as the theory of networks and systems. The papers in the proceedings bring the readers up to date on recent achievements in these areas.

This volume contains the contributions to the theory of differential and integral operators. A second volume, entitled "Recent Progress in Operator Theory", is dedicated to the other aspects of operator theory covered in the workshop.

The financial support of the following institutions for IWOTA 95 is highly appreciated:

Bayerisches Staatsministerium für Unterricht, Kultus, Wissenschaft und Kunst,
Deutsche Forschungsgemeinschaft (DFG),
Regensburger Universitätsstiftung Hans Vielberth,
Gesellschaft für Angewandte Mathematik und Mechanik (GAMM),
Akademie Verlag, Berlin,
Birkhäuser Verlag, Basel.

We also thank the University of Regensburg for supporting the workshop in many ways, in particular by providing lecture rooms and other facilities.

I.C. Gohberg, R. Mennicken and C. Tretter

List of participants

1	M.S. Agranovich	30	E. Grinshpun	59	W.N. Everitt	88	I.S. Kac
2	K. Seddighi	31	I. Koltracht	60	R. Vonhoff	89	S.G. Pyatkov
3	Ch. Davis	32	N. Suciu	61	C. Sadosky	90	J. Kos
4	B. Fritzsche	33	A. Kozhevnikov	62	M.A. Kaashoek	91	A.C.M. Ran
5	B. Kirstein	34	B. Silbermann	63	M. Moeller	92	S.N. Naboko
6	S. Roch	35	B. Gramsch	64	A.K. Motovilov	93	V. Hardt
7	S. Treil	36	M.P.H. Wolff	65	A.B. Mingarelli	94	H. Gail
8	S.M. Verduyn Lunel	37	S. Behm	66	D.Z. Arov	95	M. Weber
9	K. Lorentz	38	P. Kurasov	67	I.M. Spitkovsky	96	A. Hartmann
10	H.S.V. de Snoo	39	L. Rodman	68	J.A. Dubinskii	97	D. Gaspar
11	A. Ben-Artzi	40	H.-G. Leopold	69	V. Matsaev	98	E.W. Lytvynov
12	G. Schlüchtermann	41	E. Litsyn	70	V. Liskevich	99	W.G. Litvinov
13	R. Lauter	42	H. Langer	71	A.A. Nudelman	100	E.R. Tsekanovskii
14	S. Albeverio	43	I. Marek	72	G. Litvinchuk	101	Yu.V. Egorov
15	J. Zemanek	44	A. Gheondea	73	D. Bakic	102	L.A. Sakhnovich
16	A.S. Markus	45	H. Triebel	74	B. Najman	103	V. Peller
17	M. Solomyak	46	Ch.R. Johnson	75	V.M. Adamjan	104	M. Markin
18	Z.G. Sheftel	47	A.F. dos Santos	76	D. Pik	105	P. Gaspar
19	L.R. Volevich	48	R. Duduchava	77	G. Heinig	106	K.-H. Förster
20	M. Malamud	49	F.-O. Speck	78	V. Strauss	107	R. Gaspar
21	E. Meister	50	Ya. Roitberg	79	M. Taghavi	108	P. Jonas
22	D.V. Yakubovich	51	K. Makarov	80	V. Pivovarchik	109	Yu.M. Berezansky
23	V. Kapustin	52	B.-W. Schulze	81	F. Ali Mehmeti	110	R. Mennicken
24	E. Schrohe	53	M. Demuth	82	N. Gopal Reddy	111	I.C. Gohberg
25	R. Denk	54	H.R. Beyer	83	K. Naimark	112	C. Tretter
26	J. Brasche	55	I. Roitberg	84	B.M. Levitan	113	V. Kondratiev
27	M.A. Nudelman	56	V. Mikhailets	85	A.A. Shkalikov	114	B.A. Plamenevskii
28	A. Kulesko	57	I. Suciu	86	R.O. Griniv	115	J. Saurer
29	M. Rakowski	58	V.S. Rabinovich	87	A.I. Kozhanov		

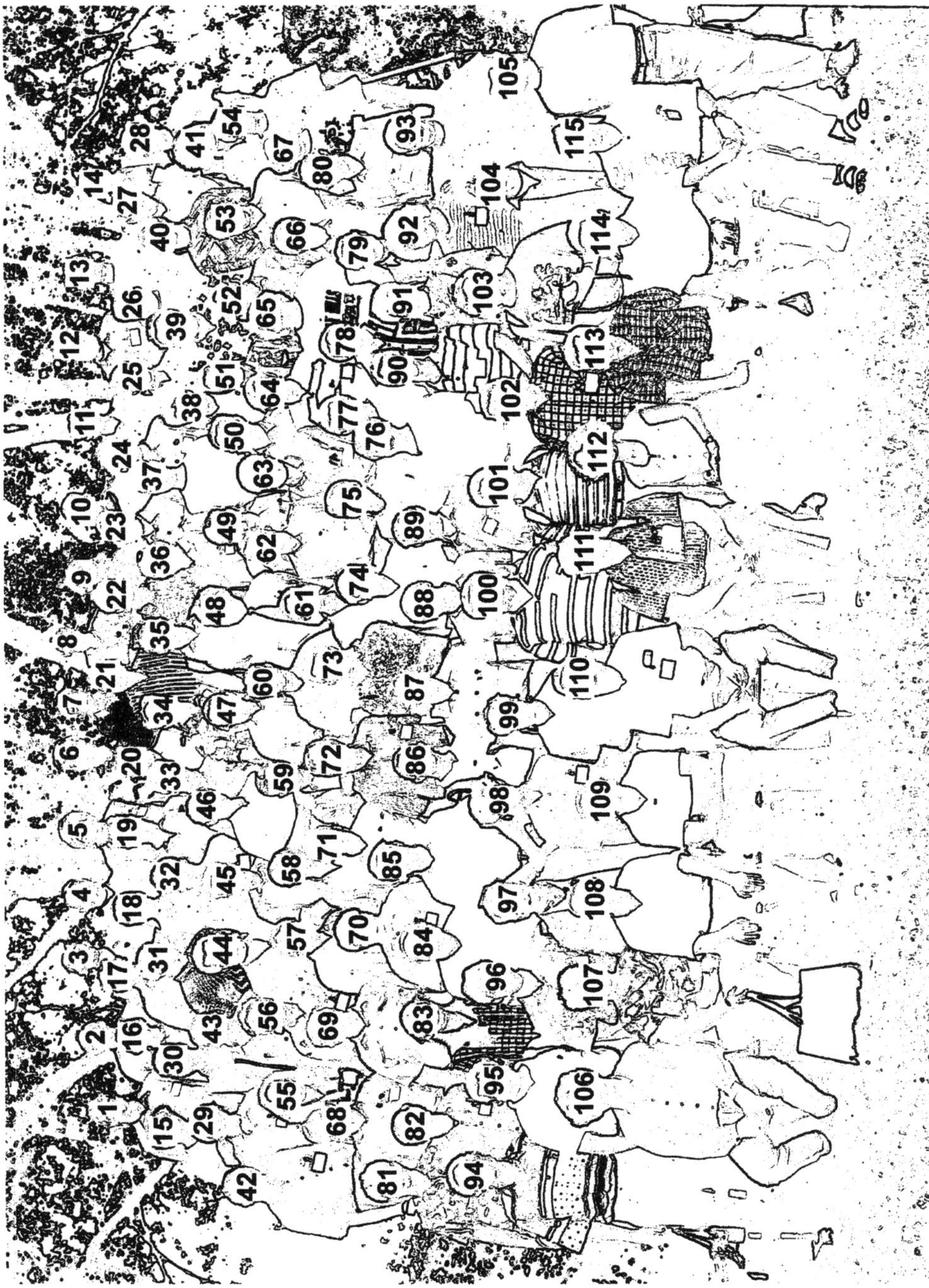

Operator Theory:
Advances and Applications, Vol. 102

Limit behaviour in a singular perturbation problem, regularized convolution operators and the three-body quantum problem

S. ALBEVERIO and K.A. MAKAROV

A model of a quantum mechanical system related to the three-body problem is studied. The model is defined in terms of a symmetric pseudodifferential operator (PDO) with unbounded symbol. The entire family of self-adjoint extensions of this operator is studied using harmonic analysis. A regularization procedure for this PDO is introduced, the limit behavior of the regularized operators when the regularization parameter is removed is analyzed and a nontrivial attractor is exhibited.

1. Introduction

In the quantum mechanical three-body problem with short range forces the phenomenon of the Efimov effect is known to arise: the presence of infinitely many three-body bound states at the lower end of the spectrum [E] (see also [AN], [OS], [S], [T], [Y]). This phenomenon is related to the one of a model three-body problem with point-like interactions, where any self-adjoint realization of the energy operator is known to be nonsemibounded from below, and it has an infinite series of negative eigenvalues tending to minus infinity ("the fall to the center" phenomenon) [MF1], [MF2], [MM] (see also the review [Fl]). A deep relationship between the asymptotics of the eigenvalues in the three-body problem with point-like interactions and the accumulation law of those to the three-body threshold in the case of the Efimov effect was discussed in [AHKW] (see also [F], [FM]).

The three-body problem both in the case of the point-like interactions and in the case of the Efimov effect are closely connected with the theory of integral equations of convolution type with meromorphic symbols.

In the simplest case of scalar identical particles the extension theory for the three-body Hamiltonian with point-interactions can be reduced to the one of a formally symmetric unbounded operator acting on the space $L_2(\mathbb{R})$:

$$\mathcal{A} = W(I - \mathcal{L})W,$$

where W is the multiplication operator by the function $\sqrt{\cosh x}$,

$$(Wf)(x) = \sqrt{\cosh x}\, f(x),$$

and $\mathcal{L}$ is a self-adjoint integral operator of convolution type,

$$(\mathcal{L}f)(x) = \int_{\mathbb{R}} L(x-y)f(y)\,dy, \tag{1.1}$$

with the kernel [MS]

$$L(x) = \frac{4}{\sqrt{3}\pi}\log\frac{2\cosh x + 1}{2\cosh x - 1}.$$

In the case of the Efimov effect the study of the first term of the spectral asymptotics of the counting function associated with the discrete spectrum can be reduced to the computation of distribution of zeros of the Fredholm determinant $\Delta(R)$ of the convolution operator $I - \mathcal{L}$ considered now in the space $L_2([-R,R])$,

$$\Delta(R) = \det{(I-\mathcal{L})}|_{L_2([-R,R])}.$$

The characteristic feature of these two problems is that the symbol $l = 1 - \widehat{L}$ (here $\widehat{L}$ denotes the Fourier transform of the kernel function L) of the convolution operator $I - \mathcal{L}$ is given by

$$l(s) = 1 - \frac{8}{\sqrt{3}}\frac{\sinh\frac{\pi s}{6}}{s\cosh\frac{\pi s}{2}} \tag{1.2}$$

and has zeros on the real axis. This leads, firstly, to the fact that the symmetric operator $\mathcal{A}$ has nontrivial deficiency indices and, in addition, it is nonsemibounded from below (the reason for the "fall to the center" phenomenon in the case of point-interactions) and, secondly, to the fact that the Fredholm determinant $\Delta(R)$ considered as a function of R has infinitely many zeros (which is equivalent to the Efimov effect).

A natural problem arises concerning the possibility of approximation of the family of self-adjoint extensions of the "limit" object $W(1-\mathcal{L})W$ by a family of rather more simple operators

$$\mathcal{B}_R = W(1 - P_R\mathcal{L}P_R)W \tag{1.3}$$

as $R \to \infty$, where P_R is a projection from $L_2(\mathbb{R})$ onto the subspace $L_2([-R,R])$.

In the present paper we consider a closely related but slightly different problem. First, instead of the multiplication operator W by the function $\sqrt{\cosh x}$ we consider the multiplication operator by the function $e^{\frac{\alpha}{2}|x|}$ in the space $L_2(\mathbb{R})$, $\alpha > 0$ being a parameter,

$$(Wf)(x) = e^{\frac{\alpha}{2}|x|}f(x).$$

Secondly, we extend the consideration to the class not only of meromorphic symbols but also of symbols admitting an analytic continuation to some strip. Namely, we suppose that the kernel L decreases exponentially at infinity in such a manner that

$$e^{\beta|x|}L(x) \in L_\infty(\mathbb{R}) \tag{1.4}$$

for some $\beta > 0$. This requirement is a natural generalization of the properties of the meromorphic symbol (1.2).

For $\beta > \alpha$ the operator $\mathcal{A}$ is a correctly defined symmetric operator on the initial domain

$$\mathcal{D}(\mathcal{A}) = \mathcal{D}(W^2),$$

where $\mathcal{D}(W^2) = \{f : W^2 f \in L_2(\mathbb{R})\}$, since in this case the Fourier transform $\widehat{L}$ of the kernel L admits a bounded analytic continuation to the strip

$$\Pi_{\frac{\alpha}{2}} = \left\{z : |\operatorname{Im} z| < \frac{\alpha}{2}\right\}$$

and therefore, by a variant of the Paley-Wiener Theorem, we have the inclusion $L\mathcal{D}(W) \subset \mathcal{D}(W)$.

In the case where the symbol $l = 1 - \widehat{L}$ of the integral operator $1 - \mathcal{L}$ vanishes at some points in the strip $\Pi_{\frac{\alpha}{2}}$ the operator $\mathcal{A}$ has nontrivial equal deficiency indices and therefore admits a family of self-adjoint extensions.

Thirdly, instead of the family of the operators $\mathcal{B}_R = W(1 - P_R\mathcal{L}P_R)W$ we consider along with $\mathcal{A}$ the one-parameter family of (unbounded) self-adjoint operators

$$\mathcal{B}_R = W(1 - \mathcal{L}_R)W, \qquad R > 0, \tag{1.5}$$

defined on the same domain $\mathcal{D}(\mathcal{B}_R) = \mathcal{D}(\mathcal{A})$, where $\mathcal{L}_R$ is the compact integral operator with kernel

$$\mathcal{L}_R(x, y) = \chi\left(\frac{x}{R}\right) L_R(x - y)\chi\left(\frac{y}{R}\right).$$

Here $L_R(x)$ denotes the periodic extension of the function $L(x)$ from the interval $[-R, R]$ to the whole real axis and χ is the indicator of the interval $[-1, 1]$. The family $\mathcal{B}_R$ can be considered as some "regularization" of a more complicated object $\mathcal{A}$. This type of regularization is chosen due to technical reasons since the part of $\mathcal{L}_R$ corresponding to the invariant subspace $L_2([-R, R])$ can be diagonalized by Fourier series on the finite interval $[-R, R]$ and in this case the problem of the inversion of the integral operator $1 - \mathcal{L}_R$ in $L_2([-R, R])$ corresponding to the finite interval $[-R, R]$ can be solved asymptotically explicitly. The replacement of "the projection method" (1.3) by (1.5) is due the fact that (1.5) is easier to handle.

The main goal of the paper is to study the attractor of the dynamical system $R \mapsto \mathcal{B}_R$ in the sense of strong resolvent convergence. It turns out that this attractor consists of a special family of self-adjoint extensions of the symmetric operator $\mathcal{A}$ and we study precisely the corresponding dynamics in the neighborhood of the attractor.

Full details of the proofs are given in [AM1]. Our results can be considered as a first step to the proof of the universality of the Efimov effect announced in [AHKW]. In fact our main result (Theorem 3.2) is an abstract version of the nontrivial behavior of the resolvents of the three-body energy-operator with pairwise cut-off interactions approximating the so-called point-interactions of δ-function type when the cut-off parameter is removed. Operators of the type $W(I-L)W$ with $(Wf)(x) = \sqrt{\cosh x}\, f(x)$ and a special type of convolution kernels (see (1.2)) were in fact introduced in the pioneering work [MF1], [MF2] by Faddeev and Minlos: their resolvents determine the singular three-body t-matrix for the three-body Hamiltonians with δ-interactions.

The method developed here is also useful in the spectral analysis of the operators discussed in [H] (especially in the nonsemibounded case) since these operators are unitary equivalent to a direct sum of self-adjoint extensions of the operators $W(I-\mathcal{L})W$ for appropriate W and convolution operators $\mathcal{L}$ (see [AM2]).

2. Some auxiliary results

Under the assumption $\beta > \frac{3}{2}\alpha$ the symbol $l(s) = 1 - \widehat{L}(s)$ of the integral operator $I - \mathcal{L}$ satisfies the following conditions:

i) *the function* l *admits an analytic continuation to the strip* $\Pi_{\frac{3}{2}\alpha}$,

ii) *the equation*

$$l(z) = 0 \tag{2.1}$$

has a finite number of solutions in the strip $\Pi_{\frac{3}{2}\alpha}$,

iii) *The function* $l^{-1}(s)$ *is bounded in a neighborhood of infinity in* $\Pi_{\frac{\alpha}{2}}$.

If we suppose in addition that

iv) *there are no solutions of equation* (2.1) *on the boundary of the strip* $\Pi_{\frac{\alpha}{2}}$,

we have the following result.

Lemma 2.1. *Let conditions* i)–iv) *be fulfilled. Then the operator* $\mathcal{A}$ *defined on the domain* $\mathcal{D}(\mathcal{A}) = \mathcal{D}(W^2)$ *is a closed symmetric operator with equal deficiency indices* (n, n), *where* n *is the total number of zeros of equation* (2.1) *in the strip* $\Pi_{\frac{\alpha}{2}}$ *counting multiplicity.*

If the symbol l of the integral operator $1 - \mathcal{L}$ does have some zeros $a_1, \dots, a_m$ in the strip $\Pi_{\frac{\alpha}{2}}$, then the adjoint operator $\mathcal{A}^*$ has a nontrivial kernel $\operatorname{Ker} \mathcal{A}^* = \operatorname{span}\{h_1, \dots, h_m\}$ generated by functions of the form

$$h_k(x) = P_k(x) e^{i a_k x} e^{-\frac{\alpha}{2}|x|}, \qquad k = 1, \dots, m, \tag{2.2}$$

where $P_k(x)$ is an arbitrary polynomial of degree $n_k - 1$, with n_k being the multiplicity of the zero a_k. Let us note that in this case the functions $g_k(x) = P_k(x) e^{i a_k x}$ appear to be solutions of the homogeneous integral equation

$$g_k(x) - \int_{\mathbb{R}} L(x-y) g_k(y)\, dy = 0.$$

Lemma 2.2. *Under the assumptions* i)–iv) *the domain of the adjoint operator* $\mathcal{A}^*$ *can be represented in the form*

$$\mathcal{D}(\mathcal{A}^*) = \mathcal{D}(\mathcal{A}) + \mathcal{P}_+ \operatorname{Ker} \mathcal{A}^* + \mathcal{P}_- \operatorname{Ker} \mathcal{A}^*,$$

where $\mathcal{P}_\pm$ *is the orthogonal projection from* $L_2(\mathbb{R})$ *onto the subspace* $L_2(\mathbb{R}_\pm)$. *The action of the adjoint operator* $\mathcal{A}^*$ *on the space* $\mathcal{P}_\pm \operatorname{Ker} \mathcal{A}^*$ *with the basis* $\{\mathcal{P}_\pm h_k\}_{k=1}^m$ *(see* (2.2)*) is given by the formula*

$$(\mathcal{A}^* \mathcal{P}_\pm f_k)(x) = e^{\frac{\alpha}{2}|x|} \int_{\mathbb{R}_\pm} L(x-y) P_k(y) e^{ia_k y}\, dy.$$

3. The main results

Here we shall not give a detailed exposition of the extension theory for $\mathcal{A}$ in the general case (see [AM1] for such an exposition) but concentrate our attention to the case where the symbol l has only real zeros $a_1, \ldots, a_n$ in the strip $\Pi_{\frac{\alpha}{2}}$. Let us note that for a given symbol l having real zeros this case is always realized for α small enough. In order to avoid some technical complications we shall assume in addition that all the zeros $a_1, \ldots, a_n$ are simple. In accordance with the von Neumann theory all self-adjoint extensions of a symmetric operator with finite deficiency indices (n, n) are in one-to-one correspondence with the elements of the group $U(n)$ of all unitary $n \times n$-matrices.

In order to describe the attractor of the dynamical system $R \mapsto \mathcal{B}_R$ we need a special subfamily of self-adjoint extensions of $\mathcal{A}$ parametrized by points of the n-dimensional torus $\mathbf{T}^n$ seen as a subgroup of $U(n)$.

Lemma 3.1. *Let* $\omega = (\omega_1, \ldots, \omega_n) \in \mathbf{T}^n$ *be a point on the torus* $\mathbf{T}^n$. *Then the restriction* $\mathcal{A}_\omega$ *of the adjoint operator* $\mathcal{A}^*$ *to the domain*

$$\mathcal{D}(\mathcal{A}_\omega) = \mathcal{D}(\mathcal{A}) + \Omega_\omega$$

is a self-adjoint operator, where $\Omega_\omega = \operatorname{span}\{f_1, \ldots, f_n\}$ *is the n-dimensional subspace generated by the functions*

$$f_k(x) = e^{ia_k x - \frac{\alpha}{2}|x|} \begin{cases} \omega_k & \text{if } x \geq 0, \\ -\overline{\omega}_k & \text{if } x < 0. \end{cases}$$

Now we are in a position to formulate the central result of the paper, which shows that the behaviour of the dynamical system $R \mapsto \mathcal{B}_R$ in the neighbourhood of the attractor is in the general case asymptotically quasiperiodic.

Theorem 3.2. *Let the equation* $l(z) = 0$ *have only simple zeros* $a_1, \ldots, a_n$ *in the strip* $\Pi_{\frac{\alpha}{2}}$, *and let* $R \mapsto \omega(R)$ *be the trajectory on the torus* $\mathbf{T}^n$ *given by*

$$\omega(R) = (e^{ia_1 R}, \ldots, e^{ia_n R}). \tag{3.1}$$

Under the condition $\beta > 4\alpha$ we have the following description for the attractor of the dynamical system $R \mapsto \mathcal{B}_R$ in the strong resolvent sense:

$$s-\lim_{R\to\infty}((\mathcal{A}_{\omega(R)} - z)^{-1} - (\mathcal{B}_R - z)^{-1}) = 0, \qquad \operatorname{Im} z \neq 0. \tag{3.2}$$

4. A sketch of the proof

The main strategy of the proof is to obtain first some results about the convergence to zero of the difference of the operators $\mathcal{A}_{\omega(R)}^{-1}$ and $\mathcal{B}_R^{-1}$ on a dense set of smooth functions h having compact support, provided that R tends to infinity outside some neighborhood of the critical set, consisting of those values of the parameter R such that either $\mathcal{A}_{\omega(R)}$ or $\mathcal{B}_R$ have zero as a point of the spectrum. In this case we have explicit representations for $\mathcal{A}_{\omega(R)}^{-1}$ as well as for $\mathcal{B}_R^{-1}h$ (in terms of Fourier series on a finite interval $[-R, R]$, R large enough). The information is sufficient in order to get the strong resolvent convergence (3.2) outside some neighborhood of the critical set of the parameter mentioned above. In this neighborhood we use some simple arguments of perturbation theory in combination with those of the extension theory.

First, we study the asymptotic behavior in the space $L_2(\mathbb{R})$ of the solutions of the equation

$$\mathcal{B}_R f_R = h \tag{4.1}$$

for fixed right hand side h taken from a dense set of smooth functions with compact support.

Let us denote by $\mathcal{Z}$ the critical set of those values of the parameter R for which the integral operator $I - \mathcal{L}_R$ has a nontrivial kernel. Outside this set the operators $1 - \mathcal{L}_R$ and $\mathcal{B}_R$ both have a bounded inverse and for the solution of equation (4.1) we have the representation

$$f_R = \mathcal{B}_R^{-1}h = W^{-1}(1 - \mathcal{L}_R)^{-1}W^{-1}h.$$

For fixed h with a compact support the solution f_R vanishes outside the interval $[-R, R]$ for R large enough, and then the search for f_R is reduced to the inversion of the integral operator $1 - \mathcal{L}_R$ on its invariant subspace $L_2([-R, R])$. The part of $\mathcal{L}_R$ on the subspace $L_2([-R, R])$ can be diagonalized by the Fourier series of orthogonal exponents on the interval $[-R, R]$,

$$\left\{\frac{1}{\sqrt{2R}} e^{\frac{i\pi n}{R}x}\right\}_{n\in\mathbb{Z}},$$

and after this diagonalization it becomes the multiplication operator by the discrete function $n \mapsto l_n(R)$,

$$l_n(R) = \int_{-R}^{R} L(x) e^{-\frac{i\pi n}{R}x} dx$$

on the l_2-space of two-sided sequences. Therefore the solution f_R inside the interval $[-R, R]$ can be given explicitly by the Fourier series

$$f_R(x) = \frac{e^{-\frac{\alpha}{2}|x|}}{2R} \sum_{n \in \mathbb{Z}} \frac{h_n(R)}{1 - l_n(R)} e^{\frac{i\pi n}{R} x}, \tag{4.2}$$

where

$$h_n(R) = \int_{-R}^{R} h(x) e^{-\frac{i\pi n}{R} x - \frac{\alpha}{2}|x|} \, dx.$$

One can study the L_2-asymptotics of the solution f_R for $R \to \infty$ outside some neighborhood of the critical set $\mathcal{Z}$, which can now be characterized in terms of the Fourier coefficients $l_n(R)$ as the set $\{R : l_n(R) \neq 1 \text{ for all } n \in \mathbb{Z}\}$.

In order to formulate the result let us specify this neighborhood of the critical set $\mathcal{Z}$. Namely, for given $\kappa > 0$ let

$$\Xi_\kappa = \bigcup_{k=1}^{n} \bigcup_{m \in \mathbb{Z}} \left(\frac{\pi m}{a_k} - e^{-\kappa m}, \frac{\pi m}{a_k} + e^{-\kappa m} \right).$$

Using (1.4) it is easy to prove that for fixed $\kappa < \beta$ we have the inclusion

$$\mathcal{Z} \cap [N, \infty) \subset \Xi_\kappa$$

for N large enough.

It appears that the leading term in the limit $R \to \infty$ of the Fourier series (4.2) can be expressed in terms of asymptotics of Riemannian integral sums corresponding to the partition of the interval of integration by the points $\{\frac{\pi n}{R}\}_{n \in \mathbb{Z}}$ for the divergent integral

$$\frac{1}{2\pi} \int \frac{\widehat{h(\cdot) e^{-\frac{\alpha}{2}|\cdot|}}(s)}{1 - l(s)} e^{ixs} \, ds,$$

having pole type singularities of the integrand at the points $a_1, \ldots, a_n$. Fortunately, the asymptotics can be computed explicitly and we have the following result.

Lemma 4.1. *Let h be smooth and have compact support, and let q_h be the only rational function vanishing at infinity with possible simple poles at the points $a_1, \ldots, a_n$ such that the function*

$$H(s) = \frac{\widehat{h(\cdot) e^{-\frac{\alpha}{2}|\cdot|}}(s)}{1 - l(s)} - q_h(s)$$

is a smooth function on the real axis. Under the condition $\kappa < \frac{\beta}{2}$ the leading term of the asymptotic behavior of the solution f_R on the finite interval $[-R, R]$ coincides with the function

$$\Phi_h(x, R) = \frac{e^{-\frac{\alpha}{2}|x|}}{2\pi} \int_{\mathbb{R}} H(s) e^{isx} \, ds - S_h(x, R)$$

where

$$S_h(x,R) = \frac{1}{2}\sum_{k=1}^{n} \operatorname{Res} q_h(z)|_{z=a_k}(i \operatorname{sign} x - \cot Ra_k)e^{ixa_k - \frac{\alpha}{2}|x|}$$

in the sense that

$$\lim_{R\to\infty, R\notin\Xi_\kappa} ||f_R - \Phi_h(\cdot,R)||_{L_2([-R,R])} = 0. \tag{4.3}$$

Using Lemma 3.1 one can see that the function $\Phi_h(x,R)$ considered as a function on the whole axis belongs to the domain of the self-adjoint extension $\mathcal{A}_{\omega(R)}$ corresponding to the point $\omega(R)$ on the trajectory on the torus $\mathbf{T}^n$ described by (3.1), and, moreover, using the result of Lemma 2.2 we get that

$$\mathcal{A}_{\omega(R)}\Phi_h(\cdot,R) = h.$$

Outside the set Ξ_κ, $\kappa < \frac{\beta}{2}$, the operator $\mathcal{A}_{\omega(R)}$ as well as $\mathcal{B}_R$ have bounded inverses and Lemma 4.1 gives a result on the triviality of the limit $\mathcal{P}_R(\mathcal{A}_{\omega(R)} - \mathcal{B}_R)h$ as $R\to\infty$, $R\notin\Xi_\kappa$, provided that h is fixed and has compact support. Here $\mathcal{P}_R$ denotes the orthogonal projector from $L_2(\mathbb{R})$ onto the subspace $L_2([-R,R])$.

The operator $\mathcal{A}^{-1}_{\omega(R)}$ is some finite-rank perturbation of the inverse of the special self-adjoint extension $\mathcal{A}_{\omega^*}$ of $\mathcal{A}$ corresponding to the point $\omega^* = (1,1,\dots,1)\in\mathbf{T}^n$. Actually we have a variant of Krein's formula:

$$\mathcal{A}^{-1}_{\omega(R)} = \mathcal{A}^{-1}_{\omega^*} + \mathcal{K}_R,$$

where

$$\mathcal{K}_R\cdot = \sum_{k=1}^{n}\mu_k(R)(\,\cdot\,,h_k)h_k.$$

Here

$$\mu_k(R) = \frac{1}{2l'(a_k)}\cot a_k R$$

and $h_k, k=1,\dots,n$, form a basis in Ker $\mathcal{A}^*$

$$h_k(x) = \exp\left(ia_k x - \frac{\alpha}{2}|x|\right).$$

For $R\to\infty$ outside the set Ξ_κ, $\kappa<\frac{\beta}{2}$, from Lemma 4.1 one can obtain the following convergence result on the dense set of smooth functions h having compact support:

$$\lim_{R\to\infty, R\notin\Xi_\kappa}(\mathcal{A}^{-1}_{\omega^*} + \mathcal{P}_R\mathcal{K}_R\mathcal{P}_R - \mathcal{B}_R^{-1})h = 0.$$

Based on this key result and using some perturbation theory arguments we can prove the assertion of Theorem 3.2 if only $R\to\infty$ outside the set Ξ_κ, $\kappa<\frac{\beta}{2}$. We can also obtain the corresponding result for $R \to \infty$ inside the set Ξ_κ, $\kappa > 2\alpha$, using the fact that the operator-valued function

$$R \mapsto (\mathcal{A}_{\omega(R)} - z)^{-1}, \qquad \operatorname{Im} z \neq 0,$$

is uniformly continuous on the torus $\mathbf{T}^n$ together with arguments of perturbation theory, which allow us to control the fluctuations, in the norm sense, of the operator-valued function $R \mapsto (\mathcal{B}_R - z)^{-1}$:

$$\|(\mathcal{B}_R - z)^{-1} - (\mathcal{B}_{R'} - z)^{-1}\| \leq C R^{\frac{1}{2}} e^{\alpha R} |R - R'|^{\frac{1}{2}}.$$

In case $\beta > 4\alpha$ and $\kappa \in (2a, \frac{\beta}{2})$, we can combine these two facts to get the convergence result for $R \to \infty$ inside Ξ_κ as well as outside Ξ_κ, which proves the assertion of the theorem.

Acknowledgements

The second author gratefully acknowledges the support of the SFB 237 (Essen-Bochum-Düsseldorf).

References

[AHKW] Albeverio, S., Høegh-Krohn, R., Wu, T.T.: A class of exactly solvable three-body quantum mechanical problems and universal low energy behavior; Phys. Lett. A 83 (1971), 105–109.

[AM1] Albeverio, S., Makarov, K.A.: Attractors in a model related to the three-body quantum problem; Preprint SFB 237, Bochum 270 (1995); Acta Appl. Math., to appear.

[AM2] Albeverio, S., Makarov, K.A.: Spectral theory of the operator $(p^2+m^2)^{\frac{1}{2}} - ze^2/r$ in the unsemibounded case; Preprint SFB 237, Bochum (1996), in preparation.

[AN] Amado, R.D., Noble, J.V.: On Efimov's effect: A new pathology of three-particle systems; Phys. Lett. B 35 (1971), 25–27; II, Phys. Lett. D (3) 5 (1972), 1992–2002.

[E] Efimov, V.: Energy levels arising from resonant two-body forces in a three-body system; Phys. Lett. B 33 (1970), 563–564.

[F] Faddeev, L.D.: The integral equations method in the scattering theory for three and more particles; Moscow, MIFI 1971 (in Russian).

[Fl] Flamand, G.: Mathematical theory of non-relativistic two- and three-particle systems with point interactions; in: Cargèse Lectures in Theoretical Physics, edited by F.Lurcat, Gordon and Breach, New York 1967, 247–287.

[FM] Faddeev, L.D., Merkuriev, S.P.: Quantum scattering theory for several particle systems; Kluwer Academic Publishers, Dordrecht 1993.

[H] Herbst, I.: Spectral theory of the operator $(p^2+m^2)^{\frac{1}{2}} - ze^2/r$; Comm. Math. Phys. 53 (1977), 285–294; Errata ibid 55 (1977), 316.

[MF1] Minlos, R.A., Faddeev, L.D.: On the point interaction for a three-particle system in quantum mechanics; Soviet Phys. Dokl. 6 (1962), 1072–1074.

[MF2] Minlos, R.A., Faddeev, L.D.: Comment on the problem of three particles with point interactions; Soviet Phys. JETP 14 (1962), 1315–1316.

[MM] Melnikov, A.M., Minlos, R.A.: On the point-like interactions of three different particles; Adv. in Soviet Math. 5 (1981), 99–112.

[MS] Minlos, R.A., Shermatov, M.Kh.: On point-like interactions of three quantum particles; Vestnik Moskov Univ. Ser.I Mat. Mekh. 6 (1989), 7–14 (in Russian).

[OS] Ovchinnikov, Yu.N, Sigal, I.M.: Number of bound states of three body systems and Efimov's effect; Ann. Physics 123 (1979), 274–295.

[S] Sobolev, A.V.: The Efimov effect. Discrete spectrum asymptotics; Comm. Math. Phys. 156 (1993), 127–168.

[T] Tamura, H.: The Efimov effect of three-body Schrödinger operator; J. Funct. Anal. 95 (1991), 433–459.

[Y] Yafaev, D.R.: On the theory of the discrete spectrum of the three-particle Schrödinger operator; Math. USSR-Sb. 23 (1974), 535–559.

Ruhr Universität Bochum
Institut für Mathematik
44780 Bochum
Germany
sergio.albeverio@rz.ruhr-uni-bochum.de

St. Petersburg University
Dept. of Computational and Math. Physics
198024 St. Petersburg
Russia
konstanin.makarov@rz.ruhr-uni-bochum.de

1991 Mathematics Subject Classification: Primary 4720; Secondary 45E10, 81V70

Submitted: June 12, 1996

Operator Theory:
Advances and Applications, Vol. 102

Banach algebras of functions on nonsmooth domains

F. ALI MEHMETI and S. NICAISE

In this paper we study spaces of functions on domains with a conical singularity, which arise naturally if we study hyperbolic and parabolic evolution equations on such domains. We state conditions under which these spaces are Banach algebras, a result which is of independent interest. As applications, we remark that this implies local existence of solutions of Cauchy problems for semilinear, hyperbolic and parabolic partial differential equations with zero Dirichlet boundary conditions and a power-nonlinearity.

1. Introduction

Consider e.g. the wave equation on a three-dimensional bounded space-domain Ω with a smooth boundary except at zero, where it coincides in a small neighbourhood of zero with an infinite cone. Consider zero Dirichlet boundary conditions. The cone's intersection with the unit sphere may be called G. Let $\mathcal{L}(\omega, r\partial r, D_\omega)$ denote the writing of Δ in spherical coordinates $r = |x|$ and $\omega = \frac{x}{|x|}$ and $\mathcal{L}(\lambda) := \mathcal{L}(\omega, \lambda, D_\omega)$. Now denote by λ_{min} the smallest real number such that the operator $\mathcal{L}(\lambda) : \overset{\circ}{H}{}^m(G) \to H^{-m}(G)$ is not invertible and $\operatorname{Re}\lambda > 1 - \frac{3}{2}$.

Now consider the operator

$$D(A) := \{u \in \overset{\circ}{H}{}^m(\Omega) : \Delta u \in L^2(\Omega)\}, \qquad Au := -\Delta u, \quad u \in D(A).$$

From the results of this paper it follows that $D(A)$ is a Banach algebra with respect to a norm which is equivalent to the graph norm if

$$\lambda_{min} > \frac{1}{4},$$

which is a geometrical condition on G.

Our proof is valid for operators of order $2m$ with real constant coefficients and has the following structure:

1. We show that $D(A)$ is embedded in weighted Sobolev spaces, which measures the singularities of its elements at 0.
2. We study a multiplication result in these weighted Sobolev spaces which implies their Banach algebra property in certain cases. The idea is that the singularity at the boundary allows the functions in usual Sobolev spaces to be more singular, which requires weight conditions for compensation, such that products remain in the spaces under consideration. Preliminary versions of these results can be found in [2] and [3].

3. We give conditions for the multiplicativity of the trace operator in the weighted spaces. This proves the conservation of the zero Dirichlet boundary condition under multiplication and thus that the products are again in $D(A)$.

Our applications to semilinear equations use an abstract iteration scheme (cf. [12]). For the example considered above, they imply existence of solutions to the initial boundary value problem (Dirichlet conditions) for the semilinear 3D wave equation

$$\ddot{u}(t,x) - \Delta_x u(t,x) = u(t,x)^k, \qquad (t,x) \in [0, t_{max}] \times \Omega.$$

Here, k is in $\mathbb{N}$, t_{max} sufficiently small and $\lambda_{min} > \frac{1}{4}$. Below, we state similar results if the spatial part is of order $2m$ and has real constant coefficients and for the parabolic case.

Our results are inspired by the investigation of quasilinear equations on networks [1], and the results on elliptic boundary value problems on nonsmooth domains (e.g. [5, 6, 7, 8, 9, 10, 11]). The question of *global* existence for locally Lipschitz continuous semilinearities under the presence of a damping term, which compensates the interfering blow-up effects of the nonlinearity and the singular boundary, is treated in [4] (cf. also [2] for preliminary results on the operator of composition). For quasilinear hyperbolic equations of second order, local existence has been proved in [13].

2. The differential operators and embeddings of their domains

Let Ω be a bounded domain of $\mathbb{R}^n$ $(n \geq 2)$ with a smooth boundary except at 0 where Ω coincides in a small neighbourhood $\mathcal{V}$ of 0 with an infinite cone C. We denote $G = C \cap S^{n-1}$.

Let

$$L = \sum_{|\alpha|=2m} a_\alpha D^\alpha \tag{2.1}$$

be a differential operator of order $2m$ with (real) constant coefficients (only for the sake of simplicity), with a fixed $m \in \mathbb{N}^* := \{1, 2, \ldots\}$, which is strongly elliptic in the usual sense. Let $\mathcal{L}$ be its writing in spherical coordinates $r = |x|$ and $\omega = \frac{x}{|x|}$:

$$\mathcal{L}(\omega, r\partial_r, D_\omega) = r^{2m} L_0(D_x).$$

The parameter depending operator $\mathcal{L}(\lambda)$ is the operator $\mathcal{L}(\omega, \lambda, D_\omega)$ acting from $\overset{\circ}{H}{}^m(G)$ into $H^{-m}(G)$. It is well-known that $\mathcal{L}(\lambda)^{-1}$ is meromorphic and that its poles determine the singularities of the boundary value problem associated with

L. Let us denote by Λ this set of poles, also called singular exponents. Let us set

$$\lambda_{min} = \min\{\Re(\lambda) : \lambda \in \Lambda \text{ such that } \Re(\lambda) > m - \frac{n}{2}\}.$$

For a nonnegative integer l and $\gamma \in \mathbb{R}$, we recall that the Sobolev space $H^l(\Omega)$ and the weighted Sobolev space $H^l_\gamma(\Omega)$ are defined by

$$H^l(\Omega) = \{v \in \mathcal{D}'(\Omega) : D^\alpha v \in L^2(\Omega), \quad |\alpha| \le l\}, \tag{2.2}$$
$$H^l_\gamma(\Omega) = \{v \in \mathcal{D}'(\Omega) : r^{\gamma+|\alpha|-l} D^\alpha v \in L^2(\Omega), \quad |\alpha| \le l\}, \tag{2.3}$$

with respective norm

$$\|v\|_l = \left(\sum_{|\alpha|\le l} \int_\Omega |D^\alpha v|^2 \, dx \right)^{1/2}, \tag{2.4}$$

$$\|v\|_{l,\gamma} = \left(\sum_{|\alpha|\le l} \int_\Omega |r^{\gamma+|\alpha|-l} D^\alpha v|^2 \, dx \right)^{1/2}. \tag{2.5}$$

Let A be the operator from $L^2(\Omega)$ into itself associated with L, i.e.,

$$D(A) = \{u \in \overset{\circ}{H}{}^m(\Omega) : Lu \in L^2(\Omega)\}, \qquad Au = (-1)^m Lu, \quad u \in D(A).$$

The results of [3] lead to the following continuous embedding:

Theorem 2.1. *For all $l \in \mathbb{N}^*$, we have the continuous embedding*

$$D(A^l) \hookrightarrow H^{2lm}_{\gamma_{2l}}(\Omega) \tag{2.6}$$

with $\gamma_{2l} > 0$ fulfilling

$$\gamma_{2l} > 2lm - n/2 - \lambda_{min} \quad \text{if } \lambda_{min} \le 2lm - \frac{n}{2}, \tag{2.7}$$

$$\gamma_{2l} = \varepsilon \quad \text{if } \lambda_{min} > 2lm - \frac{n}{2}, \tag{2.8}$$

where $\varepsilon \in (0,1)$ is fixed and arbitrarily small if n is even, and $\varepsilon = 0$ if n is odd.

Proof. The claim is a direct consequence of the Theorems 3.3 and 5.10 of [3]. □

3. Continuity of the operator of multiplication

Our next goal is to give sufficient conditions under which the domain $D(A)$ of A is a Banach algebra. In other words, we want to know when the application

$$P : (u, v) \to u \cdot v$$

is continuous from $D(A) \times D(A)$ into $D(A)$. Since P is bilinear, we simply have to show its continuity at 0. To show such a result, we first consider the continuity of P into some weighted Sobolev spaces. It is based on a mutiplicativity result in weighted Sobolev spaces proved in [3, Th.2.6] and the embedding of the weighted Sobolev spaces $H^l_\gamma(\Omega)$ into spaces of continuous functions.

Lemma 3.1. *Let $l \in \mathbb{N}^*$ satisfy*

$$l - \frac{n}{2} > 0 \quad \text{and} \quad l - \frac{n}{2} - \gamma > 0. \tag{3.1}$$

Then we have the continuous embedding

$$H^l_\gamma(\Omega) \hookrightarrow C(\bar{\Omega}). \tag{3.2}$$

Moreover, there exists a positive constant C such that

$$|u(x)| \leq C r^{l-\frac{n}{2}-\gamma} \|u\|_{l,\gamma}, \qquad u \in H^l_\gamma(\Omega), \quad x \in \bar{\Omega}. \tag{3.3}$$

Proof. We only have to prove (3.2) and (3.3) near 0; otherwise it is the usual Sobolev embedding Theorem. In the neighbourhood $\mathcal{V}$ of 0, we perform the Euler change of variable $r = e^t$, and from Lemma 2.3 of [5], we conclude that

$$e^{\sigma t}(\eta u)(e^t, \omega) \in H^l(B)$$

with $\sigma = -(l - \frac{n}{2} - \gamma)$. Applying the usual Sobolev embedding Theorem to $H^l(B)$, we conclude that

$$e^{\sigma t}(\eta u)(e^t, \omega) \in C(\bar{B})$$

since $l > \frac{n}{2}$. Moreover, there exists a $C > 0$ such that

$$\sup_{t \in \mathbb{R}, \omega \in \bar{G}} |e^{\sigma t}(\eta u)(e^t, \omega)| \leq C \|e^{\sigma t}(\eta u)(e^t, \omega)\|_{H^l(B)}.$$

Going back to Γ, we get (3.2) and (3.3). □

Theorem 3.2. *Let $l \in \mathbb{N}^*$ and $\gamma \in \mathbb{R}$ satisfy (3.1). If $l' \in \mathbb{N}^*$ and $\gamma' \in \mathbb{R}$ are such that*

$$l' - \frac{n}{2} - \gamma' \leq 2(l - \frac{n}{2} - \gamma), \tag{3.4}$$

then P is continuous from $H^l_\gamma(\Omega) \times H^l_\gamma(\Omega)$ into $H^{l'}_{\gamma'}(\Omega)$.

Proof. Fix u, v in $H^l_\gamma(\Omega)$. For all $|\alpha| \leq l'$, we shall show that

$$r^{|\alpha| - l' + \gamma'} D^\alpha (u \cdot v) \in L^2(\Omega) \tag{3.5}$$

as well as

$$\|r^{|\alpha|-l'+\gamma'}D^\alpha(u\cdot v)\|_0 \le C\|u\|_{l,\gamma}\|v\|_{l,\gamma}. \tag{3.6}$$

i) For $|\alpha|=0$, by the estimate (3.3), we may write

$$|r^{-l'+\gamma'}(u\cdot v)(x)| \le Cr^{-l'+\gamma'}r^{(l-\frac{n}{2}-\gamma)}\|u\|_{l,\gamma}|v(x)|.$$

Integrating the square of this estimate over Ω, one obtains

$$\|r^{-l'+\gamma'}(u\cdot v)\|_0 \le C\|u\|_{l,\gamma}\|v\|_{0,\delta},$$

where $\delta = -l'+\gamma'+(l-\frac{n}{2}-\gamma)$. Since $H^l_\gamma(\Omega)\hookrightarrow H^0_{\delta'}(\Omega)$, with $\delta'=\gamma-l$, we get (3.5) and (3.6) for $|\alpha|=0$ because the inequality $\delta'\le\delta$ (equivalent to (3.4)) implies the continuous embedding $H^0_{\delta'}(\Omega)\hookrightarrow H^0_\delta(\Omega)$.
ii) For $|\alpha|\ge 1$, we use Leibniz' rule

$$D^\alpha(uv) = \sum_{\eta\le\alpha}\begin{pmatrix}\alpha\\ \eta\end{pmatrix}D^\eta u D^{\alpha-\eta}v. \tag{3.7}$$

We shall show that each term of this right-hand side belongs to $H^0_\delta(\Omega)$, with $\delta=|\alpha|-l+\gamma$. Therefore let us fix $\eta\le\alpha$, then our goal reduces to prove that the function

$$D^\eta u D^{\alpha-\eta}v$$

belongs to $H^0_\delta(\Omega)$ and that

$$\|D^\eta u D^{\alpha-\eta}v\|_{0,\delta} \le C\|u\|_{l,\gamma}\|v\|_{l,\gamma}. \tag{3.8}$$

But from Theorem 2.6 of [3] (the assumptions of that theorem being equivalent to (3.1)), we know that

$$D^\eta u D^{\alpha-\eta}v \in H^0_{\delta'}(\Omega), \tag{3.9}$$

and

$$\|D^\eta u D^{\alpha-\eta}v\|_{0,\delta'} \le C\|u\|_{l-|\eta|,\gamma}\|v\|_{l-|\alpha|+|\eta|,\gamma}. \tag{3.10}$$

with $\delta' = 2\gamma-2l+|\alpha|+\frac{n}{2}$. The conclusion follows from the fact that $\delta'\le\delta$. □

A direct consequence of this result is that $H^l_\gamma(\Omega)$ is a Banach algebra. In view of the proof of Theorem 2.6 of [3], this result mainly follows from the fact that the Sobolev space of order l is a Banach algebra when $l>\frac{n}{2}$.

A second consequence of this Theorem is that $D(A)$ is also a Banach algebra.

Theorem 3.3. *Suppose that* $2m-\frac{n}{2}>0$, $\lambda_{min}>0$ *and*

$$2m-\frac{n}{2} < 2\min\left\{\lambda_{min}, 2m-\frac{n}{2}\right\}. \tag{3.11}$$

Then $D(A)$ *is a Banach algebra.*

Proof. First, by Theorem 2.1, we have

$$D(A) \hookrightarrow H^{2m}_{\gamma_2}(\Omega). \tag{3.12}$$

Applying now Theorem 3.2, for any $u, v \in D(A)$, one has $u \cdot v \in H^{2m}_0(\Omega)$ and

$$\|u \cdot v\|_{2m,0} \leq C\|u\|_{D(A)}\|v\|_{D(A)}. \tag{3.13}$$

It remains to be proved that $u \cdot v$ satisfies the boundary conditions, i.e., we need to show that

$$u \cdot v \in \overset{\circ}{H}{}^m(\Omega). \tag{3.14}$$

So we have to check that

$$\gamma D^\alpha(u \cdot v) = 0 \text{ on } \Gamma, \qquad |\alpha| < m, \tag{3.15}$$

where γ is the trace operator on the boundary Γ of Ω.
The condition (3.15) for $|\alpha| = 0$ clearly holds because the condition on λ_{min} insures that $H^{2m}_{\gamma_2}(\Omega) \hookrightarrow C(\bar{\Omega})$. Consequently, one has

$$\gamma(u \cdot v) = u_{|\Gamma} v_{|\Gamma} = 0$$

because $u, v \in \overset{\circ}{H}{}^m(\Omega)$.
To prove (3.15) for $|\alpha| \geq 1$, we still use Leibniz' rule and simply show that

$$\gamma\{D^\eta u D^{\alpha-\eta} v\} = 0 \text{ on } \Gamma \tag{3.16}$$

for all $\eta \leq \alpha$. By Theorem 3.2, $D^\eta u D^{\alpha-\eta} v$ belongs to $H^1(\Omega)$, which gives a meaning to its trace. To conclude, we need to show that

$$\gamma\{D^\eta u D^{\alpha-\eta} v\} = \gamma(D^\eta u)\gamma(D^{\alpha-\eta} v). \tag{3.17}$$

Indeed, if (3.17) holds, then (3.16) also holds due to the fact that $u \in \overset{\circ}{H}{}^m(\Omega)$ and since $|\eta| < m$.
To establish (3.17), we recall that the space

$$C^\infty_v(\bar{\Omega}) = \{v \in C^\infty(\bar{\Omega}) : v = 0 \text{ in a neighbourhood of } 0\}$$

is dense in $H^l_\gamma(\Omega)$. Consequently, there exist two sequences $(u_n)_{n\in\mathbb{N}}$ and $(v_n)_{n\in\mathbb{N}}$ of elements of $C^\infty_v(\bar{\Omega})$ such that

$$u_n \to u \quad \text{in} \quad H^{2m}_{\gamma_2}(\Omega), \qquad n \to \infty, \tag{3.18}$$

$$v_n \to v \quad \text{in} \quad H^{2m}_{\gamma_2}(\Omega), \qquad n \to \infty. \tag{3.19}$$

From Theorem 3.2, we deduce that

$$D^\eta u_n D^{\alpha-\eta} v_n \to D^\eta u D^{\alpha-\eta} v \quad \text{in} \quad H^1(\Omega), \qquad n \to \infty.$$

The usual trace theorem leads to

$$(3.20)\qquad \gamma\{D^\eta u_n D^{\alpha-\eta} v_n\} \to \gamma\{D^\eta u D^{\alpha-\eta} v\} \quad \text{in} \quad L^2(\Gamma), \qquad n\to\infty.$$

On the other hand, by the definition of $H^l_\gamma(\Omega)$, (3.18) implies that

$$(3.21)\qquad u_n \to u \quad \text{in} \quad H^{2m}(\Omega\setminus V), \qquad n\to\infty,$$

$$(3.22)\qquad v_n \to v \quad \text{in} \quad H^{2m}(\Omega\setminus V), \qquad n\to\infty,$$

for any neighbourhood $\mathcal{V}$ of 0. By the trace theorem and Theorem 1.4.4.2 of [6], we arrive at

$$(3.23)\quad \gamma(D^\eta u_n)\gamma(D^{\alpha-\eta} v_n) \to \gamma(D^\eta u)\gamma(D^{\alpha-\eta} v) \quad \text{in} \quad L^2(\Gamma\setminus V), \qquad n\to\infty.$$

Since the elements u_n, v_n clearly satisfy (3.17), u, v also satisfy (3.17) thanks to (3.20) and (3.23), and since $\mathcal{V}$ was arbitrary. □

In the following we make some comments on the condition (3.11):
a) If $\lambda_{min} \geq 2m - \frac{n}{2}$, then (3.11) always holds. This is not surprising since in that case $D(A) = H^{2m}(\Omega) \cap \overset{\circ}{H}{}^m(\Omega)$ (as in the case of smooth domains).
b) If $\lambda_{min} < 2m - \frac{n}{2}$, then (3.11) reduces to

$$(3.24)\qquad \lambda_{min} > m - \frac{n}{4}.$$

This always holds in dimension 2, since by Theorem 1 of [8], we have

$$(3.25)\qquad \lambda_{min} > m - \frac{n}{2} + \frac{1}{2},$$

which is exactly (3.24) when $n = 2$. In dimension $n \geq 3$, (3.24) is not always satified (even for the Laplace operator). Let us also notice that the condition $\lambda_{min} > 0$ always holds in dimension 2 or 3 thanks to (3.25).

Corollary 3.4. *Under the assumptions of Theorem* 3.3, *for any positive integer* k, *the operator* $u \mapsto u^k$ *is locally Lipschitz from* $D(A)$ *into* $D(A)$.

Proof. The local Lipschitz property is a direct consequence of the Theorems 2.1 and 3.3, using the well-known identity $u^k - v^k = (u-v)\{\sum_{l=0}^{k-1} u^l v^{k-1-l}\}$. □

4. Applications to semilinear evolution equations

The first application we have in mind concerns the local existence of solutions to semilinear parabolic evolution equations.

Theorem 4.1. *Under the assumptions of Theorem 3.3, for any positive integer k, and all $u_0 \in D(A)$, there exists a $t_{max} > 0$ such that the evolution equation*

$$(4.1) \qquad \begin{cases} \dfrac{\partial u}{\partial t}(t,x) + Au(t,x) = u(t,x)^k, & t > 0, \quad x \in \Omega, \\ u(t,x) = 0, \qquad x \in \Gamma, \\ u(0,x) = u_0(x), \qquad x \in \Omega, \end{cases}$$

has a unique solution $u \in C([0, t_{max}[, D(A)) \cap C^1([0, t_{max}[, L^2(\Omega))$.

Proof. It is well-known that the ellipticity of L implies that $-A$ generates a C_0 semigroup on $L^2(\Omega)$. Moreover, Corollary 3.4 insures that the nonlinearity

$$u \mapsto u^k$$

is locally Lipschitz from $D(A)$ into $D(A)$. The conclusion is now a consequence of Theorems 6.1.4 and 6.1.7 of [12]. □

The second application concerns the local existence of solutions to semilinear hyperbolic evolution equations. In that case, we require that the bilinear form a associated with L by

$$a(u,v) = (-1)^m \langle Lu, v \rangle, \qquad u, v \in \mathring{H}^m(\Omega),$$

is symmetric and strongly coercive on $\mathring{H}^m(\Omega)$, i.e., there exists an $\alpha > 0$ such that

$$a(u,u) \geq \alpha \, \|u\|_m^2, \qquad u \in \mathring{H}^m(\Omega).$$

Then A is a positive self-adjoint operator and $D(A^{1/2}) = \mathring{H}^m(\Omega)$.

Theorem 4.2. *Under the assumptions of Theorem 3.3, for any positive integer k, and all $u_0 \in D(A), u_1 \in \mathring{H}^m(\Omega)$, there exists a $t_{max} > 0$ such that the evolution equation*

$$(4.2) \qquad \begin{cases} \dfrac{\partial^2 u}{\partial t^2}(t,x) + Au(t,x) = u(t,x)^k, & t > 0, \quad x \in \Omega, \\ u(t,x) = 0, \qquad x \in \Gamma, \\ u(0,x) = u_0(x), \quad \dfrac{\partial u}{\partial t}(0,x) = u_1(x), \qquad x \in \Omega, \end{cases}$$

possesses a unique solution $u \in C([0, t_{max}[, D(A)) \cap C^1([0, t_{max}[, \mathring{H}^m(\Omega)) \cap C^2([0, t_{max}[, L^2(\Omega))$.

Proof. We use the usual trick which consists in reducing (4.2) to a first order evolution equation

$$(4.3) \qquad \begin{cases} \dot{\mathbf{u}}(t) + \mathcal{A}\mathbf{u}(t) = F(\mathbf{u}(t)) & \text{in } X, \\ \mathbf{u}(0) = \mathbf{u}_0, \end{cases}$$

where

$$\mathcal{A} = \begin{pmatrix} 0 & -I \\ A & 0 \end{pmatrix}$$

is a linear operator in the Hilbert space $X = D(A^{1/2}) \times L^2(\Omega)$,

$$\mathbf{u} = \begin{pmatrix} u \\ \dfrac{\partial u}{\partial t} \end{pmatrix}, \qquad \mathbf{u}_0 = \begin{pmatrix} u_0 \\ u_1 \end{pmatrix},$$

and the nonlinearity F is defined by

$$F\begin{pmatrix} u \\ v \end{pmatrix} = \begin{pmatrix} 0 \\ u^k \end{pmatrix}.$$

The assumptions on the bilinear form a imply that $-\mathcal{A}$ generates a C_0 semigroup on X. Moreover, Corollary 3.4 implies that F is locally Lipschitz from $D(\mathcal{A}) = D(A) \times D(A^{1/2})$ into itself. Therefore, the conclusion still follows from Theorems 6.1.4 and 6.1.7 of [12] (going back to u). □

Remark 4.3. Using this type of approach, we establish in [4] global existence results for semilinear evolution equations with dissipation in domains with conical singularities.

References

[1] ALI MEHMETI, F.: Nonlinear waves in networks; Mathematical Research 80, Akademie Verlag, Berlin 1994.

[2] ALI MEHMETI, F., NICAISE, S.: Characterization of iterated powers of operators in nonsmooth domains and Nemetskij's operators; in: G. Lumer, S. Nicaise, B.-W. Schulze (eds.), Partial Differential Equations; Models in Physics and Biology; Mathematical Research 82, Akademie Verlag, Berlin 1994, 40–55.

[3] ALI MEHMETI, F., NICAISE, S.: Non-autonomous evolution equations on nonsmooth domains; Math. Nachr. (1997), to appear.

[4] ALI MEHMETI, F., NICAISE, S.: Nemetskij's operators and global existence of small solutions of semilinear evolution equations on nonsmooth domains; TH Darmstadt preprints 1856 (1996); Comm. Partial Differential Equations (to appear).

[5] BOURLARD, M., DAUGE, M., LUBUMA, M.-S., NICAISE, S.: Coefficients des Singularités pour des problèmes aux limites elliptiques sur une Domaine à Points coniques I: Résultats généraux pour le problème de Dirichlet; RAIRO Modél. Math. Anal. Numér. 24 (1990), 27–52.

[6] GRISVARD, P.: Elliptic problems in nonsmooth Domains; Monographs and Studies in Mathematics 21, Pitman, Boston 1985.

[7] KONDRATIEV, V.A.: Boundary value problems for elliptic equations in domains with conical or angular points; Trans. Moscow Math. Soc. 16 (1967), 227–313.

[8] KOZLOV, V.A., MAZ'YA, V.G.: Spectral properties of the operator bundles generated by elliptic boundary value problems in a cone; Functional Anal. Appl. 22 (1988), 38–46.

[9] MAZ'YA, V.G., PLAMENEVSKII, B.A.: Estimates in L^p and in Hölder classes and the Miranda-Agmon maximum principle for solutions of elliptic boundary value problems in domains with singular points on the boundary; Trans. Amer. Math. Soc 123:2 (1984), 1–56.

[10] MEISTER, E., PENZEL, F., SPECK, F.-O., TEIXEIRA, F.-S.: Some interior and exterior boundary value problems for the Helmholtz equation in a quadrant; Proc. Roy. Soc. Edinburgh Sect. A 123 (1993), 275–294.

[11] NICAISE, S.: Polygonal interface problems; Series "Methoden und Verfahren der Mathematischen Physik" 39, Peter Lang Verlag, 1993.

[12] PAZY, A.: Semigroups of linear operators and applications to partial differential equations; Appl. Math. Sci. 44, Springer Verlag, New York 1983.

[13] WITT, I.: Non-linear hyperbolic equations in domains with conical points: existence and regularity of solutions; Mathematical Research 84, Akademie Verlag, Berlin 1995.

Université de Valenciennes et du Hainaut Cambrésis
LIMAV
Institut des Sciences et Techniques de Valenciennes
B.P. 311
F-59304 - Valenciennes Cedex
France
snicaise@univ-valenciennes.fr

1991 Mathematics Subject Classification: Primary 35B65, 47N20; Secondary 35K55, 35L70

Submitted: May 30, 1996

Operator Theory:
Advances and Applications, Vol. 102

A nonlinear approach to generalized factorization of matrix functions

M.C. Câmara and A.F. dos Santos

The generalized factorization of some classes of 2×2 matrix symbols is determined by reduction to the study of certain scalar non-linear Riemann-Hilbert problems. This method is applied to several types of matrix functions, whose factorization is explicitly obtained.

1. Introduction

In this paper a generalized method for studying the existence of a canonical generalized factorization of certain classes of 2×2 non-rational matrix functions is presented. The method also provides explicit formulas for the factors when the factorization is canonical and can be modified in such a way that the factors of a non-canonical factorization may also be obtained.

The method proposed in the subsequent sections envolves the derivation of two scalar Riemann-Hilbert problems by applying certain non-linear functionals to the original matrix Riemann-Hilbert problem for the factors of the factorization. From the solution of the scalar Riemann-Hilbert problems it is possible to study the existence of non-trivial solutions of the original homogeneous Riemann-Hilbert problem and to determine the factors, once the question of existence of a canonical factorization is settled.

Contrary to existing methods for the study of generalized factorization for several classes of symbols like the Daniele-Khrapkov method ([2], [5], [8]) or Wiener-Hopf-Hilbert method ([1], [6]), which apply only to very specific classes of matrix functions, the present method seems to have a much wider application and, as far as the authors could assess, applies to most classes of functions that can be factorized by existing methods. It has also been applied successfully to classes of symbols for which other methods have failed ([3],[4]).

2. Preliminaries

We denote by $L_p(\mathbb{R})$, $1 \leq p < \infty$, the Banach space of all complex-valued Lebesgue mesurable functions defined on $\mathbb{R}$, for which $|f|^p$ is integrable, with the norm

$$\|f\|_p = \left(\int_{\mathbb{R}} |f(t)|^p \, dt\right)^{1/p}. \tag{2.1}$$

The singular integral operator $S_{\mathbb{R}} : L_p(\mathbb{R}) \to L_p(\mathbb{R})$, $1 < p < \infty$, is defined as usual by

$$S_{\mathbb{R}} f(t) = \frac{1}{i\pi} \int_{\mathbb{R}} \frac{f(u)}{u-t} \, du, \qquad t \in \mathbb{R}, \tag{2.2}$$

where the integral is understood in the sense of Cauchy's principal value. Related to this operator we define two complementary projections

$$P^{\pm} = \frac{1}{2}(I \pm S_{\mathbb{R}}), \tag{2.3}$$

where I is the identity operator in $L_p(\mathbb{R})$. We denote by $L_p^+(\mathbb{R})$ and $L_p^-(\mathbb{R})$ the images of P^+ and P^-, respectively.

By $L_\infty^+(\mathbb{R})$ ($L_\infty^-(\mathbb{R})$, respectively) we denote the space of all essentially bounded functions $f \in L_\infty(\mathbb{R})$ which admit a bounded analytic extension to the halfplane $\mathbb{C}^+ = \{z \in \mathbb{C} : \text{Im } z > 0\}$ ($\mathbb{C}^- = \{z \in \mathbb{C} : \text{Im } z < 0\}$, respectively). By $\mathcal{C}(\dot{\mathbb{R}})$ we represent the algebra of all functions which are continuous on $\mathbb{R}$ and possess equal limits at $\pm\infty$, and we denote by $\mathcal{R}(\dot{\mathbb{R}})$ the class of all rational functions in $\mathcal{C}(\dot{\mathbb{R}})$.

If A is an algebra, let $\mathcal{G}(A)$ be the group of invertible elements in A.

By a *generalized factorization* of $G \in \mathcal{G}(L_\infty(\mathbb{R}))^{2\times 2}$ relative to $L_2(\mathbb{R})$ we mean a factorization of the form

$$G = G_- \, \text{diag} \, (r^{k_j})_{j=1}^2 \, G_+ \tag{2.4}$$

with $r(\xi) = (\xi - i)/(\xi + i)$, for $\xi \in \mathbb{R}$, $k_1, k_2 \in \mathbb{Z}$, $k_1 \geq k_2$, where the factors $G_\pm$ satisfy the following conditions:

(i) $r_+ G_+^{\pm 1} \in (L_2^+(\mathbb{R}))^{2\times 2}$ for $r_+(\xi) = (\xi + i)^{-1}$,

(ii) $r_- G_-^{\pm 1} \in (L_2^-(\mathbb{R}))^{2\times 2}$ for $r_-(\xi) = (\xi - i)^{-1}$,

(iii) $G_+^{-1} P^+ G_-^{-1} I$ is an operator defined on a dense subset of $(L_2(\mathbb{R}))^2$ possessing a bounded extension to $(L_2(\mathbb{R}))^2$.

The generalized factorization is said to be *canonical* if the *partial indices* k_1 and k_2 are equal to zero.

For $G \in \mathcal{G}(\mathcal{C}(\dot{\mathbb{R}}))^{2\times 2}$ there always exists a generalized factorization as defined above.

It is well-known that $G \in \mathcal{G}(L_\infty(\mathbb{R}))^{2\times 2}$ admits a generalized factorization (relative to $L_2(\mathbb{R})$) iff the operator

$$P^+ G I^+ : (L_2^+(\mathbb{R}))^2 \to (L_2^+(\mathbb{R}))^2$$

(where I^+ denotes the identity operator on $(L_2^+(\mathbb{R}))^2$) is Fredholm. If, in particular, $G \in \mathcal{G}(C(\dot{\mathbb{R}}))^{2\times 2}$, this is equivalent to

$$\det\, G(\xi) \neq 0, \qquad \xi \in \dot{\mathbb{R}},$$

and in this case the *total index* $\operatorname{ind} G = k_1 + k_2$ is equal to the index of $\det\, G$ (as a continuous function in $\dot{\mathbb{R}}$).

3. The method

Let us first consider the question of existence of a canonical factorization of a matrix-valued function $G \in (L_\infty(\mathbb{R}))^{2\times 2}$ possessing a generalized factorization relative to $L_2(\mathbb{R})$, of the form (2.4), such that $\operatorname{ind} G = 0$.

With these assumptions, the problem is reduced to investigating whether the Riemann-Hilbert problem

$$G\phi^+ = \phi^-, \qquad \phi^\pm \in (L_2^\pm(\mathbb{R}))^2, \tag{3.1}$$

has only the trivial solution.

Let $\tilde{F} : \mathbb{C}^2 \to \mathbb{C}$ be a functional such that the set of solutions of the equation

$$\tilde{F}(x_1, x_2) = 0 \tag{3.2}$$

is known. We assume moreover that

$$F(\phi_1, \phi_2)(t) = \tilde{F}(\phi_1(t), \phi_2(t)) \quad (t \in \mathbb{R}) \tag{3.3}$$

defines a function $F : (L_2(\mathbb{R}))^2 \to L_p(\mathbb{R})$, $p \geq 1$, such that, conveniently rewriting equation (3.1) in some equivalent form

$$G_1\phi^+ = G_2\phi^- \qquad (G_2^{-1}G_1 = G), \tag{3.4}$$

we have

$$F(G_1, \phi^+) = \psi^+ + q_1, \tag{3.5}$$

$$F(G_2, \phi^-) = \psi^- + q_2 \tag{3.6}$$

where $\psi^\pm \in L_p^\pm(\mathbb{R})$ and $q_1, q_2 \in \mathcal{R}(\dot{\mathbb{R}})$.

It is obvious that every solution (ϕ^+, ϕ^-) of (3.1) must also be a solution to the Riemann-Hilbert problem

$$F(G_1\phi^+) = F(G_2\phi^-) \tag{3.7}$$

which, taking (3.5) and (3.6) into account, implies that

$$F(G_1\phi^+) = F(G_2\phi^-) = q \quad \text{with } q \in \mathcal{R}(\dot{\mathbb{R}}) \cap L_p(\mathbb{R}). \tag{3.8}$$

Let us consider the following two cases separately:

Case I: $q = 0$.
Since the set of solutions of (3.2) is known, we can solve (3.8) and check which of these solutions satisfy the equality $G\phi^+ = \phi^-$, thus answering the question of existence of a canonical generalized factorization for G.

Case II: q does not vanish identically.
In this case the solution of (3.4), which is equivalent to (3.1), can be obtained from a second scalar Riemann-Hilbert problem related to that homogeneous equation. In fact $G\phi^+ = \phi^-$ implies that, defining

$$\tilde{\phi}^- = \begin{bmatrix} 0 & 1 \\ -1 & 0 \end{bmatrix} \phi^-, \tag{3.9}$$

we must have

$$(\tilde{\phi}^-)^T G\phi^+ = 0. \tag{3.10}$$

Therefore we consider the problem of solving this non-linear scalar problem, bearing in mind that we are not looking for the whole set of its solutions, but only for those which satisfy a certain condition of the form (3.8).

It turns out that, at least for a considerably large class of 2×2 matrix-valued functions G, the introduction of such a condition not only allows us to solve the non-linear problem (3.10) - as we shall see in the following sections - but guarantees moreover that any solution of (3.10) satisfying (3.8) is also a solution of (3.1). Therefore the problem of solving the homogeneous equation $G\phi^+ = \phi^-$ becomes equivalent to solving the non-linear scalar equation (3.10) submitted to the above mentioned condition.

This happens in particular if $\tilde{F}$ is a homogeneous function in (x_1, x_2) of degree one, i.e.,

$$\tilde{F}(\lambda x_1, \lambda x_2) = \lambda \tilde{F}(x_1, x_2) \quad (\lambda \in \mathbb{C}). \tag{3.11}$$

In fact we have the following:

Theorem 3.1. *Let $\tilde{F} : \mathbb{C}^2 \to \mathbb{C}$ be a functional satisfying* (3.11) *and such that, for some decomposition $G = G_2^{-1}G_1$ an equality of the form* (3.8) *holds. Then any solution of* (3.10) *satisfying condtion* (3.8) *is also a (non-trivial) solution to the homogeneous equation $G\phi^+ = \phi^-$.*

Proof. Since equality (3.10) means that, for each value of the variable t in $\mathbb{R}$, $G\phi^+(t)$ is orthogonal (with respect to the usual inner product in $\mathbb{C}^2$) to $\phi^-(t)$, there is a scalar function $\lambda(t)$ such that

$$G\phi^+(t) = \lambda(t)\phi^-(t) \quad (t \in \mathbb{R}). \tag{3.12}$$

Taking (3.11) into account, it follows from (3.8) that $\phi^+, \phi^- \neq 0$ and

$$\lambda F(G_2\phi^-) = F(G_2\phi^-) = q \tag{3.13}$$

where q has at most a finite number of zeros. Therefore $\lambda = 1$ which means, according to (3.12), that (ϕ_+, ϕ_-) is a solution to the homogeneous equation $G\phi^+ = \phi^-$. □

The result of this theorem still holds if the function F (in the appropriate condition of the form (3.8) associated with the non-linear equation) is not homogeneous of degree one, as long as we can derive from this condition an equality of the form $F_1(G_1\phi^+) = F_1(G_2\phi^-) = \psi$ where ψ does not vanish (in $\mathbb{R}$) and F_1 is homogeneous of degree one.

In any case, if the generalized factorization of G is not canonical, by solving (3.1) we can determine the kernel of the operator $P^+GI^+ : (L_2^+(\mathbb{R}))^2 \to (L_2^+(\mathbb{R}))^2$ and thus determine the partial indices in the generalized factorization of G.

Considering now the determination of the factors $G_\pm$ in a canonical generalized factorization of G (which exists if the equation $G\phi^+ = \phi^-$ has only the zero solution) we have the following: Let $\Phi^\pm \in (L_2^\pm(\mathbb{R}))^{2\times 2}$ be such that

$$G\Phi^+ = r\Phi^- \tag{3.14}$$

where $r(\xi) = (\xi - \xi^+)/(\xi - \xi^-)$, $\xi^\pm \in \mathbb{C}^\pm$ $(\xi \in \mathbb{R})$. Then a canonical generalized factorization of G is $G = G_-G_+$ where

$$\begin{aligned} G_- &= r_-^{-1}\Phi^-, & G_+^{-1} &= r_+^{-1}\Phi^+, \\ r_-(\xi) &= \frac{1}{\xi - \xi^+}, & r_+(\xi) &= \frac{1}{\xi - \xi^-}, \end{aligned} \tag{3.15}$$

if $\Phi^-(\xi^-)$ or $\Phi^+(\xi^+)$ is an invertible matrix (cf. [4]).

Therefore we can obtain the two columns in G_- and G_+^{-1} separately by solving a homogeneous equation $r^{-1}G\phi^+ = \phi^-$, using the method presented above as well as convenient normalizing conditons (which are chosen in order to simplify the resulting non-linear scalar problem (3.10) and ensure that the invertibility condition regarding the factors is satisfied).

In the next sections we show how this method can be applied to some classes of functions which are of interest in applications.

4. The *RW* class

We consider now a class of 2×2 matrix functions first proposed by Rawlins and Williams in a somewhat different form.

Let G be a 2×2 matrix function of the form

$$G = \begin{bmatrix} 1 & \rho\alpha \\ -\rho^{-1}\alpha^{-1} & 1 \end{bmatrix} \tag{4.1}$$

where $\alpha \in \mathcal{G}(L_\infty^+(\mathbb{R}))$ is continuous in $\dot{\mathbb{R}}$ and $\rho^2 = q \in \mathcal{R}(\dot{\mathbb{R}})$.

Let moreover q be a quotient of two first degree polynomials with different zeros,

$$q(\xi) = \rho^2(\xi) = \frac{\xi - \xi_1}{\xi - \xi_2} \tag{4.2}$$

with $\xi_1, \xi_2 \in \mathbb{C} \setminus \mathbb{R}$. Since our main purpose is to show how the method proposed applies to this class of matrix functions (which we shall denote by RW), we will only study an example where $\alpha = \alpha_+ \in \mathcal{G}(L_\infty^+(\mathbb{R}))$ and ξ_1, ξ_2 in (4.2) belong to the upper half-plane $\mathbb{C}^+$. Any other case can be treated analogously. Let, for instance

$$\rho^2(\xi) = \frac{\xi - 2i}{\xi - i} \qquad (\xi \in \mathbb{R}). \tag{4.3}$$

Regarding the question of existence of a canonical generalized factorization for G we have the following result.

Theorem 4.1. *Let $G \in RW$ be of the form*

$$G = \begin{bmatrix} 1 & \rho\alpha_+ \\ -\rho^{-1}\alpha_+^{-1} & 1 \end{bmatrix} \tag{4.4}$$

where $\alpha_+ \in \mathcal{G}(L_\infty^+(\mathbb{R}))$ is continuous in $\dot{\mathbb{R}}$ and ρ^2 is defined by (4.3). Then G possesses a canonical generalized factorization.

Proof. Since G is continuous in $\dot{\mathbb{R}}$ and its total index is zero, it is enough to show that the equation

$$G\phi^+ = \phi^-, \qquad \phi^\pm \in (L_2^\pm(\mathbb{R}))^2, \tag{4.5}$$

admits only the solution $\phi^+ = \phi^- = 0$.
Defining

$$\tilde{F} : \mathbb{C}^2 \to \mathbb{C}\,, \quad \tilde{F}(x_1, x_2) = x_1 x_2 \tag{4.6}$$

we have (see (3.3))

$$\begin{aligned} &F : (L_2(\mathbb{R}))^2 \to L_1(\mathbb{R}) \\ &F(\phi) = \phi_1\phi_2 \quad \text{where } \phi = (\phi_1, \phi_2), \end{aligned} \tag{4.7}$$

and, for $\phi^\pm = (\phi_1^\pm, \phi_2^\pm)$,

$$F(G\phi^+) = F(\phi^-) \iff -\rho^{-1}\alpha_+^{-1}[(\phi_1^+)^2 - \rho^2\alpha_+^2(\phi_2^+)^2] = \phi_1^-\phi_2^-\,. \tag{4.8}$$

Thus any solution to (4.5) must satisfy

$$-\alpha_+^{-1}[(\phi_1^+)^2 - \rho^2\alpha_+^2(\phi_2^+)^2] = \rho\phi_1^-\phi_2^- = \frac{K}{\xi - i} \tag{4.9}$$

but, since both sides of (4.8) represent functions belonging to $L_1(\mathbb{R})$, we must have $K = 0$. Therefore $\phi_1^- = 0$ or $\phi_2^- = 0$ and in either case it follows from (4.5) that $\phi^+ = \phi^- = 0$. □

Remark 4.2. If ξ_1 and ξ_2 in (4.2) do not belong to the same half-plane $\mathbb{C}^+$ or $\mathbb{C}^-$ and assuming, for example, that $\xi_1 \in \mathbb{C}^+$ and $\xi_2 \in \mathbb{C}^-$, we get (instead of (4.9)):

$$-\frac{(\xi-\xi_2)^{1/2}}{\xi-\xi_2}\,\alpha_+^{-1}\,[(\phi_1^+)^2 - \rho^2\alpha_+^2(\phi_2^+)^2] = \frac{(\xi-\xi_1)^{1/2}}{\xi-\xi_2}\,\phi_1^-\phi_2^- \tag{4.10}$$

and thus

$$\frac{(\xi-\xi_1)^{1/2}}{\xi-\xi_2}\,\phi_1^-\phi_2^- = \frac{K}{\xi-\xi_2} \qquad (K \in \mathbb{C}).$$

Since the left-hand side of this equality represents a function in $L_1(\mathbb{R})$ we must have $K = 0$ and we conclude, as in Theorem 4.1, that $\phi^+ = \phi^- = 0$.

Next we obtain a canonical generalized factorization for G, using the procedure presented in Section 3.

The two columns of the factors G_+^{-1}, G_- can be obtained, as mentioned in Section 3, by solving the homogeneous equation

$$r^{-1}G\phi^+ = \phi^-, \qquad r(\xi) = \frac{\xi-i}{\xi-i} \qquad (\xi \in \mathbb{R}). \tag{4.11}$$

In this case, condition (3.8) takes the form

$$-\alpha_+^{-1}[(\phi_1^+)^2 - \rho^2\alpha_+^2(\phi_2^+)^2] = r^2\rho\phi_1^-\phi_2^- = q \tag{4.12}$$

where

$$q(\xi) = \frac{P_1(\xi)}{(\xi-i)(\xi+i)^2} \qquad (\xi \in \mathbb{R}) \tag{4.13}$$

for some polynomial P_1 with degree not greater than one.

For $q = 0$ it follows from (4.14) and (4.11) that $\phi^+ = \phi^- = 0$. Since we are looking for non-trivial solutions to (4.11) we consider only the case where q does not vanish identically (cf. Section 3).

The associated non-linear scalar equation (3.10) can be written in the form

$$\frac{\phi_1^+ + \rho\alpha_+\phi_2^+}{\phi_1^+ - \rho\alpha_+\phi_2^+} = -\alpha_+^{-1}\rho^{-1}\frac{\phi_1^-}{\phi_2^-}. \tag{4.14}$$

It is clear from (4.12) that the most convenient normalizing conditions to be satisfied by the solutions of (4.11), from which we obtain the two columns in G_+^{-1}, G_-, are

$$\phi_2^+(i) = 0, \qquad \phi_1^+(i) \neq 0 \tag{4.15}$$

and

$$\phi_1^+(2i) = 0, \qquad \phi_2^+(2i) \neq 0. \tag{4.16}$$

These, moreover, cannot be satisfied simultaneously since it follows from (4.12) that if (4.15) is satisfied then we must have $q(\xi) = K/(\xi+i)^2$, $K \neq 0$, which implies that $\phi_1^+(2i) \neq 0$.

First we look for solutions to (4.11) satisfying (4.15). Thus we have

$$\phi_2^+ = r\tilde{\phi}_2^+, \qquad \tilde{\phi}_2^+ \in L_2^+(\mathbb{R}); \tag{4.17}$$

condition (4.12) can be expressed in the form

$$\begin{cases} (\phi_1^+)^2 - \dfrac{(\xi-2i)(\xi-i)}{(\xi+i)^2}\alpha_+^2(\tilde{\phi}_2^+)^2 = -\dfrac{C_1\alpha_+}{(\xi+i)^2}, \\ \phi_1^-\phi_2^- = \dfrac{C_1\rho^{-1}}{(\xi-i)^2} \end{cases} \tag{4.18}$$

where C_1 is a constant and, regarding (4.14), we have

$$\frac{\phi_1^+ + \rho\alpha_+ r\tilde{\phi}_2^+}{\phi_1^- - \rho\alpha_+ r\tilde{\phi}_2^+} = -\rho^{-1}\alpha_+^{-1}\frac{\phi_1^-}{\phi_2^-}. \tag{4.19}$$

It follows from the first condition in (4.18) that neither the nominator nor the denominator on the left-hand side of (4.19) vanishes in the upper half-plane $\mathbb{C}^+$. Analogously, the second condition in (4.18) implies that neither ϕ_1^- nor ϕ_2^- vanishes in the lower half plane $\mathbb{C}^-$. Taking this into account we rewrite equation (4.19) as

$$q_-^{-1}\log\frac{\phi_1^+ + q_- r_+\alpha_+\tilde{\phi}_2^+}{\phi_1^+ - q_- r_+\alpha_+\tilde{\phi}_2^+} = q_-^{-1}\log(-\rho^{-1}\alpha_+^{-1}) + q_-^{-1}\log\frac{\phi_1^-}{\phi_2^-} \tag{4.20}$$

where

$$q_-(\xi) = (\xi-i)^{1/2}(\xi-2i)^{1/2}, \quad r_+(\xi) = (\xi+i)^{-1}. \tag{4.21}$$

Since

$$q_-^{-1}\log\frac{\phi_1^+ + q_- r_+\alpha_+\tilde{\phi}_2^+}{\phi_1^- - q_- r_+\alpha_+\tilde{\phi}_2^+} \in L_2^+(\mathbb{R}), \tag{4.22}$$

$$q_-^{-1}\log\frac{\phi_1^-}{\phi_2^-} \in L_2^-(\mathbb{R}), \tag{4.23}$$

(cf. [4]) we have

$$q_-^{-1}\log\frac{\phi_1^+ + q_- r_+\alpha_+\tilde{\phi}_2^+}{\phi_1^+ - q_- r_+\alpha_+\tilde{\phi}_2^+} = P^+ q_-^{-1}\log(-\rho^{-1}\alpha_+^{-1}), \tag{4.24}$$

$$q_-^{-1}\log\frac{\phi_1^-}{\phi_2^-} = -P^- q_-^{-1}\log(-\rho^{-1}\alpha_+^{-1}). \tag{4.25}$$

The last two equalities, together with (4.18), yield

$$(4.26)\quad \phi^+ = \begin{bmatrix} \phi_1^+ \\ \phi_2^+ \end{bmatrix} = K_1 r_+ \begin{bmatrix} \alpha_+^{1/2} \operatorname{ch}\left[\frac{1}{2} q_- P^+(q_-^{-1} \log(-\rho^{-1}\alpha_+^{-1}))\right] \\ \alpha_+^{-1/2}\rho^{-1} \operatorname{sh}\left[\frac{1}{2} q_- P^+(q_-^{-1} \log(-\rho^{-1}\alpha_+^{-1}))\right] \end{bmatrix}$$

$$(4.27)\quad \phi^- = \begin{bmatrix} \phi_1^- \\ \phi_2^- \end{bmatrix} = iK_1 r_- \begin{bmatrix} \rho^{-1/2} \exp\left[-\frac{1}{2} q_- P^-(q_-^{-1} \log(-\rho^{-1}\alpha_+^{-1}))\right] \\ \rho^{1/2} q_-^{-1} \exp\left[\frac{1}{2} q_- P^-(q_-^{-1} \log(-\rho^{-1}\alpha_+^{-1}))\right] \end{bmatrix}$$

with $r_-(\xi) = \xi - i$ $(\xi \in \mathbb{R})$.

We obtain analogously the solutions to (4.11) satisfying (4.16), from which we determine the second columns of G_+^{-1} and G.

The canonical generalized factorization for G which is obtained in this way is explicitly given in the following theorem.

Theorem 4.3. *With the assumptions of Theorem* 4.1, *a canonical generalized factorization for G is $G = G_- G_+$ with*

$$G_+^{-1} = \begin{bmatrix} \alpha_+^{1/2} \operatorname{ch}\left[\frac{1}{2} q_- P^+(q_-^{-1} \log(-\eta))\right] & i\alpha_+^{1/2}\rho \operatorname{sh}\left[\frac{1}{2} q_- P^+(q_-^{-1} \log \eta)\right] \\ \alpha_+^{-1/2}\rho^{-1} \operatorname{sh}\left[\frac{1}{2} q_- P^+(q_-^{-1} \log(-\eta))\right] & i\alpha_+^{-1/2} \operatorname{ch}\left[\frac{1}{2} q_- P^+(q_-^{-1} \log(\eta))\right] \end{bmatrix}$$

$$G_- = \begin{bmatrix} i\rho^{-1/2} \exp\left[-\frac{1}{2} q_- P^-(q_-^{-1} \log(-\eta))\right] & i\rho^{1/2} \exp\left[-\frac{1}{2} q_- P^-(q_-^{-1} \log \eta)\right] \\ i\rho^{-1/2} \exp\left[\frac{1}{2} q_- P^-(q_-^{-1} \log(-\eta))\right] & i\rho^{1/2} \exp\left[\frac{1}{2} q_- P^-(q_-^{-1} \log \eta)\right] \end{bmatrix}$$

for $q_-(\xi) = (\xi - 2i)^{1/2}(\xi - i)^{1/2}$ and $\eta = \rho^{-1}\alpha_+^{-1}$.

Proof. To see that $G = G_- G_+$ is indeed a canonical generalized factorization of G (relative to $L_2(\mathbb{R})$) it is enough to verify that and $GG_+^{-1} = G_-$ and

$$(4.28)\qquad r_+ G_+^{-1} \in L_2^+((\mathbb{R}))^{2\times 2}, \qquad r_- G_- \in (L_2^-(\mathbb{R}))^{2\times 2},$$

$$(4.29)\qquad \det\, G_+(2i) \neq 0$$

where $r_\pm(\xi) = (\xi \pm i)^{-1}$.

As to the first two conditions, they can be verified directly, taking into account that

$$q_- P^\pm(q_-^{-1} \log\, \eta) = \rho r_-^{-1} P^\pm(r_- \rho^{-1} \log(\rho^{-1}\alpha_+^{-1})).$$

Condition (4.29) is also satisfied. In fact the first column in $r_+G_+^{-1}$ satisfies (4.15) and therefore $\phi_1^+(2i) \neq 0$, while the second column in $r_+G_+^{-1}$, $[\phi_3^+\phi_4^+]^T$, is such that $\phi_3^+(2i) = 0$, $\phi_4^+(2i) \neq 0$. Thus det $G_+^{-1}(2i) = (2i)^2\phi_1^+(2i)\phi_4^+(2i) \neq 0$. □

Remark 4.4. The canonical generalized factorization of matrix functions G of the form

$$G = \begin{bmatrix} 1 & a \\ b & -ab \end{bmatrix},$$

where a, b admit a canonical bounded factorization, can be determined analogously, as long as we assume that either $a = a_-a_+$ with a_-^2 of the form (4.3) or $b = b_-b_+$ with b_+^2 of the same type.

The class of matrix functions which we considered in this section has also been studied, in order to obtain conditions for the existence of a canonical generalized factorization and formulas for its factors, in [1]. However, a comparison between the results obtained in that paper with those obtained here clearly shows the advantages of the method proposed in this paper in terms of simplicity of the method itself and of the resulting formulas for the factors.

5. The class $\mathcal{N}$

A class of symbols involving two rationally-independent scalar functions which appears in connection with some eigenvalue problems is the class of matrix functions of the form $G = \alpha I_2 + \beta N$ where I_2 is the identity matrix of order 2, α, β are in $C(\dot{\mathbb{R}})$ and N is a nilpotent rational matrix. Taking for N, for example, the matrix

$$N = \begin{bmatrix} 1 & r \\ -r^{-1} & -1 \end{bmatrix}, \qquad r(\xi) = \frac{\xi - i}{\xi + i}, \tag{5.1}$$

we get

$$G = \begin{bmatrix} \alpha + \beta & \beta r \\ -\beta r^{-1} & \alpha - \beta \end{bmatrix}. \tag{5.2}$$

It should be remarked that although there is some formal similarity between this class and the Daniele-Khrapkov class of matrix functions (cf. Section 6) they have different properties, namely, the conditions for existence of canonical factorization, as we shall see, are quite different in the two classes.

In fact, as far as the method of solution of the equation $G\phi^+ = \phi^-$ is concerned, there is a fundamental difference between the two classes; in the class $\mathcal{N}$ the functional $\tilde{F}$ is linear whereas in the Daniele class it is quadratic just as in the Rawlins-Williams class dealt with in Section 4.

Let us assume that α admits a bounded canonical factorization $\alpha = \alpha_-\alpha_+$ and, therefore ind $G = 0$ (since det $G = \alpha^2$).

In this case the homogeneous equation $G\phi^+ = \phi^-$ is equivalent to

$$(5.3)\qquad \begin{cases} (\alpha+\beta)\phi_1^+ + \beta r\phi_2^+ = \phi_1^- \\ -\beta\phi_1^+ + (\alpha-\beta)r\phi_2^+ = r\phi_2^- \end{cases}$$

where $(\phi_1^\pm, \phi_2^\pm) \in (L_2^\pm(\mathbb{R}))^2$.

It is easy to see that $\tilde{F}(x_1, x_2) = x_1 + x_2$ satisfies the assumptions of Theorem 3.1. Rewriting (5.3) in the form $G_1\phi^+ = G_2\phi^-$ with

$$(5.4)\qquad G_1 = \alpha_-^{-1}\begin{bmatrix} 1 & 0 \\ 0 & r \end{bmatrix} G, \qquad G_2 = \alpha_-^{-1}\begin{bmatrix} 1 & 0 \\ 0 & r \end{bmatrix}$$

and defining F as in (3.3), we have

$$(5.5)\qquad F(G_1\phi^+) = F(G_2\phi^-) \iff \alpha_+(\phi_1^+ + r\phi_2^+) = \alpha_-^{-1}(\phi_1^- + r\phi_2^-),$$

from which we derive

$$(5.6)\qquad \alpha_+(\phi_1^+ + r\phi_2^+) = \alpha_-^{-1}r(r^{-1}\phi_1^- + \phi_2^-) = \frac{K}{r_+},$$

K being a constant in $\mathbb{C}$.

On the other hand, the non-linear scalar equation (3.10) takes the form

$$(5.7)\qquad 1 + \frac{\beta}{\alpha} = \frac{r^{-1}\phi_1^-}{r^{-1}\phi_1^- + \phi_2^-} + \frac{r\phi_2^+}{\phi_1^+ + r\phi_2^+}.$$

By solving this equation subject to condition (5.6) we get the following.

Theorem 5.1. *Let* $G = \alpha I + \beta N$, *with* $\alpha, \beta \in C(\dot{\mathbb{R}})$, *be such that* ind (det G) $= 0$. *Then* G *possesses a canonical generalized factorization iff*

$$(5.8)\qquad \left(P^- \frac{1+\beta/\alpha}{r_-}\right)(-i) \neq 0$$

with $r_-(\xi) = \xi - i$. *If condition* (5.8) *is not satisfied, the solutions of* (3.1) *are of the form*

$$(5.9)\qquad \begin{cases} \phi_1^+ = -K\alpha_+^{-1}P^+ \dfrac{\beta/\alpha}{r_+}, \\ \phi_2^+ = K\alpha_+^{-1}r^{-1}P^+ \dfrac{1+\beta/\alpha}{r_+}, \\ \phi_1^- = Kr\alpha_-P^- \dfrac{1+\beta/\alpha}{r_-}, \\ \phi_2^- = -K\alpha_-P^- \dfrac{\beta/\alpha}{r_-}, \end{cases}$$

where K is a constant and

$$r_-(\xi) = \xi - i, \qquad r_+(\xi) = \xi + i \quad (\xi \in \mathbb{R}). \tag{5.10}$$

In this case the partial indices in the non-canonical generalized factorization of G are $k_1 = 1$, $k_2 = -1$.

Proof. G possesses a canonical generalized factorization iff the homogeneous equation $G\phi^+ = \phi^-$, $\phi^\pm \in (L_2^\pm(\mathbb{R}))^2$, admits only the solution $\phi^+ = \phi^- = 0$. Now, if (ϕ^+, ϕ^-) satisfies (3.1), it also satisfies (5.6). If $K = 0$ then, from

$$\phi_1^+ + r\phi_2^+ = 0 = r^{-1}\phi_1^- + \phi_2^- \tag{5.11}$$

it follows that (ϕ^+, ϕ^-) is a solution to (3.1) iff $\phi^+ = \phi^- = 0$. Therefore a non-trivial solution of the homogeneous equation exists only if (5.6) holds with $K \neq 0$. In that case we have, from Theorem 3.1, that (ϕ^+, ϕ^-) satisfies $G\phi^+ = \phi^-$ iff it is a solution to equation (5.7).

Taking (5.6) into account we see that neither of the denominators on the right-hand side of (5.7) vanishes and therefore this equation can be easily solved. We have, from (5.6) and (5.7),

$$K(1 + \beta/\alpha) = r_+\alpha_-^{-1}\phi_1^- + r_-\alpha_+\phi_2^+ \tag{5.12}$$

and therefore

$$K(P^- \frac{1 + \beta/\alpha}{r_-}) = r^{-1}\alpha_-^{-1}\phi_1^-. \tag{5.13}$$

$$K(P^+ \frac{1 + \beta/\alpha}{r_+}) = r\alpha_+\phi_2^+. \tag{5.14}$$

Due to the presence of the factors r^{-1} and r on the right-hand sides of (5.13) and (5.14), respectively, non-trivial solutions of (5.7) (satisfying (5.6) with $K \neq 0$) exist iff

$$\left(P^- \frac{1 + \beta/\alpha}{r_-}\right)(-i) = 0 = \left(P^+ \frac{1 + \beta/\alpha}{r_+}\right)(i). \tag{5.15}$$

It can be seen, however, that

$$\left(P^+ \frac{1 + \beta/\alpha}{r_+}\right)(i) = \frac{1}{2\pi i}\int_{\mathbb{R}} \frac{(1 + \beta/\alpha)(t)}{t^2 + 1}\,dt = \left(P^- \frac{1 + \beta/\alpha}{r_-}\right)(-i). \tag{5.16}$$

Therefore we conclude that G possesses a canonical generalized factorization unless

$$\left(P^- \frac{1 + \beta/\alpha}{r_-}\right)(-i) = 0, \tag{5.17}$$

and in this case we have, from (5.13), (5.14) and (5.6):

$$\phi_1^- = Kr\alpha_- P^- \frac{1+\beta/\alpha}{r_-},$$
$$\phi_2^+ = Kr^{-1}\alpha_+^{-1}P^+ \frac{1+\beta/\alpha}{r_+},$$
$$\phi_2^- = \left(\frac{K\alpha_-}{r_+} - \phi_1^-\right) r^{-1} = -K\alpha_- P^- \frac{\beta/\alpha}{r_-},$$
$$\phi_1^+ = \frac{K\alpha_+^{-1}}{r_+} - r\phi_2^+ = -K\alpha_+^{-1}P^+ \frac{\beta/\alpha}{r_+}.$$

Thus we also see that if (5.17) holds then dim (Ker $(P^+GP^+)) = 1$ and, since the total index k of G is zero, we have the partial indices $k_1 = 1$, $k_2 = -1$. □

If condition (5.8) is satisfied, a canonical generalized factorization for G can be determined, as mentioned in Section 4, by solving the equation

$$r^{-1}G\phi^+ = \phi^-, \qquad \phi^\pm \in (L_2^\pm(\mathbb{R}))^2, \tag{5.18}$$

subject to convenient normalizing conditions. The procedure is entirely analogous to that used in the proof of the previous theorem in order to solve (3.1).

Equation (5.18) is equivalent to

$$\tilde{G}_1\phi^+ = \tilde{G}_2\phi^- \tag{5.19}$$

where $\tilde{G}_1 = r^{-1}G_1$ and $\tilde{G}_2 = G_2$ (see (5.4)). For $\tilde{F}$ and F defined as above, we get

$$F(\tilde{G}_1\phi^+) = F(\tilde{G}_2\phi^-) \iff r^{-1}\alpha_+(\phi_1^+ + r\phi_2^+) = \alpha_-^{-1}(\phi_1^- + r\phi_2^-) \tag{5.20}$$

It follows that

$$\alpha_+(\phi_1^+ + r\phi_2^+) = r^2\alpha_-^{-1}(r^{-1}\phi_1^- + \phi_2^-) = \frac{p_1}{r_+^2} \tag{5.21}$$

where p_1 is a polynomial of the form

$$p_1(\xi) = K_1\xi + K_2 \quad (K_1, K_2 \in \mathbb{C}). \tag{5.22}$$

Since $p_1 = 0$ yields, as in the proof of Theorem 5.1, $\phi^+ = \phi^- = 0$, the non-trivial solutions of (5.18) according to Theorem 3.1 are the solutions of the non-linear scalar equation (3.10) or, equivalently (5.7).

It should be remarked that the same non-linear scalar equation of the form (3.10) is associated to both homogeneous equations (3.1) and (5.18), the difference being established by different conditions ((5.6) and (5.21), respectively) which must be imposed on the solutions.

From (5.21) we see that, by imposing certain zeros to p_1, we can simplify this condition and, at the same time, guarantee that the two denominators on the right-hand side of (5.7) do not vanish. This can be done through a convenient choice of normalizing conditions for the factors G_+ and G_-.

In fact, imposing

$$\phi_2^-(-i) = 0\,, \quad \phi_1^-(-i) \neq 0 \tag{5.23}$$

we see that we must have $p_1(-i) = 0$ and therefore condition (5.21) takes the form

$$\alpha_+(\phi_1^+ + r\phi_2^+) = \frac{K_1}{r_+}, \qquad \alpha_-^{-1}(\phi_1^- + \tilde{\phi}_2^-) = \frac{K_1}{r_-} \tag{5.24}$$

with

$$\tilde{\phi}_2^- = r\phi_2^- \quad \text{and} \quad K_1 \neq 0.$$

In this case (5.7) becomes, in terms of $\phi_1^\pm$, ϕ_1^- and $\tilde{\phi}_2^-$,

$$1 + \frac{\beta}{\alpha} = \frac{\phi_1^-}{\phi_1^- + \tilde{\phi}_2^-} + \frac{r\phi_2^+}{\phi_1^+ + r\phi_2^+}. \tag{5.25}$$

If, alternatively, we impose

$$\phi_1^+(i) = 0\,, \quad \phi_2^+(i) \neq 0, \tag{5.26}$$

we must have $p_1(i) = 0$, in which case we have, for (5.21),

$$\alpha_+\ (\tilde{\phi}_1^+ + \phi_2^+) = \frac{K_1}{r_+}, \qquad \alpha_-^{-1}(r^{-1}\phi_1^- + \phi_2^-) = \frac{K_1}{r_-} \tag{5.27}$$

where

$$\tilde{\phi}_1^+ = r^{-1}\phi_1^+ \text{ and } K_1 \neq 0.$$

In this case (5.7) takes the form

$$1 + \frac{\beta}{\alpha} = \frac{r^{-1}\phi_1^-}{r^{-1}\phi_1^- + \phi_2^-} + \frac{\phi_2^+}{\tilde{\phi}_1^+ + \phi_2^+}. \tag{5.28}$$

Both equations (5.25) and (5.28) can be solved as in the proof of Theorem 5.1.

Now, if (ϕ_1^+, ϕ_2^+), $(\phi_1^-, \tilde{\phi}_2^-)$ is a solution of (5.25) satisfying (5.24) and $(\tilde{\phi}_3^+, \phi_4^+)$, (ϕ_3^-, ϕ_4^-) is a solution of (5.28) satisfying (5.27), we see that, for $\phi_2^- = r^{-1}\phi_2^-$ and $\phi_3^+ = r\phi_3^+$, the following holds:

(i) (ϕ_1^+, ϕ_2^+), (ϕ_1^-, ϕ_2^-) is a solution of (5.18) satisfying (5.23);

(ii) (ϕ_3^+, ϕ_4^+), (ϕ_3^-, ϕ_4^-) is a solution of (5.18) satisfying (5.26) (with $\phi_3^\pm, \phi_4^\pm$ corresponding to $\phi_1^\pm, \phi_2^\pm$, respectively);

(iii) $\phi_1^+(i) \neq 0$ (and, similarly, $\phi_4^-(-i) \neq 0$), since condition (5.24) excludes this possibility.

Thus, defining

$$\Phi^- = \begin{bmatrix} \phi_1^- & \phi_3^- \\ \phi_2^- & \phi_4^- \end{bmatrix}, \qquad \Phi^+ = \begin{bmatrix} \phi_1^+ & \phi_3^+ \\ \phi_2^+ & \phi_4^+ \end{bmatrix}$$

we have (see (3.14)) $G\Phi^+ = r\Phi^-$ where, as a consequence of (iii), $\Phi^+(i)$ is invertible. Therefore $G = G_-G_+$, with $G_\pm$ as in (3.15), is a canonical generalized factorization for G.

This factorization is explicitly given as follows.

$$G_+^{-1} = \begin{bmatrix} \alpha_+^{-1} - \alpha_+^{-1}r_-^{-1}P^+(\frac{\beta}{\alpha}r_-) & -r_-^{-1}\alpha_+^{-1}P^+(\frac{\beta}{\alpha}r_+) \\ \alpha_+^{-1}r_+^{-1}P^+(\frac{\beta}{\alpha}r_-) & \alpha_+^{-1} + \alpha_+^{-1}r_+^{-1}P^+(\frac{\beta}{\alpha}r_+) \end{bmatrix},$$

$$G_- = \begin{bmatrix} \alpha_- + \alpha_-r_-^{-1}P^-(\frac{\beta}{\alpha}r_-) & r_-^{-1}\alpha_-P^-(\frac{\beta}{\alpha}r_+) \\ -r_+^{-1}\alpha_-P^-(\frac{\beta}{\alpha}r_-) & \alpha_- - \alpha_-r_+^{-1}P^-(\frac{\beta}{\alpha}r_+) \end{bmatrix}$$

with $r_\pm(\xi) = (\xi \pm i)^{-1}$.

6. Some other classes of interest

Another class of 2×2 symbols involving two rationally-independent scalar functions, which is particularly interesting from the point of view of applications, is the so-called Daniele-Khrapkov class $\mathcal{D}$. It consists of all 2×2 matrix-valued functions of the form $G = \alpha I_2 + \beta R$ where α, β are scalar functions (in $\mathcal{C}(\dot{\mathbb{R}})$) and R is a rational 2×2 matrix function such that

(6.1)
$$R^2 = qI_2,$$

where I_2 is the identity matrix of order 2 as before.

If we take R in the form

(6.2)
$$R = \begin{bmatrix} 0 & 1 \\ q & 0 \end{bmatrix}, \qquad q^{\pm 1} \in \mathcal{R}(\dot{\mathbb{R}}),$$

then we have

(6.3)
$$G = \begin{bmatrix} \alpha & \beta \\ \beta q & \alpha \end{bmatrix}.$$

The problem of determining the generalized factorization of such a matrix function has been addressed in a number of papers (cf. [2], [7], [8]), regarding the case where q is a quotient of two first degree polynomials. The more difficult case where q is a quotient of two second degree polynomials with simple zeros has been studied by a method which essentially corresponds to the one presented in Section 3 (cf. [3], [4]). Necessary and sufficient conditions for existence of a canonical generalized factorization have been established and the factors in this factorization, as well as the partial indices in the case of non-canonical factorization, have been determined explicitly.

Both classes $\mathcal{D}$ and $\mathcal{N}$ can be considered as subclasses of a larger class defined as follows.

Let R and N be 2×2 rational matrices such that R is of the form (6.2) and

$$N = \begin{bmatrix} 1 & \rho \\ -\rho^{-1} & -1 \end{bmatrix}, \qquad \rho^2 = q. \tag{6.4}$$

We denote by $\mathcal{D} - \mathcal{N}$ the class of the all 2×2 matrices

$$G = \alpha I_2 + \beta R + \gamma N \tag{6.5}$$

where α, β, γ are scalar functions. This is, therefore, a class of 2×2 symbols involving three rationally-independent entries.

The generalized factorization of matrices in this class can still be studied using the method presented in Section 3, at least in the case where $\alpha, \beta, \gamma \in \mathcal{C}(\dot{\mathbb{R}})$ and $\rho \in \mathcal{R}(\dot{\mathbb{R}})$.

Let, for instance, $\rho(\xi) = r^{-1}(\xi) = (\xi + i)/(\xi - i)$ for $\xi \in \mathbb{R}$. In this case det $G = (\alpha r^{-1} + \beta)(\alpha r^{-1} - \beta) r^2$ and we define $\delta = \alpha r^{-1} + \beta$, $\tilde{\delta} = (\alpha r^{-1} - \beta) r^2$, so that det $G = \delta\, \tilde{\delta}$.

Then we have the following.

Theorem 6.1. *Let $G \in \mathcal{D} - \mathcal{N}$ with $\alpha, \beta, \gamma \in \mathcal{C}(\dot{\mathbb{R}})$ and $\rho = r^{-1}$ be such that δ and $\tilde{\delta}$ have a canonical bounded factorization. Then G possesses a canonical generalized factorization $G = G_- G_+$ where*

$$G_+^{-1} = \begin{bmatrix} \tilde{\delta}_+^{-1} & r_+^{-1} \tilde{\delta}_+^{-1} P^+[(\alpha - \gamma)\delta_+^{-1} \tilde{\delta}_-^{-1} r_+] \\ -r \tilde{\delta}_+^{-1} & \delta_+^{-1} - r_-^{-1} \tilde{\delta}_+^{-1} P^+[(\alpha - \gamma)\delta_+^{-1/2} \tilde{\delta}_-^{-1} r_+] \end{bmatrix},$$

$$G_- = \begin{bmatrix} r^{-1} \tilde{\delta}_- & \delta_- - r^{-1} r_+^{-1} \tilde{\delta}_- P^-[(\alpha - \gamma)\delta_+^{-1/2} \tilde{\delta}_-^{-1} r_+] \\ -\tilde{\delta}_- & r_+^{-1} \tilde{\delta}_- P^-[(\alpha - \gamma)\delta_+^{-1/2} \tilde{\delta}_-^{-1} r_+] \end{bmatrix}$$

with $r_\pm(\xi) = (\xi \pm i)^{-1}$ and $\delta = \delta_- \delta_+$, $\tilde{\delta} = \tilde{\delta}_- \tilde{\delta}_+$.

References

[1] BASTOS, M. A., DOS SANTOS, A. F.: Generalized factorization for a class of 2 × 2 matrix functions with rationally-independent entries; Complex Variables Theory Appl. 22 (1993), 153–174.

[2] CÂMARA, M. C.: Factorization in a Banach algebra and the Gelfand transform; Math. Nachr. 176 (1995), 17–37.

[3] CÂMARA, M. C., DOS SANTOS, A. F., BASTOS, A.: Generalized factorization for Daniele-Khrapkov matrix functions-Partial indices; J. Math. Anal. Appl. 190 (1995), 142–164.

[4] CÂMARA, M. C., DOS SANTOS, A. F., BASTOS, A.: Generalized factorization for Daniele Khrapkov matrix functions-Explicit formulas; J. Math. Anal. Appl. 190 (1995), 295–328.

[5] DANIELE, V. G.: On the solution of two coupled Wiener-Hopf equations; SIAM J. Appl. Math. 44 (1984), 667–680.

[6] HURD, R. A.: The Wiener-Hopf-Hilbert method for diffraction problems; Canad. J. Phys. 54 (1976), 775–780.

[7] LEBRE, A., DOS SANTOS, A. F.: Generalized factorization for a class of non-rational 2 × 2 matrix functions; Integral Equations Operator Theory 13 (1990), 671–700.

[8] PRÖSSDORF, S., SPECK, F.-O.: A factorization procedure for two by two matrix functions on the circle with two rationally independent entries; Proc. Roy. Soc. Edinburgh Sect. A 115 (1990), 119–138.

Departamento de Matemática
Instituto Superior Técnico
Av. Rovisco Pais
1096 Lisboa Codex
Portugal
d394@beta.ist.utl.pt

1991 Mathematics Subject Classification: 45E10, 47B35

Submitted: May 31, 1996

Operator Theory:
Advances and Applications, Vol. 102

Completeness of scattering systems with obstacles of finite capacity

J. VAN CASTEREN and M. DEMUTH

Let K_0 be a free Feller operator and let K_Σ be the corresponding operator with Dirichlet boundary conditions on $\Gamma = \mathbb{R}^n \setminus \Sigma$. The scattering system established by $\{K_0, K_\Sigma\}$ is complete, i.e., the wave operators exist and are complete if the singularity region Γ has finite capacity. One consequence is the stability of the absolutely continuous spectra of K_0 and K_Σ, respectively. Such sets can be unbounded, which yields a non-local freedom of perturbation in the scattering theory. Kato-Feller potentials can be included.

1. Assumptions and results

A free Feller operator is a generator K_0 of a semigroup given via a transition density function satisfying the basic assumptions of stochastic spectral analysis (see Demuth, van Casteren [7], [8]). They are called BASSA and repeated here.

Let $p_0(t,x,y)$ be a continuous function $(t,x,y) \to p_0(t,x,y)$ on $(0,\infty) \times \mathbb{R}^n \times \mathbb{R}^n$ satisfying the following assumptions:

A1: p_0 is non-negative and verifies the Chapman-Kolmogorov equation

$$\int_{\mathbb{R}^n} p_0(s,x,u)p_0(t,u,y)du = p_0(s+t,x,y)$$

for $s,t > 0$; $x,y \in \mathbb{R}^n$, where du is the Lebesgue measure; and its total mass is

$$\int_{\mathbb{R}^n} p_0(t,x,y)dy = 1$$

for $t > 0$, $x \in \mathbb{R}^n$.

A2: The function p_0 is symmetric, i.e.,

$$p_0(t,x,y) = p_0(t,y,x)$$

for $t > 0$; $x,y \in \mathbb{R}^n$.

A3: Let $f \in C_\infty(\mathbb{R}^n)$, the set of continuous functions vanishing at infinity. Then it is assumed that

$$\lim_{t\to 0} \int_{\mathbb{R}^n} f(y)p_0(t,x,y)dy = f(x).$$

A4: Let again $f \in C_\infty(\mathbb{R}^n)$. Then the function p_0 has to have the Feller property, i.e.,

$$x \to \int_{\mathbb{R}^n} f(y) p_0(t,x,y) dy,$$

belongs to $C_\infty(\mathbb{R}^n)$.

Under these conditions p_0 is a kernel of a semigroup, called Feller semigroup. Its generator is denoted by K_0 and one has

$$e^{-tK_0} C_\infty(\mathbb{R}^n) \subset C_\infty(\mathbb{R}^n).$$

K_0 is called the free Feller operator. It is the L^2-generator of a Markov process

$$((\Omega, \mathcal{F}, P_x), X(t), (\mathbb{R}^n, \mathcal{B}^n)).$$

For any $f \in L^2$ we have

$$(1.1) \qquad (e^{-tK_0} f)(x) = \int_{\mathbb{R}^n} p_0(t,x,y) f(y) dy$$

$$(1.2) \qquad \&= \quad E_x\{f(X(t))\}.$$

A series of examples for K_0 are given in [7] or [8]. The simplest case is the Laplacian. But also pseudo-differential operators or relativistic Hamiltonians are included.

Perturbations by obstacles can be introduced stochastically. Let Γ be a closed set in $\mathbb{R}^n$. Its first hitting time is defined as

$$(1.3) \qquad T_\Gamma = \inf_s \{s > 0, X(s) \in \Gamma\}.$$

We set $\Sigma = \mathbb{R}^n \setminus \Gamma$ and introduce a family of operators by

$$(1.4) \qquad (U(t)f)(x) := E_x\{f(X(t)), T_\Gamma > t\}.$$

Then $U(t) \mid_{L^2(\Sigma)}$ forms a strongly continuous semigroup on $L^2(\Sigma)$. Its generator, denoted by K_Σ, corresponds to K_0 with Dirichlet boundary condition on $\partial\Gamma$. K_Σ is a self-adjoint operator in $L^2(\Sigma)$.

In order to study the scattering between K_0 and K_Σ one has to consider the two space wave operators

$$(1.5) \qquad \Omega_\pm(K_\Sigma, J, K_0) := \operatorname*{s-lim}_{t\to\pm\infty} e^{itK_\Sigma} J e^{-itK_0} P_{ac}(K_0)$$

and

$$(1.6) \qquad \Omega_\pm(K_0, J^*, K_\Sigma) := \operatorname*{s-lim}_{t\to\pm\infty} e^{itK_0} J^* e^{-itK_\Sigma} P_{ac}(K_\Sigma).$$

$P_{ac}(\cdot)$ denotes the projection operator onto the absolutely continuous subspace. The identification operator J is given by

(1.7) $$Jf := f{\restriction}_{\Sigma}, \quad f \in L^2(\mathbb{R}^n).$$

One has

(1.8) $$J^*J = 1 - 1_\Gamma = 1_\Sigma$$

where 1_Σ is the projection operator in $L^2(\mathbb{R}^n)$ given by the indicator function on Σ, and

(1.9) $$JJ^* = 1_{L^2(\Sigma)},$$

the identity in $L^2(\Sigma)$. Because of (1.8) and (1.9) the scattering system $\{K_0, K_\Sigma\}$ is complete if

(1.10) $$\operatorname*{s-lim}_{t\to\pm\infty} 1_\Gamma e^{-itK_0} P_{ac}(K_0) = 0.$$

This entails us with the stability of the absolutely continuous spectra,

(1.11) $$\sigma_{ac}(K_0) = \sigma_{ac}(K_\Sigma).$$

If $K_0 = -\Delta$ and if Γ is compact it was shown by Arsenev [1] or again by Deift, Simon [5] that the scattering system is complete. More general K_0 were studied by Demuth [6]. For generators of diffusion semigroups Stollmann [10] has considered also more general Γ, but he could not include Γ of finite capacity.

Now we define the capacity of a set Γ by

(1.12) $$\operatorname{cap}(\Gamma) := \inf\{(K_0^{\frac12} f, K_0^{\frac12} f) + (f,f), f \in \operatorname{dom} K_0^{\frac12}, f \geq 1_U, U \text{ open}, \Gamma \subset U\}.$$

The 1-equilibrium potential of Γ, defined as

$$E_x\{e^{-T_\Gamma}, T_\Gamma < \infty\},$$

is the unique minimizing element of (1.12) and the following inequality (see e.g. Fukushima [9], Chapter 3) is valid:

(1.13) $$\|E_\bullet\{e^{-T_\Gamma}, T_\Gamma < \infty\}\|_{L^1} \leq \operatorname{cap}(\Gamma).$$

Often the Newton capacity is defined via the carré du champ (or squared gradient operator) Γ_1, which is given by

$$[\Gamma_1(f,g)](x) = \operatorname*{s-lim}_{s\to 0} \frac{1}{2s} E_x\{[f(X(s)) - f(x)][g(X(s)) - g(x)]\}$$

for all f, g as long as this limit makes sense. If the process is a diffusion with infinite lifetime we have for $f \in \operatorname{dom} K_0^{\frac12}$

$$\int_{\mathbb{R}^n} [\Gamma_1(f,f)](x)dx = \int_{\mathbb{R}^n} |(K_0^{\frac12} f)(x)|^2 dx$$

(see e.g. Bakry [2]).

Now we are able to formulate the results in this note.

Theorem 1.1.

a) *Let K_0 be a self-adjoint free Feller operator in $L^2(\mathbb{R}^n)$ given by $p_0(t,x,y)$. Assume that p_0 satisfies BASSA. Moreover we assume that e^{-tK_0} is $L^1 - L^\infty$ smoothing, i.e.,*

$$\sup_{x,y\in\mathbb{R}^n} p_0(t,x,y) < \infty. \tag{1.14}$$

Let Γ be a closed set in $\mathbb{R}^n$ and let $\Sigma = \mathbb{R}^n \setminus \Gamma$ be the complement of Γ. Denote by K_Σ the Friedrichs extension of $K_0 \restriction (\text{dom } K_0 \cap L^2(\Sigma))$, i.e., K_Σ is the operator corresponding to K_0 with Dirichlet boundary condition on $\partial\Gamma$.
If Γ has finite capacity then the wave operators $\Omega_\pm(K_\Sigma, J, K_0)$ and $\Omega_\pm(K_0, J^*, K_\Sigma)$ exist and are complete, implying

$$\sigma_{ac}(K_0) = \sigma_{ac}(K_\Sigma). \tag{1.15}$$

b) *Assume additionally a Kato-Feller potential $V : \mathbb{R}^n \to \mathbb{R}$, i.e.,*

$$\lim_{\tau\to 0} \sup_x \int_0^\tau ds \int_{\mathbb{R}^n} dy \; p_0(s,x,y) \mid V(y) \mid \; = 0. \tag{1.16}$$

Then $K_0 \dot{+} V$ is a well-defined self-adjoint operator in $L^2(\mathbb{R}^n)$. Correspondingly we denote by $(K_0 \dot{+} V)_\Sigma$ the Friedrichs extension of $K_0 + V \restriction [\text{dom } (K_0+V) \cap L^2(\Sigma)]$. Corresponding to (1.13) we assume in this perturbed case

$$\int_{\mathbb{R}^n} dx E_x \left\{ e^{-\int_0^{T_\Gamma} [a+V(X(u))]du}, \; T_\Gamma < \infty \right\} < \infty, \tag{1.17}$$

where $a > 0$ is chosen large enough.
Then the scattering system $\{(K_0 \dot{+} V)_\Sigma, K_0 \dot{+} V\}$ is complete. Among others this implies

$$\sigma_{ac}(K_0 \dot{+} V) = \sigma_{ac}((K_0 \dot{+} V)_\Sigma). \tag{1.18}$$

If in addition the potential V is in $L^1(\mathbb{R}^n)$ the scattering systems $\{K_0 \dot{+} V, K_0\}$ and $\{(K_0 \dot{+} V)_\Sigma, K_\Sigma\}$ are complete, implying the stability of the absolutely continuous spectra:

$$\begin{aligned} \sigma_{ac}(K_0) &= \sigma_{ac}(K_0 \dot{+} V) \\ &= \sigma_{ac}((K_0 \dot{+} V)_\Sigma) \\ &= \sigma_{ac}(K_\Sigma). \end{aligned} \tag{1.19}$$

Remark 1.2. Standard examples of sets of finite capacity are unions of balls with decreasing radii. Let $K_0 = -\Delta$, the Laplacian in $L^2(\mathbb{R}^n)$. Let $\Gamma = \bigcup B_m$, where

B_m are balls of radius r_m. Take $r_m < 1$. Then for $n \geq 5$ the capacity of Γ can be estimated by

$$\text{cap } (\Gamma) \leq c \sum_m r_m^{n-2}.$$

In particular, if $n \geq 5$ there are Γ of finite capacity consisting of balls B_m for which

$$\sum_m r_m = \infty.$$

That means we have included star-shaped or hedgehog-shaped regions with finitely many unbounded peaks.

Moreover the result shows that the behaviour of Kato-Feller potentials on sets of finite capacity is irrelevant for the scattering. The existence of $\Omega_+(K_0 \dot{+} V, K_0)$ implies the existence of $\Omega_+((K_0 \dot{+} V)_\Sigma, J, K_0)$ where $\Sigma = \mathbb{R}^n \setminus \Gamma$ and if Γ has finite capacity. It is already known that in scattering the results are not affected by the local behaviour of the perturbation. However here we have a non-local freedom of the potentials because sets of finite capacity can be unbounded.

2. Proof of the results

Denote by $\mathcal{L}_1$, $\mathcal{L}_2$, $\mathcal{L}_\infty$ the set of trace class, Hilbert-Schmidt, and compact operators, respectively. The proof of Theorem 1.1 is based on an abstract criterion for the completeness of two-space scattering systems. This criterion has its own interest, because it is more general than a trace class condition for differences of semigroups or resolvents.

Lemma 2.1. *Let K_1, K_2 be self-adjoint semibounded operators in different Hilbert spaces $\mathcal{H}_1$ and $\mathcal{H}_2$. Let J be a bounded identification operator between $\mathcal{H}_1$ and $\mathcal{H}_2$. Then the wave operators*

$$\Omega_\pm(K_2, J, K_1) := \operatorname*{s-lim}_{t \to \pm\infty} e^{itK_2} J e^{-itK_1} P_{ac}(K_1), \tag{2.1}$$

$$\Omega_\pm(K_1, J^*, K_2) := \operatorname*{s-lim}_{t \to \pm\infty} e^{itK_1} J^* e^{-itK_2} P_{ac}(K_2) \tag{2.2}$$

exist if

$$e^{-K_2}(e^{-K_2} J - J e^{-K_1}) e^{-K_1} \in \mathcal{L}_1(\mathcal{H}_1, \mathcal{H}_2) \tag{2.3}$$

and

$$e^{-K_2} J - J e^{-K_1} \in \mathcal{L}_\infty(\mathcal{H}_1, \mathcal{H}_2). \tag{2.4}$$

The wave operators are complete, if additionally

$$(J^* J - 1_{\mathcal{H}_1}) e^{-K_1} \in \mathcal{L}_\infty(\mathcal{H}_1, \mathcal{H}_1) \tag{2.5}$$

and

$$(J J^* - 1_{\mathcal{H}_2}) e^{-K_2} \in \mathcal{L}_\infty(\mathcal{H}_2, \mathcal{H}_2). \tag{2.6}$$

Proof. Because of (2.3) (compare also [3], p. 347) the strong limits

$$\underset{t\to\pm\infty}{\text{s-lim}}\ e^{it(e^{-K_2})}e^{-K_2}Je^{-K_1}e^{-it(e^{-K_1})}P_{ac}(e^{-K_1})$$

exist, implying the existence of

$$\underset{t\to\pm\infty}{\text{s-lim}}\ e^{it(e^{-K_2})}e^{-K_2}Je^{-it(e^{-K_1})}P_{ac}(e^{-K_1}).$$

From (2.4) the existence of the limits

$$\underset{t\to\pm\infty}{\text{s-lim}}\ e^{it(e^{-K_2})}Je^{-K_1}e^{-it(e^{-K_1})}P_{ac}(e^{-K_1})$$

then follows.
Again a density argument and the invariance principle yield the existence of $\Omega_\pm(K_2, J, K_1)$. The same is true for $\Omega_\pm(K_1, J^*, K_2)$.
The wave operator $\Omega_+(K_2, J, K_1)$ is called complete if

$$\underset{t\to\infty}{\text{s-lim}}(J^*J - 1_{\mathcal{H}_1})e^{-itK_1}P_{ac}(K_1) = 0.$$

The Riemann-Lebesgue lemma in conjunction with (2.5) suffices to conclude this. Condition (2.6) provides the completeness of $\Omega_\pm(K_1, J^*, K_2)$. □

Proof (of Theorem 1.1). a) Now we are able to prove the main theorem. One has to show that cap $(\Gamma) < \infty$ is sufficient for the conditions in (2.3)–(2.6). In our situation $JJ^* - 1_{L^2(\Sigma)} = 0$ and hence (2.6) holds trivially. The condition in (2.5) will follow from

(2.7) $$1_\Gamma e^{-K_0} \in \mathcal{L}_2(L^2(\mathbb{R}^n), L^2(\mathbb{R}^n)).$$

But $1_\Gamma e^{-K_0}$ is a Hilbert-Schmidt operator if the Lebesgue measure of Γ, denoted by meas Γ, is finite.

(2.8) $$\int_\Gamma dx \int_{\mathbb{R}^n} dy\ p_0(1,x,y)^2 \le \text{meas}\,\Gamma \cdot \sup_{x\in\mathbb{R}^n} p_0(2,x,x).$$

Finally notice that finite capacity implies finite measure.
In order to show (2.4) we prove

(2.9) $$e^{-K_\Sigma}J - Je^{-K_0} \in \mathcal{L}_2(L^2(\mathbb{R}^n), L^2(\mathbb{R}^n))$$

if

(2.10) $$\int_{\mathbb{R}^n} dx\ [E_x(T_\Gamma < 1)]^2 < \infty.$$

This follows from the stochastic representation of the kernel $(e^{-K_\Sigma})(x,y)$. Let $E_x^{y,t}(\cdot)$ be the measure which pins the motion $\{X(t), t \ge 0\}$ at x at time 0 and at y at time t. It is given by

$$E_x^{y,t}(A) = E_x\{p_0(t-s, X(s), y)1_A\}$$

whenever A is an event in the field $\mathcal{F}_s$, $s < t$. For any $\lambda > 0$ the kernel of the semigroup $e^{-\lambda K_\Sigma}$ is given as

$$(e^{-\lambda K_\Sigma})(x,y) = E_x^{y,\lambda}\{T_\Gamma \geq \lambda\}.$$

Hence

$$(e^{-\lambda K_0} - J^* e^{-\lambda K_\Sigma} J)(x,y) = E_x^{y,\lambda}\{T_\Gamma < \lambda\}.$$

This kernel is symmetric in x and y. On account of

$$\begin{aligned}&\|e^{-2K_0} - J^* e^{-2K_\Sigma} J\|_{HS}\\ &\leq \|e^{-K_0}(e^{-K_0} - J^* e^{-K_\Sigma} J)\|_{HS} + \|(e^{-K_0} - J^* e^{-K_\Sigma} J) J^* e^{-K_\Sigma} J\|_{HS},\end{aligned}$$

it suffices to estimate the integrals

$$\int_{\mathbb{R}^n} dx \int_{\mathbb{R}^n} dy \left| \int_{\mathbb{R}^n} du\ p_0(1,x,u)\ E_u^{y,1}\{T_\Gamma < 1\} \right|^2$$

and

$$\int_{\mathbb{R}^n} dx \int_{\mathbb{R}^n} dy \left| \int_{\mathbb{R}^n} du\ E_x^{u,1}\{T_\Gamma < 1\}\ E_u^{y,1}\{T_\Gamma \geq 1\} \right|^2 .$$

But both expressions are smaller than

$$\sup_{x,y \in \mathbb{R}^n} p_0(2,x,y) \int_{\mathbb{R}^n} dx\ (E_x\{T_\Gamma < 1\})^2,$$

and hence (2.10) yields (2.9).
It remains to verify (2.3). For showing

$$e^{-K_\Sigma}(e^{-K_\Sigma} J - J e^{-K_0}) e^{-K_0} \in \mathcal{L}_1(L^2(\mathbb{R}^n), L^2(\Sigma)) \tag{2.11}$$

we use a trace class theorem developed in [4]. This is a general trace class criterion for products of integral operators. Assume A and B to be integral operators with measurable kernels $A(\cdot,\cdot)$, $B(\cdot,\cdot) : \mathbb{R}^n \times \mathbb{R}^n \to \mathbb{C}$. The product AB is a trace class operator if $A(\cdot,x)$ and $B(x,\cdot)$ are in $L^2(\mathbb{R}^n, dy)$ and if

$$\int_{\mathbb{R}^n} dx \left(\int_{\mathbb{R}^n} dy \mid A(y,x) \mid^2 \right)^{\frac{1}{2}} \left(\int_{\mathbb{R}^n} dy \mid B(x,y) \mid^2 \right)^{\frac{1}{2}} = K < \infty. \tag{2.12}$$

In that case the trace norm of AB is smaller than K. For the special integral operators in (2.11) we have

$$\left(\int_{\mathbb{R}^n} dy \mid (e^{-K_\Sigma})(y,x) \mid^2 \right)^{\frac{1}{2}} \leq \sup_{x,y} p_0(2,x,y)^{\frac{1}{2}}.$$

Hence the condition in (2.11) is satisfied if

$$\int_{\mathbb{R}^n} dx \left[\int_{\mathbb{R}^n} dy \left| \int_{\mathbb{R}^n} du\ E_x^{u,1}\{T_\Gamma < 1\}\ p_0(1,u,y) \right|^2 \right]^{\frac{1}{2}} \tag{2.13}$$

$$\leq \sup_{x,y\in\mathbb{R}^n} p_0(2,x,y)^{\frac{1}{2}} \int_{\mathbb{R}^n} dx\ E_x\{T_\Gamma < 1\}$$

is finite.
Together with (2.10) part a) of the theorem is satisfied if

$$\int_{\mathbb{R}^n} dx\ E_x\{T_\Gamma < 1\} < \infty. \tag{2.14}$$

This holds because the capacity of Γ is finite, i.e.,

$$\int_{\mathbb{R}^n} dx\ E_x\{T_\Gamma < 1\} \ \leq\ e\cdot \int_{\mathbb{R}^n} dx\ E_x\{e^{-T_\Gamma}, T_\Gamma < \infty\} \tag{2.15}$$

$$\leq\ e \cdot \text{cap}\,(\Gamma).$$

b) The completeness of the scattering system $\{(K_0\dot{+}V)_\Sigma, K_0\dot{+}V\}$ follows in the same way as that for $\{K_\Sigma, K_0\}$ using (1.17) and the estimate of the perturbed kernel:

$$(e^{-\lambda(K_0\dot{+}V)})(x,y) \leq Me^{\lambda a} \sup_{x,y\in\mathbb{R}^n} p_0(\frac{\lambda}{2},x,y)^{\frac{1}{2}}\ p_0(\lambda,x,y)^{\frac{1}{2}} \tag{2.16}$$

(M, a, λ positive constants).
In order to study the system $\{K_0 + V, K_0\}$ we use again Lemma 2.1. For this one space situation one has to verify that the operator

$$e^{-(K_0\dot{+}V)} - e^{-K_0} \text{ belongs to } \mathcal{L}_\infty \tag{2.17}$$

and that

$$e^{-(K_0\dot{+}V)}(e^{-(K_0\dot{+}V)} - e^{-K_0})e^{-K_0} \text{ belongs to } \mathcal{L}_1. \tag{2.18}$$

The operator in (2.18) has a kernel which can be written as

$$\int_0^1 ds(e^{-(2-s)(K_0\dot{+}V)}Ve^{-(1+s)K_0})(x,y) \tag{2.19}$$

$$= \int_0^1 ds \int_{\mathbb{R}^n} du(e^{-\frac{2-s}{2}(K_0\dot{+}V)})(x,u) \tag{2.20}$$

$$\int_{\mathbb{R}^n} dv(e^{-\frac{2-s}{2}(K_0\dot{+}V)})(u,v)V(v)p_0(1+s,v,y).$$

The CDSS [4] trace class theorem is used for the factors (see (2.12))

$$e^{-\frac{2-s}{2}(K_0\dot{+}V)} \circ e^{-\frac{2-s}{2}(K_0\dot{+}V)} V\, e^{-(1+s)K_0}. \tag{2.21}$$

On account of (2.16) we have

$$\left\{ \int_{\mathbb{R}^n} dy \ \mid (e^{-\frac{2-s}{2}(K_0\dot{+}V)})(y,x) \mid^2 \right\}^{\frac{1}{2}} \leq ((e^{-(2-s)(K_0\dot{+}V)})(x,x)]^{\frac{1}{2}} \tag{2.22}$$
$$\leq M e^{\frac{2-s}{2}a} \sup_{x,y;\ \frac{1}{4}\leq\lambda\leq 1} p_0(\lambda,x,y)^{\frac{1}{2}}$$

For the other kernel in (2.21) we get

$$\int_{\mathbb{R}^n} dx \left\{ \int_{\mathbb{R}^n} dy \left| \int_{\mathbb{R}^n} dv \ (e^{-\frac{2-s}{2}(K_0\dot{+}V)})(x,v)V(v)p_0(1+s,v,y) \right|^2 \right\}^{\frac{1}{2}} \tag{2.23}$$
$$\leq \sup_{x,y\in\mathbb{R}^n;\ 1\leq\lambda\leq 4} p_0(\lambda,x,y)^{\frac{1}{2}} \int_{\mathbb{R}^n} dx \int_{\mathbb{R}^n} dv (e^{-\frac{2-s}{2}(K_0\dot{+}V)})(x,v)V(v)$$
$$\leq M\ e^{2a} \sup_{x,y\in\mathbb{R}^n;\ \frac{1}{2}\leq\lambda\leq 4} p_0(\lambda,x,y)^{\frac{3}{2}} \cdot \|V\|_{L^1}.$$

The inequalities in (2.22) and (2.23) lead to

$$\int_0^1 ds \|e^{-(2-s)(K_0\dot{+}V)} V\, e^{-(1+s)K_0}\|_{\text{trace}} < \infty. \tag{2.24}$$

In order to show that

$$\int_0^1 ds\ e^{-(2-s)(K_0\dot{+}V)} V\, e^{-(1+s)K_0} \in \mathcal{L}_1, \tag{2.25}$$

it is sufficient if the function

$$s \longrightarrow e^{-(2-s)(K_0\dot{+}V)} V\, e^{-(1+s)K_0}$$

is continuous in the trace norm. This turns out to be true because V belongs to L^1, because $p_0(t,x,y)$ and $(e^{-t(K_0\dot{+}V)})(x,y)$ are continuous in t, and because $e^{-t(K_0\dot{+}V)}$ is a strongly continuous semigroup in $L^1(\mathbb{R}^n)$. This will be sketched shortly.

The continuity of $e^{-(2-s)(K_0\dot{+}V)} V\, e^{-(1+s)K_0}$ in the trace norm follows if

$$\left\| \left[e^{-\frac{2-s+k}{2}(K_0\dot{+}V)} - e^{-\frac{2-s}{2}(K_0\dot{+}V)} \right] e^{-\frac{2-s+k}{2}(K_0\dot{+}V)} V\, e^{-(1+s+k)K_0} \right\|_{\text{trace}} \tag{2.26}$$
$$\stackrel{k\rightarrow 0}{\longrightarrow} 0,$$

$$\text{(2.27)}\qquad \left\| e^{-\frac{2-s}{2}(K_0 \dot{+} V)} \left[e^{-\frac{2-s+k}{2}(K_0 \dot{+} V)} - e^{-\frac{2-s}{2}(K_0 \dot{+} V)} \right] V\, e^{-(1+s+k)K_0} \right\|_{\text{trace}} \overset{k\to 0}{\longrightarrow} 0,$$

and

$$\text{(2.28)}\qquad \| e^{-(2-s)(K_0 \dot{+} V)} V \left[e^{-(1+s+k)K_0} - e^{-(1+s)K_0} \right] \|_{\text{trace}} \overset{k\to 0}{\longrightarrow} 0.$$

Using always the CDSS [4] trace class theorem (2.12) we show (2.26)–(2.28). For (2.26) we have

$$\int_{\mathbb{R}^n} dy \mid e^{-\frac{2-s+k}{2}(K_0 \dot{+} V)} - e^{-\frac{2-s}{2}(K_0 \dot{+} V)})(y,x) \mid^2$$
$$= (e^{-(2-s+k)(K_0 \dot{+} V)})(x,x) - 2(e^{-(2-s+\frac{k}{2})(K_0 \dot{+} V)})(x,x) + e^{-(2-s)(K_0 \dot{+} V)}(x,x).$$

For $k \to 0$ this tends to zero because $t \to (e^{-t(K_0 \dot{+} V)})(x,y)$ is continuous for Kato-Feller potentials V. The last expression can be estimated by the free density. Dominated convergence implies (2.26).

For (2.27) we notice

$$\int_{\mathbb{R}^n} dx \left\{ \int_{\mathbb{R}^n} dy \left| \int_{\mathbb{R}^n} dv \right.\right.$$
$$\left.\left. \left[(e^{-\frac{2-s+k}{2}(K_0 \dot{+} V)})(x,v) - e^{-\frac{2-s}{2}(K_0 \dot{+} V)}(x,v) \right] V(v)\, p_0(1+s+k,v,y) \right|^2 \right\}^{\frac{1}{2}}$$
$$\leq \sup_{x,y\in\mathbb{R}^n;\, \frac{1}{2}\leq\lambda\leq 3} p_0(\lambda,x,y) \| (e^{-\frac{2-s+k}{2}(K_0 \dot{+} V)} - e^{-\frac{2-s}{2}(K_0 \dot{+} V)}) V \|_{L^1}$$

The convergence follows because $e^{-t(K_0 \dot{+} V)}$ is a strongly continuous semigroup also from $L^1(\mathbb{R}^n) \to L^1(\mathbb{R}^n)$.

The proof of (2.28) is similar to that of (2.26). One has

$$\int_{\mathbb{R}^n} dx \left\{ \int_{\mathbb{R}^n} dy \left| \int_{\mathbb{R}^n} dv \right.\right.$$
$$\left.\left. (e^{-\frac{2-s}{2}(K_0 \dot{+} V)})(x,v) V(v) (p_0(1+s+k,v,y) - p_0(1+s,v,y)) \right|^2 \right\}^{\frac{1}{2}}$$
$$\leq \int_{\mathbb{R}^n} dx \left\{ \int_{\mathbb{R}^n} dv \int_{\mathbb{R}^n} dv' \right.$$
$$(e^{-\frac{2-s}{2}(K_0 \dot{+} V)})(x,v) \mid V(v) \mid (e^{-\frac{2-s}{2}(K_0 \dot{+} V)})(x,v') \mid V(v') \mid$$
$$\left. \Big(p_0(2+2s+2k,v,v') - 2p_0(2+2s+k,v,v') + p_0(2+2s,v,v') \Big) \right\}^{\frac{1}{2}}.$$

The integrand tends to zero because $t \to p_0(t,x,y)$ is continuous. The whole expression can be estimated by $\|V\|_{L^1}$.
This proves (2.18), the trace class property of $e^{-(K_0 \dot{+} V)}(e^{-(K_0 \dot{+} V)} - e^{-K_0})e^{-K_0}$.
It remains to prove (2.17), the compactness of $e^{-(K_0 \dot{+} V)} - e^{-K_0}$. We will sketch it here. Let $V = V_+ - V_-$. Then

$$(2.29) \quad e^{-(K_0 \dot{+} V)} - e^{-K_0} = e^{-(K_0 \dot{+} V)} - e^{-(K_0 \dot{+} V_+)} + e^{-(K_0 \dot{+} V_+)} - e^{-K_0}.$$

The kernels can be estimated by

$$\begin{aligned}
&\left|(e^{-(K_0 \dot{+} V)})(x,y) - (e^{-(K_0 \dot{+} V_+)})(x,y)\right|^2 \\
&= \left((e^{-(K_0 \dot{+} V)})(x,y)\right)^2 + \left((e^{-(K_0 \dot{+} V_+)})(x,y)\right)^2 \\
&\quad -2(e^{-(K_0 \dot{+} V)})(x,y) \cdot (e^{-(K_0 \dot{+} V_+)})(x,y) \\
&\leq \left((e^{-(K_0 \dot{+} V)})(x,y)\right)^2 - \left((e^{-(K_0 \dot{+} V_+)})(x,y)\right)^2.
\end{aligned}$$

Therefore the Hilbert-Schmidt norm of $e^{-(K_0 \dot{+} V)} - e^{-(K_0 \dot{+} V_+)}$ is smaller than

$$\begin{aligned}
&\int_{\mathbb{R}^n} \left[(e^{-2(K_0 \dot{+} V)})(x,x) - (e^{-2(K_0 \dot{+} V_+)})(x,x)\right] dx \\
&= \int_{\mathbb{R}^n} dx \int_0^2 ds \left(e^{-(2-s)(K_0 \dot{+} V)} \mid V_- \mid e^{-s(K_0 \dot{+} V_+)}\right)(x,x) \\
&\leq \int_0^2 ds \int_{\mathbb{R}^n} dx \int_{\mathbb{R}^n} dv (e^{-(2-s)(K_0 \dot{+} V)})(x,v) \mid V_-(v) \mid (e^{-s(K_0 \dot{+} V)})(v,x) \\
&\leq 2 \sup_{x,y} (e^{-2(K_0 + V)})(x,y) \cdot \|V_-\|_{L^1}.
\end{aligned}$$

In the same way $e^{-(K_0 \dot{+} V_+)} - e^{-K_0} \in \mathcal{L}_2$ if V_+ belongs to $L^1(\mathbb{R}^n)$. □

References

[1] ARSENEV, A.A.: Singular potentials and resonances (in Russian); Publishing house of Moscow University, Moscow 1974.

[2] BAKRY, D.: L' hypercontractivité et son utilisation en théorie de semigroups; Course Saint-Flour, 1992, Lecture Notes in Math. (to appear).

[3] BAUMGÄRTEL, H.; WOLLENBERG, M.: Mathematical scattering theory; Birkhäuser, Basel 1983.

[4] CASTEREN, J. VAN; DEMUTH, M.; STOLLMANN, P.; STOLZ, G.: Trace norm estimates for products of integral operators and diffusion semigroups; Integral Equations Operator Theory 23 (1995), 145–153.

[5] DEIFT, P.; SIMON, B.: On the decoupling of finite singularities from the question of asymptotic completeness in two body quantum systems; J. Funct. Anal. 23 (1976), 218–238.

[6] DEMUTH, M.: Zeitabhängige Streutheorie für Erzeuger von Markov–Prozessen mit anisotropen und singulären Potentialen; Report Karl–Weierstraß–Institut, R–Math–04/85, Berlin 1985.

[7] DEMUTH, M.; VAN CASTEREN, J.: Perturbations of generalized Schrödinger operators in stochastic spectral analysis; Lecture Notes in Phys. 403 (1992), 1–15.

[8] DEMUTH, M.; VAN CASTEREN, J.: Framework and results of stochastic spectral analysis; Operator Theory: Adv. Appl. 70 (1994), 123–132.

[9] FUKUSHIMA, M.: Dirichlet forms and Markov processes; North Holland, Amsterdam 1980.

[10] STOLLMANN, P.: Scattering by obstacles of finite capacity; J. Funct. Anal. 121 (1994), 416–425.

Technische Universität Clausthal
Institut für Mathematik
Erzstraße 1
38678 Clausthal-Zellerfeld
Germany
demuth@math.tu-clausthal.de

Dept. of Math. and Comp. Science
University of Antwerp (UIA)
Universitetsplein 1
2610 Antwerp-Wilrijk
Belgium
vcaster@wins.uia.ac.de

1991 Mathematics Subject Classification: Primary 47A40; Secondary 60J35

Submitted: March 11, 1996

Operator Theory:
Advances and Applications, Vol. 102

Examples of positive operators in Krein space with 0 a regular critical point of infinite rank

B. ĆURGUS and B. NAJMAN[†]

It is shown that the operators associated with the perturbed wave equation in $\mathbb{R}^n$ and with elliptic operators with an indefinite weight function and mildly varying coefficients on $\mathbb{R}^n$ are similar to a self-adjoint operator in a Hilbert space. These operators have the whole $\mathbb{R}$ as the spectrum. It is shown that they are positive operators in corresponding Krein spaces, and the whole problem is reduced to showing that 0 is not a singular critical point.

1. Introduction

Let $\mathcal{K}$ be a Krein space, A a positive operator in $\mathcal{K}$ with nonempty resolvent set. Then A has a spectral function with the only possible critical points being 0 and ∞. In [3] we found sufficient conditions for a perturbation B in order that $A_1 = A + B$ is also a positive operator with nonempty resolvent set and that the nonsingularity of 0 and/or ∞ persists under this perturbation. We refer to [8] for the definitions and properties of Krein space operators.

In this note we give examples of an operator A and a perturbation B such that both 0 and ∞ are regular critical points of $A_1 = A+B$ and hence A_1 is similar to a self-adjoint operator in a Hilbert space. Note that in these examples both 0 and ∞ are critical points of infinite rank, i.e., there does not exist a neighbourhood Δ of one of these two points such that $E(\Delta)\mathcal{K}$ is a Pontryagin space. The examples we consider are the operator associated with the perturbed wave equation and an elliptic operator with an indefinite weight. The wave equation example implies a well-posedness result which seems to be difficult to prove without the Krein space theory. For other examples of 0 being a regular critical point of a positive operator we refer to [5].

In [3] we have proved the following result:

Theorem 1.1. *Let $(\mathcal{K}, [\,\cdot\,|\,\cdot\,])$ be a Krein space, and let a, b be two symmetric forms in $\mathcal{K}$. Assume that a is closed, symmetric and positive (that is, $a(x) > 0$ for all $x \in \mathcal{D}(a)$, $x \neq 0$). Further assume that $\mathcal{D}(a) \subseteq \mathcal{D}(b)$ and that there exist real numbers α and β such that*

$$(1.1) \qquad \alpha \leq \frac{b(x)}{a(x)} \leq \beta \qquad \text{for all} \quad x \in \mathcal{D}(a).$$

[†]Branko Najman died unexpectedly in August 1996.

Let

$$a_\kappa = a + \kappa b, \qquad \kappa \in \mathbb{R}.$$

For $\kappa\alpha > -1$ *the form* a_κ *is also a closed positive symmetric form on* $\mathcal{D}(a)$. *Let* A *and* A_κ *be the positive self-adjoint operators associated in* $(\mathcal{K}, [\,\cdot\,|\,\cdot\,])$ *with* a *and* a_κ, *respectively (see* [7]). *Assume that the operator* A *has nonempty resolvent set and that* ∞ *is not a singular critical point of* A.
There exist real numbers $\kappa^\pm$ *such that* $\kappa^- < 0 < \kappa^+$, *for* $\kappa_- < \kappa < \kappa_+$ *the operator* A_κ *has nonempty resolvent set and such that* ∞ *is not a singular critical point of* A_κ. *Moreover, following statements are equivalent:*

(i) 0 *is not a singular critical point of* A.
(ii) 0 *is not a singular critical point of* A_κ.
(iii) A *is similar to a self-adjoint operator in* $(\mathcal{K}, (\,\cdot\,|\,\cdot\,))$.
(iv) A_κ *is similar to a self-adjoint operator in* $(\mathcal{K}, (\,\cdot\,|\,\cdot\,))$.

2. Perturbed wave equation

The example to be described is an extension of the example in [2, 6].

Let $\mathcal{G}$ be a Hilbert space with a scalar product $(\,\cdot\,|\,\cdot\,)$, H a nonnegative injective self-adjoint operator in $\mathcal{G}$. For $\alpha \in \mathbb{R}$ let $\mathcal{G}_\alpha$ be the Hilbert space completion of $(\mathcal{D}(H^\alpha), (H^\alpha\cdot\,|H^\alpha\cdot))$. Denote by $\|\cdot\|_\alpha$ the norm of this Hilbert space. The operator H^β can be extended to an isometry between $\mathcal{G}_\alpha$ and $\mathcal{G}_{\alpha-\beta}$. Denote by $\mathcal{H}$ the Hilbert space $\mathcal{G}_{1/4} \oplus \mathcal{G}_{-1/4}$ and by $\langle\,\cdot\,|\,\cdot\,\rangle$ its natural scalar product. If $x \in \mathcal{G}_{1/4}$ then $|(x|y)| \le \|x\|_{1/4}\|y\|_{-1/4}$, $y \in \mathcal{G}$. Therefore the scalar product $(\,\cdot\,|\,\cdot\,)$ can be extended by continuity from $\mathcal{G}_{1/4} \times \mathcal{G}$ to $\mathcal{G}_{1/4} \times \mathcal{G}_{-1/4}$ and similarly from $\mathcal{G} \times \mathcal{G}_{1/4}$ to $\mathcal{G}_{-1/4} \times \mathcal{G}_{1/4}$. Define an indefinite scalar product on $\mathcal{H}$ by

$$[x|y] = (x_1|y_2) + (x_2|y_1), \qquad x = (x_1, x_2),\ y = (y_1, y_2) \in \mathcal{H}.$$

The space $\mathcal{H}$ with the indefinite scalar product $[\,\cdot\,|\,\cdot\,]$ is a Krein space. The fundamental symmetry is

$$\mathbf{J} = \begin{bmatrix} 0 & H^{-1/2} \\ H^{1/2} & 0 \end{bmatrix}.$$

Define the operator $\mathbf{A}$ in $\mathcal{H}$ on $\mathcal{D}(\mathbf{A}) = \mathcal{G}_{3/4} \oplus \mathcal{G}_{1/4}$ by

$$\mathbf{A} = \begin{bmatrix} 0 & I \\ H & 0 \end{bmatrix}.$$

The operator $\mathbf{A}$ is a self-adjoint operator in $(\mathcal{H}, [\,\cdot\,|\,\cdot\,])$. Since

$$[\mathbf{A}x|x] = (Hx_1|x_1) + (x_2|x_2), \qquad x = (x_1, x_2) \in \mathcal{D}(\mathbf{A}), \tag{2.1}$$

the operator $\mathbf{A}$ is positive in $(\mathcal{H}, [\,\cdot\,|\,\cdot\,])$. The form $[\mathbf{A}x|y]$, $x, y \in \mathcal{D}(\mathbf{A})$ is closable. Let $\mathbf{a}$ be its closure. It follows from (2.1) that the domain of $\mathbf{a}$ is $\mathcal{D}(\mathbf{a}) = \mathcal{G}_{1/2} \oplus \mathcal{G}$

and that

$$\mathbf{a}(x,y) = (H^{1/2}x_1|H^{1/2}y_1) + (x_2|y_2), \qquad x=(x_1,x_2),\ y=(y_1,y_2)\in\mathcal{D}(\mathbf{a}).$$

Since the operators **A** and **J** commute we have:

Lemma 2.1. *The operator* **A** *is similar to a self-adjoint operator in* $\mathcal{H}$. *In particular, neither* ∞ *nor* 0 *is a singular critical point of* **A**.

Let q and V be symmetric $H^{1/2}$-bounded operators in $\mathcal{G}$. We define the form **b** on $\mathcal{D}(\mathbf{b}) = \mathcal{D}(\mathbf{a})$,

$$\mathbf{b}(x,y) = (qx_1|qy_1) + (Vx_1|y_2) + (x_2|Vy_1), \quad x=(x_1,x_2),\ y=(y_1,y_2)\in\mathcal{D}(\mathbf{a}).$$

The operator formally associated with the form **b** in $(\mathcal{H},[\cdot|\cdot])$ is

$$\mathbf{B} = \begin{bmatrix} V & 0 \\ q^2 & V \end{bmatrix}.$$

Lemma 2.2. *Under the above assumptions*

$$\alpha \le \frac{\mathbf{b}(x)}{\mathbf{a}(x)} \le \beta \qquad \text{for all} \quad x\in\mathcal{D}(\mathbf{a}). \tag{2.2}$$

where

$$\begin{aligned} \alpha &= \frac{1}{2}\left(\|qH^{-1/2}\|^2 - (\|qH^{-1/2}\|^4 + 4\|VH^{-1/2}\|^2)^{1/2}\right), \\ \beta &= \frac{1}{2}\left(\|qH^{-1/2}\|^2 + (\|qH^{-1/2}\|^4 + 4\|VH^{-1/2}\|^2)^{1/2}\right). \end{aligned}$$

Proof. Let $x=(x_1,x_2)\in\mathcal{D}(\mathbf{a}) = \mathcal{G}_{1/2}\oplus\mathcal{G}$ and let $\mathbf{r}(x) = \dfrac{\mathbf{b}(x)}{\mathbf{a}(x)}$. Then

$$\mathbf{r}(x) = \frac{\|qx_1\|^2 + 2\,\Re(Vx_1|x_2)}{\|H^{1/2}x_1\|^2 + \|x_2\|^2}.$$

Set $y_1 = H^{1/2}x_1$, $x_2 = y_2$. Note that the mapping $x\mapsto y=(y_1,y_2)$ is a bijection of $\mathcal{D}(\mathbf{a})$ onto $\mathcal{G}\oplus\mathcal{G}$. Then

$$\mathbf{r}(y) = \frac{\|qH^{-1/2}y_1\|^2 + 2\,\mathrm{Re}(VH^{-1/2}y_1|y_2)}{\|y_1\|^2+\|y_2\|^2},$$

hence for every $\gamma > 0$

$$\mathbf{r}(y) \le \frac{(\|qH^{-1/2}\|^2 + \gamma\|VH^{-1/2}\|^2)\|y_1\|^2 + \frac{1}{\gamma}\|y_2\|^2}{\|y_1\|^2+\|y_2\|^2},$$

$$\mathbf{r}(y) \geq \frac{(\|qH^{-1/2}\|^2 - \gamma\|VH^{-1/2}\|^2)\|y_1\|^2 - \frac{1}{\gamma}\|y_2\|^2}{\|y_1\|^2 + \|y_2\|^2}.$$

Picking first

$$\gamma = \frac{-\|qH^{-1/2}\|^2 + (\|qH^{-1/2}\|^4 + 4\|VH^{-1/2}\|^2)^{1/2}}{2\|VH^{-1/2}\|^2}$$

and then

$$\gamma = \frac{\|qH^{-1/2}\|^2 + (\|qH^{-1/2}\|^4 + 4\|VH^{-1/2}\|^2)^{1/2}}{2\|VH^{-1/2}\|^2},$$

we find $\alpha \leq \mathbf{r}(y) \leq \beta$. □

Corollary 2.3. *There exist numbers $\kappa^- < 0 < \kappa^+$ such that for $\kappa \in (\kappa^-, \kappa^+)$ the form $\mathbf{a}_\kappa = \mathbf{a} + \kappa\mathbf{b}$ defined on $\mathcal{G}_{1/2} \oplus \mathcal{G}$ is closed, symmetric and bounded from below.*

Let $\mathbf{A}_\kappa$ be the associated operator in the Krein space $(\mathcal{H}, [\cdot|\cdot])$. From the Lemmas 2.1, 2.2, and a result of P. Jonas (see [3, Proposition 6]) it follows that $\mathbf{A}_\kappa$ is a positive operator with nonempty resolvent set. From Theorem 1.1 (iii) and Lemma 2.1 we conclude (compare also [2, Theorem 3.5])

Theorem 2.4. *Let q and V be symmetric $H^{1/2}$-bounded operators in $\mathcal{G}$. Then for real κ with $|\kappa|$ sufficiently small, the operator $\mathbf{A}_\kappa$ is similar to a self-adjoint operator in the Hilbert space $(\mathcal{H}, (\cdot|\cdot))$.*

It follows from Theorem 2.4 that the operator $i\mathbf{A}_\kappa$ generates a uniformly bounded C_0 group of operators in $\mathcal{H}$. Since the Cauchy problem

$$\left(\frac{d}{dt} - i\kappa V\right)^2 u + (H + \kappa q^2)u = 0, \qquad u(0) = u_0, \qquad \frac{du}{dt}(0) = u_1, \tag{2.3}$$

can be written as

$$\frac{dU}{dt} = i\mathbf{A}_\kappa U, \;\; U(0) = U_0, \quad \text{where} \quad U = \begin{bmatrix} u \\ \left(-i\frac{d}{dt} - \kappa V\right) u \end{bmatrix},$$

it follows that the Cauchy problem (2.3) is well-posed in $\mathcal{G}_{1/4} \oplus \mathcal{G}_{-1/4}$.

In particular, if $\mathcal{G} = L^2(\mathbb{R}^n)$ and H is the self-adjoint realization of the Laplace operator in $\mathcal{G}$, then we obtain a well-posedness result for the perturbed wave equation in $\mathbb{R}^n$. Note that the boundedness of $fH^{-1/2}$ in this case amounts to the inequality

$$\int_{\mathbb{R}^n} |fu|^2 \leq \int_{\mathbb{R}^n} |\nabla u|^2 \qquad \text{for all} \quad u \in C_o^\infty(\mathbb{R}^n).$$

We refer to the inequality (IV.4.6) in [7] for sufficient conditions to satisfy this inequality.

3. Elliptic operators with mildly varying coefficients on $\mathbb{R}^n$

For simplicity, we study only second order operators. Consider the form

$$a_1(x,y) = \sum_{|\alpha|+|\beta|\le 2} \int a_{\alpha\beta} D^\alpha x D^\beta y$$

on $\mathcal{D}(a_1) = H^1(\mathbb{R}^n)$ where we assume:

(i) for all α, β we have $a_{\alpha\beta} \in L^\infty(\mathbb{R}^n)$,

(ii) for some weakly mixed elliptic polynomial (see [1]) $p(x) = \sum_{|\alpha|+|\beta|\le 2} a^o_{\alpha\beta} x^{\alpha+\beta}$ we have

$$\max_{\alpha,\beta}\{\|a_{\alpha\beta} - a^o_{\alpha\beta}\|_{L^\infty(\mathbb{R}^n)}\} < r,$$

where r is a number to be specified later.

We are interested in the operator $A_1 = (\operatorname{sgn} x_n) \sum_{|\alpha|+|\beta|\le 2} D^\beta a_{\alpha\beta} D^\alpha$. Denote

$$a(x,y) = \sum_{|\alpha|+|\beta|\le 2} \int a^o_{\alpha\beta} D^\alpha x D^\beta y, \quad b(x,y) = \sum_{|\alpha|+|\beta|\le 2} \int (a_{\alpha\beta} - a^o_{\alpha\beta}) D^\alpha x D^\beta y$$

for $x, y \in \mathcal{D}(a) = H^1(\mathbb{R}^n)$.

From the ellipticity of p we find the constant $C > 0$ such that $a(x) \ge C\|u\|^2_{H^1(\mathbb{R}^n)}$. From our assumption $|b(x)| \le Kra(x), x \in \mathcal{D}(a)$, with K depending on a only. Hence for r sufficiently small, all our assumptions are satisfied. Since the operator JA, $J = \operatorname{sgn} x_n$, is similar to a self-adjoint operator in $L^2(\mathbb{R}^n)$ by [1], if r is sufficiently small the operator A_1 has the same property.

Similar results hold for the operator $A = \frac{1}{w} L$ where L is a second order positive elliptic operator in $L^2(\mathbb{R}^n)$ (or, more generally, in $L^2(\Omega)$ with appropriate boundary conditions) with constant coefficients and w is a function which vanishes on a set of measure zero and which attains positive and negative values on sets of positive measure. With its natural domain, the operator A is a self-adjoint operator in the Krein space L^2_w. We consider the perturbations of the form $B = \frac{1}{w} q$, where q is a function. Then the forms a and b are given by

$$a(x,y) = (Lx, y), \qquad b(x,y) = \left((\operatorname{sgn} q)|q|^{1/2} x, |q|^{1/2} y\right) .$$

Then $\frac{b}{a}$ is bounded if and only if $\frac{\| |q|^{1/2} x\|^2_{L^2}}{a(x)}$ is bounded. This is equivalent to $|q|^{1/2} L^{-1/2}$ being a bounded operator in L^2. Sufficient conditions for that can be found in the literature; sufficient conditions for the boundedness of this operator in the case of finite domain Ω can be found in [4].

References

[1] ĆURGUS, B., NAJMAN, B.: Positive differential operators in Krein space $L^2(\mathbb{R}^n)$; Preprint.

[2] ĆURGUS, B., NAJMAN, B.: Quasi-uniformly positive operators in Krein spaces; Operator Theory: Adv. Appl. 80 (1995), 90–99.

[3] ĆURGUS, B., NAJMAN, B.: Perturbations of range; Proc. Amer. Math. Soc., to appear.

[4] EDMUNDS, D.E., TRIEBEL, H.: Eigenvalue distributions of some degenerate elliptic operators: an approach via entropy numbers; Math. Ann. 299 (1994), 311–340.

[5] FLEIGE, A., NAJMAN, B.: Nonsingularity of critical points of some differential and difference operators; Operator Theory: Adv. Appl., this volume, (1997), 85–95.

[6] JONAS, P.: On the spectral theory of operators associated with perturbed Klein-Gordon and wave type equations; J. Operator Theory 29 (1993), 207–224.

[7] KATO, T.: Perturbation Theory of Linear Operators; Springer Verlag, Berlin 1966.

[8] LANGER, H.: Spectral function of definitizable operators in Krein spaces; Functional Analysis, Proceedings, Dubrovnik 1981, Lecture Notes in Math. 948 (1982), 1-46.

Department of Mathematics
Western Washington University
Bellingham, WA 98225
USA
curgus@cc.wwu.edu

Department of Mathematics
University of Zagreb
Bijenička 30
10000 Zagreb
Croatia

1991 Mathematics Subject Classification: Primary 47B50; Secondary 47F05, 35P05, 47D03, 34G10

Submitted: March 18, 1996

Operator Theory:
Advances and Applications, Vol. 102

On Hilbert-Schmidt operators and determinants corresponding to periodic ODE systems

R. DENK

In this paper the structure of infinite determinants corresponding to linear periodic ODE systems is investigated. Making use of the theory of Hilbert-Schmidt operators and their determinants it can be shown that the infinite determinant characterizing the stability of such an ODE system has polynomial structure. In the proof we use the fact that the trace of the commutator of two specific operators vanishes. The knowledge of the asymptotic structure of the finite section determinants enables us to improve the convergence of the infinite determinant which is the basis for numerical applications.

1. Introduction

The aim of this paper is to describe the structure and convergence of the infinite determinant corresponding to a linear periodic ODE system of the form

$$P(D,x)\,y(x) = 0 \tag{1.1}$$

with $P(D,x) := D^m + A^{(m-1)}(x)D^{m-1} + \ldots + A^{(1)}(x)D + A^{(0)}(x)$ where $A^{(j)}$ are 1-periodic matrix valued functions and $D := \frac{d}{dx}$. Determinants connected with periodic ordinary differential equations have a long history, starting with the famous work of G.W. HILL [9]. Whereas for Hill's equation a number of results about the structure and convergence of the determinant is known (see [11], [12], [13], for instance), the general form of system (1.1) has not yet been investigated. From an operator theoretical point of view the determinant of system (1.1) appears as a regularized determinant of some Hilbert-Schmidt operator. The close connection between the stability of (1.1) and the regularized determinant is known even for partial differential equations (see [10]). In order to make this connection useful for applications one has to know more about the structure of the regularized determinant. As we will see in Sections 2 and 3, in the case considered here the structure is very simple and allows us to improve the convergence of the corresponding finite section determinants. Thus the determinantal approach is not only interesting in a theoretical sense but also for applications.

In order to simplify the notation we will not consider system (1.1) but the corresponding equivalent first order system which we will write in the form

$$y'(x) = A(x) \cdot y(x) \tag{1.2}$$

where $A(\cdot) \in L^\infty(\mathbb{R}, \mathbb{C}^{n\times n})$ is 1-periodic. (We will return to system (1.1) at the end of Section 2.) First we want to fix some notations. I_n stands for the unit matrix in

$\mathbb{C}^{n\times n}$ and $Y(x)$ for the fundamental solution of (1.2), i.e. the matrix solution with the initial value $Y(0) = I_n$. In the following we will deal with functions on the torus $\mathbb{T} := \mathbb{R}/\mathbb{Z}$ with values in $\mathbb{C}^n$, the corresponding L^2-space $L^2(\mathbb{T},\mathbb{C}^n)$ and the Sobolev spaces $W_p^1(\mathbb{T},\mathbb{C}^n) := \{f \in L^2(\mathbb{T},\mathbb{C}^n) : f \text{ absolutely continous}, f' \in L^p(\mathbb{T},\mathbb{C}^n)\}$. As usual, we set $H_1(\mathbb{T},\mathbb{C}^n) := W_2^1(\mathbb{T},\mathbb{C}^n)$.

Instead of $L^2(\mathbb{T},\mathbb{C}^n)$ we will frequently consider the isometrically isomorphic Hilbert space $\ell^2(\mathbb{Z},\mathbb{C}^n) =: H$, making use of the Fourier transform. Operators in H will be written as infinite block matrices (with respect to the standard basis in H). The operator of multiplication with a function Z is denoted by M_Z. For $Z \in L^\infty(\mathbb{T},\mathbb{C}^{n\times n})$ this operator is an element of the class $\mathcal{L}(H)$ of all bounded linear operators in H and has the form $M_Z = (Z_{k-l})_{k,l\in\mathbb{Z}}$ with

$$Z_k := \int_{\mathbb{T}} Z(t)\exp(-2\pi \mathrm{i} kt)\,dt \in \mathbb{C}^{n\times n}. \tag{1.3}$$

The symbol $\mathcal{S}_p(H)$ stands for the Neumann-Schatten class of order p in $\mathcal{L}(H)$. For a Hilbert-Schmidt operator $B \in \mathcal{S}_2(H)$ we will consider the regularized determinant $\Delta_2(1-B)$ (as a standard reference for Hilbert-Schmidt operators and regularized determinants, we mention [8]).

The operator $P_N \in \mathcal{L}(H)$ is defined as the orthogonal projection in H onto the $(2N+1)n$-dimensional subspace

$$\{(c_k)_{k\in\mathbb{Z}} \in H : c_k = 0 \text{ for } |k| > N\}.$$

For $B = (B_{kl})_{k,l\in\mathbb{Z}} \in \mathcal{L}(H)$ we set $\det(1-B) := \lim_N \det(P_N(1-B)P_N)$, provided the limit exists.

We now return to equation (1.2). The stability of this equation is characterized by the so-called Floquet exponents which can be defined as the complex numbers ν for which $\exp(\nu)$ is an eigenvalue of $Y(1)$. The following lemma summarizes different possibilities to describe the Floquet exponents.

Lemma 1.1. *For any $\nu \in \mathbb{C}$ the following statements are equivalent:*

(i) *ν is a Floquet exponent of (1.2).*

(ii) *$-\nu$ is an eigenvalue of the operator L in $L^2(\mathbb{T},\mathbb{C}^n)$ defined by*

$$D(L) := H_1(\mathbb{T},\mathbb{C}^n) \quad \text{and} \quad Lf := f' - Af.$$

(iii) *The regularized determinant $\Delta_2(1 - B_L(\nu))$ is equal to zero, where B_L is defined by*

$$B_L(\nu) := 1 - (L+\nu)F \in \mathcal{S}_2(H)$$

with

$$F := \operatorname{diag}((2\pi \mathrm{i} l + \delta_{0,l})^{-1} I_n)_{l\in\mathbb{Z}} \in \mathcal{S}_2(H).$$

Whereas the equivalence of (i) and (ii) is obvious, condition (iii) can be derived from the more general considerations in [10], p. 110. Indeed, the explicit formula for F is not crucial for the equivalence but for the following calculations.

Condition (i) of Lemma 1.1 leads in a direct way to a numerical method to compute the Floquet exponents. Here we want to concentrate on condition (iii) and study the properties of the regularized determinant mentioned there. As we will see, $\Delta_2(1 - B_L(\nu))$ has a very simple structure which enables us to use the determinantal approach in applications. Thus generalizations of well-known algorithms in the context of Hill's equation ([4], [11], [12], [16]) are obtained.

2. The structure of the regularized determinant

In order to obtain information about the regularized determinant $\Delta_2(1-B_L(\nu))$ which appears in Lemma 1.1 (iii) we investigate the behaviour of this determinant under transformations of the operator L. First we will prove the following lemma which is connected with the question under which conditions the commutator of a bounded operator and a Hilbert Schmidt operator is a trace class operator and seems to be of interest for itself (cf. also Remark 2.3).

Lemma 2.1. *Let $Z \in W^1_\infty(\mathbb{T}, \mathbb{C}^{n\times n})$ with $\det Z(x) \neq 0$ for all $x \in \mathbb{T}$. Then $F - M_Z F M_Z^{-1}$ is a trace class operator with vanishing trace.*

Proof. It is enough to show that $P_N F M_Z - M_Z P_N F$ converges (for $N \to \infty$) to $C := FM_Z - M_Z F$ in $\mathcal{S}_1(H)$. Indeed, in that case we have $P_N F - M_Z P_N F M_Z^{-1} \to F - M_Z F M_Z^{-1}$ in $\mathcal{S}_1(H)$ ([7], p. 107). But

$$\operatorname{tr}(P_N F - M_Z P_N F M_Z^{-1}) = \operatorname{tr}(P_N F) - \operatorname{tr}(M_Z P_N F M_Z^{-1}) = 0$$

for every N ([8], p. 100) which shows that $F - M_Z F M_Z^{-1}$ has vanishing trace.

Let $C = (C_{kl})_{k,l\in\mathbb{Z}}$. For $k \neq 0 \neq l$ we obtain

$$C_{kl} = \left(\frac{1}{2\pi ik} - \frac{1}{2\pi il}\right) Z_{k-l} = -\frac{1}{2\pi il}\cdot\frac{1}{2\pi ik} Z'_{k-l}$$

where $M_{Z'} = (Z'_{k-l})$ is bounded because the derivative Z' is an element of $L^\infty(\mathbb{T}, \mathbb{C}^{n\times n})$ ([7], p. 567). Therefore, $(1-P_0)C(1-P_0) = -(1-P_0)FM_{Z'}F(1-P_0)$ is an element of $\mathcal{S}_1(H)$. The same is true for C which differs from this operator only by a finite rank operator.

Now consider

$$D^N := (D^N_{kl})_{k,l\in\mathbb{Z}} := C - (P_N F M_Z - M_Z P_N F).$$

We still have to show that $\|D^N\|_{\mathcal{S}_1} \to 0$. To this end we decompose

$$\begin{aligned} D^N &= P_N D^N P_N + (1-P_N)D^N(1-P_N) \\ &\quad + P_N D^N (1-P_N) + (1-P_N)D^N P_N. \end{aligned} \tag{2.1}$$

The first term in this sum is equal to zero for all N. As we have seen above, $C \in \mathcal{S}_1(H)$ and thus $D^N \in \mathcal{S}_1(H)$. Therefore, the second term tends to zero in $\mathcal{S}_1(H)$ ([7], p. 107).

In order to estimate the third term in (2.1) we use ([14], p. 239)

$$\|P_N D^N (1-P_N)\|_{\mathcal{S}_1} \leq \sum_{k=-N}^{N} \Big(\sum_{|l|>N} |D^N_{kl}|^2 \Big)^{1/2}. \tag{2.2}$$

We estimate the sum

$$\sum_{k=1}^{N} \left(\sum_{l=N+1}^{\infty} |D^N_{kl}|^2 \right)^{1/2}, \tag{2.3}$$

the remaining parts of the sum in (2.2) can be treated analogously. Direct calculations show that for $|k| \leq N$ and $|l| > N$ we have $D^N_{kl} = -(2\pi i l)^{-1} Z_{k-l}$. So (2.3) can be estimated by

$$\begin{aligned}
\sum_{k=1}^{N} \left(\sum_{l=N+1}^{\infty} \left| \frac{Z_{k-l}}{2\pi l} \right|^2 \right)^{1/2} &\leq \frac{1}{2\pi(N+1)} \sum_{k=1}^{N} \left(\sum_{l=N+1}^{\infty} |Z_{k-l}|^2 \right)^{1/2} \\
&= \frac{1}{2\pi(N+1)} \sum_{k=1}^{N} \left(\sum_{l=N+1-k}^{\infty} \left| \frac{Z'_{-l}}{2\pi l} \right|^2 \right)^{1/2} \\
&\leq \left(\frac{1}{2\pi(N+1)} \sum_{k=1}^{N} \frac{1}{2\pi(N+1-k)} \right) \cdot \|Z'\|_{L_2} \\
&\leq \frac{1}{2\pi(N+1)} \frac{1}{2\pi} (1 + \ln N) \cdot \|Z'\|_{L_2} \\
&\to 0.
\end{aligned}$$

Therefore, also the third term on the right-hand side of (2.1) converges to 0 in $\mathcal{S}_1(H)$. Instead of the last term of this sum we consider the adjoint operator $P_N(D^N)^*(1-P_N)$. The block coefficient at the position $(k,l) \in \mathbb{Z}^2$ of this operator is given by $-(2\pi i l)^{-1} Z^*_{l-k}$ if $|k| \leq N$ and $|l| > N$ and by 0 else. (Z^*_k denotes the adjoint matrix of Z_k.) From this we see that the estimate for the third term of (2.1) is also valid for the last term if Z is replaced by Z^*. Therefore, we obtain $\|D^N\|_{\mathcal{S}_1} \to 0$ which finishes the proof of this lemma. □

Corollary 2.2. *Let Z be as in Lemma 2.1. Then $\Delta_2(1 - B_L(\nu))$ and $\Delta_2(1 - B_{M_Z^{-1} L M_Z}(\nu))$ are equal up to a constant nonvanishing factor which does not depend on ν.*

Proof. Taking the definitions for $B_L(\nu)$ and $B_{M_Z^{-1} L M_Z}(\nu)$, respectively, and applying the product theorem for regularized determinants ([8], p. 169) we immediately obtain

$$
\begin{aligned}
\Delta_2(1 - B_{M_Z^{-1}LM_Z}(\nu)) &= \Delta_2(1 - B_L(\nu)) \cdot \Delta_2(F^{-1}M_Z F M_Z^{-1}) \\
&\quad \cdot \exp(-\operatorname{tr}[(1 - LF)(1 - F^{-1}M_Z F M_Z^{-1})]) \\
&\quad \cdot \exp(-\operatorname{tr}[\nu(F - M_Z F M_Z^{-1})]).
\end{aligned} \tag{2.4}
$$

In order to see that the right-hand side of (2.4) is well-defined, we note that $1 - F^{-1}M_Z F M_Z^{-1} = (M_Z F^{-1} - F^{-1}M_Z)FM_Z^{-1}$. But the commutator of M_Z and F^{-1} is (up to a finite rank operator) equal to $-M_{Z'}$ and thus bounded. So $1 - F^{-1}M_Z F M_Z^{-1} \in \mathcal{S}_2(H)$. Lemma 2.1 tells us that the last factor in (2.4) is equal to 1 while the second and third factors do not depend on ν. That the second factor does not vanish follows from the invertibility of $F^{-1}M_Z F M_Z^{-1}$. □

Remark 2.3. The proof of Lemma 2.1 uses the close connection between F and the derivation. The results of this lemma are obvious if F is replaced by any $\tilde{F} \in \mathcal{S}_1(H)$. In general, however, it is not clear under which conditions on $\tilde{F}$ and Z the lemma remains true. At least it is not valid if F is replaced by any $\tilde{F} \in \mathcal{S}_2(H)$ and M_Z by an arbitrary invertible $B \in \mathcal{L}(H)$, as the following simple example shows. Decompose $H = H_+ \oplus H_-$ with $H_+ = \ell^2(\mathbb{N} \cup \{0\}, \mathbb{C}^n)$, $H_- = \ell^2(-\mathbb{N}, \mathbb{C}^n)$ and define B and F with respect to this decomposition as the operator matrices

$$
B = \begin{pmatrix} \frac{1}{2}\mathrm{id}_{H_-} & 0 \\ 0 & \mathrm{id}_{H_+} \end{pmatrix}, \quad F = \begin{pmatrix} 0 & F_{12} \\ 0 & 0 \end{pmatrix}
$$

with $F_{12} \in \mathcal{S}_2(H_+, H_-)\backslash\mathcal{S}_1(H_+, H_-)$. Then $F - B^{-1}FB = -F \notin \mathcal{S}_1(H)$.

Now we want to investigate not the regularized determinant $\Delta_2(1 - B_L(\nu))$ but the matrix determinant $\det(1 - B_L(\nu))$ as defined in Section 1. The well-known formula which connects these determinants ([8], p. 169) leads (after straightforward calculations) to the existence of $\det(1 - B_L(\nu))$. We obtain the relation

$$
\det(1 - B_L(\nu)) = \exp(-n(1 - \nu) - \operatorname{tr} A_0) \cdot \Delta_2(1 - B_L(\nu)) \tag{2.5}
$$

where A_0 is defined analogously to (1.3). As an immediate consequence of this equation, the Floquet exponents are exactly the zeros of $\det(1 - B_L(\nu))$. The following modification of this infinite determinant will turn out to be useful not only in the proof of Theorem 2.5 but also in Section 3.

Lemma 2.4. *For* $\nu \in \Lambda := \{z \in \mathbb{C} : \det \sinh \frac{z - A_0}{2} \neq 0\}$ *we set*

$$
\overline{B}_L(\nu) := \left((1 - \delta_{kl})A_{k-l}(2\pi i l + \nu - A_0)^{-1}\right)_{k,l \in \mathbb{Z}}.
$$

Then $\det(1 - \overline{B}_L(\nu))$ *exists for* $\nu \in \Lambda$, *and we obtain*

a) $\det(1 - B_L(\nu)) = \det(1 - \overline{B}_L(\nu)) \cdot \det(2 \sinh \frac{\nu - A_0}{2})$ *for* $\nu \in \Lambda$,

b) $\det(1 - \overline{B}_L(\nu)) \to 1$ *for* $|\operatorname{Re} \nu| \to \infty$.

Proof. Comparing the definitions of $B_L(\nu)$ and $\overline{B}_L(\nu)$ we see for $k,l \in \mathbb{Z}$ and $\nu \in \Lambda$

$$(1 - B_L(\nu))_{kl} = (1 - \overline{B}_L(\nu))_{kl} \cdot \frac{2\pi il + \nu - A_0}{2\pi il + \delta_{0,l}}. \tag{2.6}$$

Therefore the finite section determinants fulfill

$$\begin{aligned} &\det(P_N(1 - B_L(\nu))P_N) \\ &= \det(P_N(1 - \overline{B}_L(\nu))P_N) \cdot \prod_{l=-N}^{N} \det\left(\frac{2\pi il + \nu - A_0}{2\pi il + \delta_{0,l}}\right) \\ &= \det(P_N(1 - \overline{B}_L(\nu))P_N) \cdot \det\left[(\nu - A_0)\prod_{l=1}^{N}\left(1 + \left(\frac{\nu - A_0}{2\pi l}\right)^2\right)\right]. \end{aligned}$$

For $N \to \infty$ the last determinant converges to $\det(2\sinh\frac{\nu - A_0}{2}) \neq 0$ as we can see from the product formula for the sinh-function applied to matrices. Thus $\det(1 - \overline{B}_L(\nu))$ exists for $\nu \in \Lambda$ and equality a) holds. To obtain b), we use the estimation

$$\begin{aligned} \|\overline{B}_L(\nu)\|_{\mathcal{S}_2}^2 &= \sum_{k,l} |(\overline{B}_L(\nu))_{kl}|^2 \\ &\leq \sum_{k\neq l} |A_{k-l}|^2 \cdot |(2\pi il + \nu - A_0)^{-1}|^2 \\ &\leq \|A\|_{L_2}^2 \cdot \sum_l |(2\pi il + \nu - A_0)^{-1}|^2 \end{aligned}$$

which shows $\overline{B}_L(\nu) \in \mathcal{S}_2(H)$ and $\|\overline{B}_L(\nu)\|_{\mathcal{S}_2} \to 0$ for $|\mathrm{Re}\,\nu| \to \infty$. From the continuity of the regularized determinant we see

$$\det(1 - \overline{B}_L(\nu)) = \Delta_2(1 - \overline{B}_L(\nu)) \to 1 \text{ for } |\mathrm{Re}\,\nu| \to \infty. \qquad \square$$

Theorem 2.5. *The determinant* $\det(1 - B_L(\nu))$ *is (up to normalization) a polynomial in* $\exp(\nu)$. *More precisely, the following equality holds for every* $\nu \in \mathbb{C}$:

$$\det(1 - B_L(\nu)) = (-1)^n \exp(-\tfrac{1}{2}(n\nu + \operatorname{tr} A_0)) \cdot \det(Y(1) - \exp(\nu)I_n).$$

Proof. Due to the theorem of Floquet-Lyapunov there exists a $Z \in W_\infty^1(\mathbb{T}, \mathbb{C}^{n\times n})$ with $\det Z(x) \neq 0$ which transforms (1.2) to a constant system, i.e. we have $(M_Z^{-1} L M_Z)f = f' - Kf$ for $f \in H_1(\mathbb{T}, \mathbb{C}^n)$ where $K \in \mathbb{C}^{n\times n}$ is a constant matrix with $\exp K = Y(1)$. From Corollary 2.2 we obtain the existence of some constant $c \neq 0$, not depending on ν, with

$$\begin{aligned} \det(1 - B_L(\nu)) &= \exp(-n(1-\nu) - \operatorname{tr} A_0) \cdot \Delta_2(1 - B_L(\nu)) \\ &= c \cdot \exp(-n(1-\nu) - \operatorname{tr} A_0) \cdot \Delta_2(1 - B_{M_Z^{-1}LM_Z}(\nu)) \\ &= c \cdot \exp(\operatorname{tr} K - \operatorname{tr} A_0) \cdot \det(1 - B_{M_Z^{-1}LM_Z}(\nu)). \end{aligned}$$

We calculate the last determinant explicitly ($B_{M_Z^{-1}LM_Z}(\nu)$ is block diagonal). Similarly to the proof of Lemma 2.4 (or using this lemma) we get

$$\begin{aligned} \det(1 - B_{M_Z^{-1}LM_Z}(\nu)) &= \det(2\sinh\tfrac{\nu - K}{2}) \\ &= \det\left[\exp(-\tfrac{\nu+K}{2}) \cdot (\exp(\nu)I_n - \exp K)\right] \\ &= (-1)^n \exp(-\tfrac{1}{2}(n\nu + \operatorname{tr} K)) \cdot \det(Y(1) - \exp(\nu)I_n), \end{aligned}$$

and therefore

$$\begin{aligned} &\exp(\tfrac{1}{2}(n\nu + \operatorname{tr} A_0)) \cdot \det(1 - B_L(\nu)) \\ &= (-1)^n c \exp(\tfrac{1}{2}(\operatorname{tr} K - \operatorname{tr} A_0)) \cdot \det(Y(1) - \exp(\nu)I_n) \qquad (2.7)\\ &= (-1)^n \tilde{c} \cdot \det(Y(1) - \exp(\nu)I_n). \end{aligned}$$

Here $\tilde{c} := c \exp(\frac{1}{2}(\operatorname{tr} K - \operatorname{tr} A_0))$ is independent of ν.

It remains to compute $\tilde{c}$. The left-hand side of (2.7) can be written as (cf. Lemma 2.4 a))

$$\begin{aligned} &\exp(\tfrac{1}{2}(n\nu + \operatorname{tr} A_0)) \cdot \det(2\sinh\tfrac{\nu - A_0}{2}) \cdot \det(1 - \overline{B}_L(\nu)) \\ &= (-1)^n \det(\exp A_0 - \exp(\nu)I_n) \cdot \det(1 - \overline{B}_L(\nu)). \end{aligned}$$

Due to Lemma 2.4 b) this expression tends to $(-1)^n \det(\exp A_0)$ for $\operatorname{Re}\nu \to -\infty$. Using the formula of Liouville we get

$$\det(\exp A_0) = \exp\left(\operatorname{tr}\int_0^1 A(t)\,dt\right) = \det Y(1).$$

Comparing the limits of both sides of (2.7) for $\operatorname{Re}\nu \to -\infty$ the constant $\tilde{c}$ is seen to be equal to 1 which finishes the proof of the theorem. □

Remark 2.6. a) In the proof of Theorem 2.5 the equivalence of (i) and (iii) in Lemma 1.1 was not used. On the other hand, this equivalence follows immediately from the formula of Theorem 2.5.

b) Due to Theorem 2.5 the Floquet exponents of (1.2) can be computed if the value of $\det(1 - B_L(\nu))$ is known for $n - 1$ different values of ν. (The leading coefficient and the constant term of the polynomial appearing at the right-hand side of Theorem 2.5 are known.) Therefore, it is important to investigate the convergence of this infinite determinant for fixed $\nu \in \mathbb{C}$. This will be done in Section 3. In particular, in the case of Hill's equation where $n = 2$ we obtain the classical result that the Floquet exponents can be computed from the value of $\det(1 - B_L(0))$, for instance.

We now return to equation (1.1) and assume the dimension of the matrices $A^{(j)}(x)$ to be equal to $\tilde{n}$. If we transform this system to the equivalent first order system (1.2) and apply the results above we obtain the determinant of an infinite block matrix whose coefficients have dimension $m\tilde{n}$. This dimension, however, can be reduced to $\tilde{n}$ (what means an important improvement with respect to computational aspects) as we can see from the following lemma. In this lemma the Fourier coefficients $P_k(t)$ of the polynomial $P(D,x)$ (cf. equation (1.1)) are defined by

$$P_k(t) := \delta_{0,k} I_n t^m + A_k^{(m-1)} t^{m-1} + \ldots + A_k^{(1)} t + A_k^{(0)}, \qquad k \in \mathbb{Z}.$$

Lemma 2.7. *Define* $1 - B_L^{(m)}(\nu) := \left((2\pi i l + \delta_{0,l})^{-m} P_{k-l}(2\pi i l + \nu)\right)_{k,l\in\mathbb{Z}}$. *Then the Floquet exponents of* (1.1) *are exactly the zeros of* $\det(1 - B_L^{(m)}(\nu))$. *The function* $\exp(\frac{1}{2} m\tilde{n}\nu) \cdot \det(1 - B_L^{(m)}(\nu))$ *is a polynomial in* $\exp(\nu)$ *of degree* $m\tilde{n}$ *with constant term* $(-1)^{m\tilde{n}} \exp(\frac{1}{2} \operatorname{tr} A_0^{(m-1)})$ *and leading coefficient* $\exp(-\frac{1}{2} \operatorname{tr} A_0^{(m-1)})$.

Proof. Transforming (1.1) to a first order system and applying Theorem 2.5 we obtain an infinite block matrix whose (k,l)-coefficient is equal to

$$\frac{1}{2\pi i l + \delta_{0,l}} \begin{bmatrix} \alpha_l \delta_{kl} I_n & -\delta_{kl} I_n & & \\ & \ddots & \ddots & \\ & & \alpha_l \delta_{kl} I_n & -\delta_{kl} I_n \\ A_{k-l}^{(0)} & A_{k-l}^{(1)} & \cdots & \alpha_l \delta_{kl} I_n + A_{k-l}^{(m-1)} \end{bmatrix}$$

where we have set $\alpha_l := 2\pi i l + \nu$. Straightforward calculations show that this determinant can be reduced by elementary column transformations to $\det(1 - B_L^{(m)}(\nu))$ as defined in the lemma. □

The determinant of Lemma 2.7 is important for applications of the determinantal approach to the mechanics of vibrations ([1], [2]). For classical Hill systems we have $A^{(1)}(x) = 0$ and no complex computation is necessary (if the input data are real) because in this case

$$1 - B_L^{(2)}(\nu) = \left(\frac{1}{(2\pi l)^2 - \delta_{0,l}} \left[(2\pi l + i\nu)^2 \delta_{kl} I_n - A_{k-l}^{(0)} \right] \right)_{k,l\in\mathbb{Z}}$$

which is a real matrix function for $\nu \in i\mathbb{R}$ (for a more detailed analysis of Hill systems, cf. also [5]).

3. On the convergence of the infinite determinant

In Section 2 the calculation of the Floquet exponents of (1.2) was reduced to the evaluation of $\det(1 - B_L(\nu))$ for a finite number of different $\nu \in \mathbb{C}$. In this section we want to investigate the convergence of the finite section determinants appearing in the definition of $\det(1 - B_L(\nu))$. From now on we will restrict ourselves to the case where the matrix function $A(\cdot)$ is a trigonometric polynomial, i.e. we have $A_k = 0$ for $|k| > b$ with some $b \in \mathbb{N}_0$. In the following let $\nu \in \mathbb{C}$ be fixed. We tacitly assume that all factors and determinants which appear in the formulas below are different from zero. We will use the abbreviations B_{kl} and $\overline{B}_{kl}$ instead of $(B_L(\nu))_{kl}$ and $(\overline{B}_L(\nu))_{kl}$, respectively, and set $\delta_N := \det(P_N(1 - B_L(\nu))P_N)$ and $\overline{\delta}_N := \det(P_N(1-\overline{B}_L(\nu))P_N)$. The first and second lemma of this section deal with the asymptotic behaviour of the sequence $(\overline{\delta}_N)_N$ and $(\delta_N)_N$, respectively.

Lemma 3.1. *Define the complex numbers $\overline{\gamma}_N$ for $N \in \mathbb{N}$ by*

$$
\begin{aligned}
\overline{\gamma}_N := \quad & \det\left[I_n - \sum_{p=1}^{b} \overline{B}_{-N,-N+p}\overline{B}_{-N+p,-N} \right. \\
& \left. - \sum_{\substack{p,q=1\\ p\neq q}}^{b} \overline{B}_{-N,-N+p}\overline{B}_{-N+p,-N+q}\overline{B}_{-N+q,-N} \right] \\
& \cdot \det\left[I_n - \sum_{p=1}^{b} \overline{B}_{N,N-p}\overline{B}_{N-p,N} \right. \\
& \left. - \sum_{\substack{p,q=1\\ p\neq q}}^{b} \overline{B}_{N,N-p}\overline{B}_{N-p,N-q}\overline{B}_{N-q,N} \right].
\end{aligned}
\tag{3.1}
$$

Then $\overline{\delta}_N - \overline{\gamma}_N\overline{\delta}_{N-1} = O(N^{-4})$ for $N \to \infty$.

Proof. We make use of the transformation of $\overline{B}_L(\nu)$ to a one-sided infinite matrix $C := (C_{kl})_{k,l=0}^{\infty}$ given by

$$
C_{kl} := \begin{pmatrix} \overline{B}_{-k,-l} & \overline{B}_{-k,l} \\ \overline{B}_{k,-l} & \overline{B}_{kl} \end{pmatrix}, \quad k, l \in \mathbb{N},
$$

(and obvious modifications for $k = 0$ or $l = 0$), cf. [12], p. 16. Now we will use the fact that

$$
\begin{aligned}
& \det(1 - C)_N - \det\Big(I_{2n} - \sum_{p=1}^{b} C_{N,N-p}C_{N-p,N} \\
& - \sum_{\substack{p,q=1\\ p\neq q}}^{b} C_{N,N-p}C_{N-p,N-q}C_{N-q,N} \Big) \det(1 - C)_{N-1} = O(N^{-4}),
\end{aligned}
\tag{3.2}
$$

where $(1 - C)_N := (\delta_{kl}I_n - C_{kl})_{k,l=0}^{N}$. To prove (3.2), one has to generalize Satz 6.11 in [13] where the analogue of (3.2) for scalar-valued C_{kl} can be found. The generalization to matrix-valued C_{kl} can be made using the main ideas from [13] and some technical estimates for submatrices and subdeterminants of $(1-C)_N$. We want to omit the complicated but straightforward calculations; the details can be found in [3]. From (3.2) the desired result follows, because $\det(1-C)_N = \overline{\delta}_N$ for all N, and for N large enough the second determinant in (3.2) is equal to $\overline{\gamma}_N$. □

Remark 3.2. The convergence order of N^{-4} appearing in Lemma 3.1 can be improved if the definition of $\overline{\gamma}_N$ is modified by additional sums. In principle it is possible to describe the asymptotics of the sequence $(\overline{\delta}_N)$ up to an arbitrary order. This can be seen from a generalization of Satz 5.11 in [13] to the matrix case; again we refer to [3] for the details. For the application to the methods of convergence improvement which will be discussed later in this section, the order given in Lemma 3.1 is sufficient.

Lemma 3.3. *We have* $\delta_N - \gamma_N \delta_{N-1} = \mathrm{O}(N^{-4})$ *for*

$$\gamma_N := 1 + \frac{\operatorname{tr}(\nu - A_0)^2}{(2\pi N)^2} + 2\sum_{p=1}^{b} \frac{\operatorname{tr}(A_p A_{-p})}{(2\pi)^2 N(N-p)}\,.$$

Proof. Substituting the definition of $\overline{B}_{kl}$ into the expression for $\overline{\gamma}_N$ as given in Lemma 3.1 we see that the first and second factor in (3.1) is equal to

$$\begin{aligned}(3.3)\qquad &\det\left(\frac{\mp 2\pi \mathrm{i} N + \nu - A_0}{\mp 2\pi \mathrm{i} N}\right)^{-1} \cdot \det\Bigg[\frac{\mp 2\pi \mathrm{i} N + \nu - A_0}{\mp 2\pi \mathrm{i} N} \\ &- \sum_p \frac{A_{\mp p}}{\mp 2\pi \mathrm{i} N}\left(\frac{\mp 2\pi \mathrm{i}(N-p) + \nu - A_0}{\mp 2\pi \mathrm{i}(N-p)}\right)^{-1} \frac{A_{\pm p}}{\mp 2\pi \mathrm{i}(N-p)} \\ &+ \sum_{p,q} \frac{A_{\mp p}}{\mp 2\pi \mathrm{i} N}\left(\frac{\mp 2\pi \mathrm{i}(N-p) + \nu - A_0}{\mp 2\pi \mathrm{i}(N-p)}\right)^{-1} \frac{A_{\mp(q-p)}}{\mp 2\pi \mathrm{i}(N-p)} \\ &\left(\frac{\mp 2\pi \mathrm{i}(N-q) + \nu - A_0}{\mp 2\pi \mathrm{i}(N-q)}\right)^{-1} \frac{A_{\pm q}}{\mp 2\pi \mathrm{i}(N-q)}\Bigg],\end{aligned}$$

where the upper sign corresponds to the first and the lower sign to the second factor in (3.1). First we want to rewrite the product of the second factors in (3.3) with different signs up to an accuracy of $\mathrm{O}(N^{-4})$. We make use of $\det(I_n + A) = 1 + \operatorname{tr} A + \mathrm{O}(N^{-4})$ for $A = \mathrm{O}(N^{-2})$ and of

$$\left(I_n \mp \frac{\nu - A_0}{2\pi \mathrm{i}(N-p)}\right)^{-1} = I_n \pm \frac{\nu - A_0}{2\pi \mathrm{i}(N-p)} + \mathrm{O}(N^{-2}).$$

Elementary calculations show

$$\overline{\gamma}_N \;=\; \det\left(I_n + \left(\frac{\nu - A_0}{2\pi N}\right)^2\right)^{-1}.$$

$$
\begin{aligned}
&\cdot\Bigg[1+\frac{\operatorname{tr}(\nu-A_0)^2}{(2\pi N)^2}+\sum_p\frac{2\operatorname{tr}(A_pA_{-p})}{2\pi N\cdot 2\pi(N-p)}\\
&+\sum_p\left(\frac{1}{(2\pi\mathrm{i}N)^2 2\pi\mathrm{i}(N-p)}-\frac{1}{2\pi iN(2\pi\mathrm{i}(N-p))^2}\right)\\
&\qquad\cdot\operatorname{tr}\Big[(\nu-A_0)(A_pA_{-p}-A_{-p}A_p)\Big]\\
&+\sum_{p\neq q}\frac{\operatorname{tr}(A_pA_{q-p}A_{-q})-\operatorname{tr}(A_{-p}A_{p-q}A_q)}{2\pi\mathrm{i}N\cdot 2\pi\mathrm{i}(N-p)\cdot 2\pi\mathrm{i}(N-q)}\Bigg]+\mathrm{O}(N^{-4}).
\end{aligned}
\tag{3.4}
$$

For $\alpha,\beta\in\mathbb{C}$ one obviously has

$$
\frac{1}{N(N-\alpha)(N-\beta)}-\frac{1}{N^3}=\mathrm{O}(N^{-4}).
$$

Therefore, the first sum in (3.4) is of order N^{-4} and can be omitted. The second sum is equal to

$$
\frac{1}{(2\pi\mathrm{i}N)^3}\sum_{p\neq q}\Big[\operatorname{tr}(A_pA_{q-p}A_{-q})-\operatorname{tr}(A_{-p}A_{p-q}A_q)\Big]+\mathrm{O}(N^{-4}),
$$

and a simple change of the summation index shows that the value of this sum is equal to zero. So we have shown

$$
\overline{\gamma}_N=\det\left(I_n+\left(\frac{\nu-A_0}{2\pi N}\right)^2\right)^{-1}\cdot(\gamma_N+\mathrm{O}(N^{-4})).
$$

But from the connection between δ_N and $\overline{\delta}_N$ (see (2.6)) and Lemma 3.1 we can conclude $\delta_N-\gamma_N\delta_{N-1}=\mathrm{O}(N^{-4})$, and the proof is complete. □

The preceding lemmas allow us to describe the order of convergence for the determinants $\det(1-B_L(\nu))$ and $\det(1-\overline{B}_L(\nu))$ (Theorem 3.4) and to improve this order (Theorem 3.5).

Theorem 3.4. *For $N\to\infty$ we have $\delta_N-\delta_{N-1}=\mathrm{O}(N^{-2})$ and $\overline{\delta}_N-\overline{\delta}_{N-1}=\mathrm{O}(N^{-2})$. In general, the exponent -2 cannot be replaced by any smaller number.*

Proof. From Lemma 3.3 and $1-\gamma_N=\mathrm{O}(N^{-2})$ we know $\delta_N-\delta_{N-1}=(\delta_N-\gamma_N\delta_{N-1})-(1-\gamma_N)\delta_{N-1}=\mathrm{O}(N^{-2})$. We write

$$
\overline{\delta}_N-\overline{\delta}_{N-1}=\prod_{l=-N}^{N}\det\left(\frac{2\pi\mathrm{i}l+\nu-A_0}{2\pi\mathrm{i}l+\delta_{0,l}}\right)^{-1}\cdot\left[\delta_N-\det\left(I_n+\left(\frac{\nu-A_0}{2\pi N}\right)^2\right)\delta_{N-1}\right].
$$

Due to the general assumptions at the beginning of this section, the product $\prod_{l=-N}^{N}\cdots$ remains bounded for $N\to\infty$ whereas the last factor is equal to $\delta_N-\delta_{N-1}+\mathrm{O}(N^{-2})$ and thus of order N^{-2}.

The following examples show that the estimation of the theorem cannot be improved without additional assumptions. If $B_L(\nu)$ is block diagonal, i.e. $A_k = 0$ for $k \neq 0$ then

$$\delta_N - \delta_{N-1} = \det\left[(\nu - A_0)\prod_{l=1}^{N-1}\left(I_n + \left(\frac{\nu - A_0}{2\pi l}\right)^2\right)\right] \cdot \left(\det\left(I_n + \left(\frac{\nu - A_0}{2\pi N}\right)^2\right) - 1\right)$$

has exactly convergence order N^{-2}. To see the same for $\overline{\delta}_N$ is more complicated. We take $n = 1$, $A_0 = 0$, $A_1 = A_{-1} = -1$. Then direct calculations show

$$\begin{aligned}
\overline{\delta}_N &= \overline{\delta}_{N-1} - \Big(\frac{1}{2\pi \mathrm{i} N - \nu}\,\frac{1}{2\pi \mathrm{i}(N-1) - \nu} \\
&\quad + \frac{1}{2\pi \mathrm{i} N + \nu}\,\frac{1}{2\pi \mathrm{i}(N-1) + \nu}\Big)\overline{\delta}_{N-2} + \mathrm{O}(N^{-3}) \\
&= \overline{\delta}_{N-1} + \frac{1}{2\pi^2 N(N-1)}\,\overline{\delta}_{N-2} + \mathrm{O}(N^{-3}).
\end{aligned}$$

If we take ν with $|\det(1 - \overline{B}_L(\nu))| > 2\varepsilon$, the finite section determinant $\overline{\delta}_{N-2}$ fulfills $|\overline{\delta}_{N-2}| > \varepsilon$ for N large enough and thus

$$|\overline{\delta}_N - \overline{\delta}_{N-1}| \geq \frac{\varepsilon}{2\pi^2 N(N-1)} + \mathrm{O}(N^{-3})$$

which shows that the estimation for $\overline{\delta}_N$ cannot be improved. □

Theorem 3.5. *Let $f_0(\operatorname{tr}(\nu - A_0)^2) \neq 0$ and $f_p(2\operatorname{tr}(A_p A_{-p})) \neq 0$ for $p = 1, \ldots, b$ where the auxiliary functions f_p are defined by*

$$f_p(z) := \begin{cases} \sinh\left(\frac{\sqrt{z}}{2}\right) \cdot \left(\frac{2}{\sqrt{z}}\right) & \text{if } p \text{ even,} \\ \cosh\left(\frac{\sqrt{z}}{2}\right) & \text{if } p \text{ odd.} \end{cases}$$

Let the modified sequence $(\tilde{\delta}_N)_N$ be given by

$$\tilde{\delta}_N := \delta_N \cdot \prod_{m=1}^{N}\left[\left(1 + \frac{\operatorname{tr}(\nu - A_0)^2}{(2\pi m)^2}\right)\prod_{\substack{p=1\\ p<2m}}^{b}\left(1 + \frac{2\operatorname{tr}(A_p A_{-p})}{\pi^2(2m-p)^2}\right)\right]^{-1}.$$

Then $\tilde{\delta}_N - \tilde{\delta}_{N-1} = \mathrm{O}(N^{-4})$ and

$$\det(1 - B_L(\nu)) = \lim_{N\to\infty}\tilde{\delta}_N \cdot f_0(\operatorname{tr}(\nu - A_0)^2)\prod_{p=1}^{b} f_p(2\operatorname{tr}(A_p A_{-p})).$$

Proof. From the definition of $\tilde{\delta}_N$ we immediately see $\tilde{\delta}_N - \tilde{\delta}_{N-1} = (\delta_N - \tilde{\gamma}_N \delta_{N-1}) \prod_{m=1}^{N}[\dots]^{-1}$ where the product is the same as in the theorem and

$$\tilde{\gamma}_N := \left(1 + \frac{\operatorname{tr}(\nu - A_0)^2}{(2\pi N)^2}\right) \prod_{p=1}^{b} \left(1 + \frac{2\operatorname{tr}(A_p A_{-p})}{\pi^2 (2N-p)^2}\right).$$

It is easy to see that $\tilde{\gamma}_N - \gamma_N = \mathrm{O}(N^{-4})$ and thus $\tilde{\delta}_N - \tilde{\delta}_{N-1} = \mathrm{O}(N^{-4})$. On the other hand,

$$\det(1 - B_L(\nu)) = \lim_N \delta_N = \lim_N \tilde{\delta}_N \cdot \prod_{m=1}^{\infty} \tilde{\gamma}_m ,$$

and the value for the infinite product can be calculated from the well-known product formulas for the sinh- and cosh-function. □

Remark 3.6. The possibility to apply the determinantal approach to numerical problems always depends on methods of convergence acceleration. From this point of view Theorem 3.5 is important. While in this theorem the convergence order is N^{-4}, for Hill systems the order can be improved up to N^{-8} and in special cases even N^{-12} ([5], [15], [16]). In this sense Hill's equation was not only the first equation for which infinite determinants have been defined but also the equation for which this method works best. But even the order N^{-4} which can be achieved for general systems of the form (1.1) is enough to ensure the comparability of the determinantal method with standard methods. This can be seen from numerical examples. Because we do not want to go into details concerning numerical aspects we just want to state one typical result. The following table contains a comparison of computing time and relative error for the determinantal method with and without acceleration of convergence. The system considered in this example was some model problem of the form (1.1) with m equal to 2 and the dimension of the matrices $A^{(j)}(x)$ equal to 2. So there are four essentially different Floquet exponents, and the relative error stated in the table is the maximum of the relative errors of these exponents. The calculation was done in Fortran 77 on a SUN workstation, and the computing time is given in CPU-seconds.

Block dimension N	Using (δ_N)		Using $(\tilde{\delta}_N)$	
	error	time	error	time
5	$1.6 \cdot 10^{-1}$	0.03	$2.2 \cdot 10^{-3}$	0.03
10	$8.7 \cdot 10^{-2}$	0.05	$2.9 \cdot 10^{-4}$	0.05
20	$4.5 \cdot 10^{-2}$	0.10	$3.7 \cdot 10^{-5}$	0.11
40	$2.3 \cdot 10^{-2}$	0.27	$4.6 \cdot 10^{-6}$	0.27

Table 1: Relative error and computing time for the determinantal method.

As we can see from the table, the acceleration of convergence as described in Theorem 3.5 has almost no influence on the computational time but is crucial for the accuracy of the method. That the determinantal approach is considerably faster than numerical integration can be seen from the corresponding data for the solution of the initial value problem: The computing time needed to obtain a relative error of $2.7 \cdot 10^{-2}$, $2.4 \cdot 10^{-3}$ and $1.0 \cdot 10^{-5}$ was in this example 1.55, 2.02 and 3.13 CPU-seconds, respectively! Thus using infinite determinants is more than ten times faster than solving the initial value problem. This comparison confirms earlier results on the determinantal method vs. numerical integration, see [5], [15] and others. Finally, we want to remark that the determinantal approach which was discussed here can be used as a first step in a two-step algorithm (where the second step is an eigenvalue method). For Hill systems, first results in this direction can be found in [6].

References

[1] ADAMS, E., KEPPLER, H., SCHULTE, U.: On the simulation of vibrations of industrial gear drives (complex interaction of pyhsics, mathematics, numerics, and experiments); Archive of Applied Mechanics 65 (1995), 142–160.

[2] BOLOTIN, W.W.: The dynamic stability of elastic systems; Holden-Day, San Francisco 1964.

[3] DENK, R.: Die Determinantenmethode zur Bestimmung der charakteristischen Exponenten von Hillschen Differentialgleichungs-Systemen; Thesis Universität Regensburg, Regensburg 1993.

[4] ———: Hill's equation systems and infinite determinants; Math. Nachr. 175 (1995), 47–60.

[5] ———: Convergence improvement for the infinite determinants of Hill systems, Z. angew. Math. Mech. 75 (1995), 463–470.

[6] ———: The determinantal method for Hill systems; Z. angew. Math. Mech. 76 (1996) S2, 509–510.

[7] GOHBERG, I.C., GOLDBERG, S., KAASHOEK, M.A.: Classes of linear operators, Vol. 1/2; Birkhäuser Verlag, Basel 1990/1993.

[8] GOHBERG, I.C., KREIN, M.G.: Introduction to the theory of linear nonselfadjoint operators; Transl. Math. Monogr. 18, Amer. Math. Soc., Providence, R. I., 1969.

[9] HILL, G. W.: On the part of the motion of the lunar perigee which is a function of the mean motions of the sun and moon; Acta Math. 8 (1886), 1–36.

[10] KUCHMENT, P.: Floquet theory for partial differential equations; Birkhäuser Verlag, Basel 1993.

[11] MAGNUS, W., WINKLER, S.: Hill's equation; Interscience Publishers, New York 1966.

[12] MENNICKEN, R.: On the convergence of infinite Hill-type determinants, Arch. Rational Mech. Anal. 30 (1968), 12–37.

[13] MENNICKEN, R., WAGENFÜHRER, E.: Über die Konvergenz verallgemeinerter Hillscher Determinanten; Math. Nachr. 72 (1976), 21–49.

[14] PIETSCH, A.: Eigenvalues and s-numbers; Cambridge University Press, Cambridge 1987.

[15] WAGENFÜHRER, E.: Ein Verfahren höherer Konvergenzordnung zur Berechnung des charakteristischen Exponenten der Mathieuschen Differentialgleichung; Numer. Math. 27 (1976), 53–65.

[16] ———: Die Determinantenmethode zur Berechnung des charakteristischen Exponenten der endlichen Hillschen Differentialgleichung; Numer. Math. 35 (1980), 405–420.

Universität Regensburg
NWF I - Mathematik
93040 Regensburg
Germany
Robert.Denk@mathematik.uni-regensburg.de

1991 Mathematics Subject Classification: Primary 34D20; Secondary 47B10, 15A15

Submitted: November 29, 1995

Operator Theory:
Advances and Applications, Vol. 102

On estimates of the first eigenvalue in some elliptic problems

Yu.V. Egorov and V.A. Kondratiev

The first eigenvalue for the Dirichlet problem is often a very important characterization of a system in Physics, Elasticity, Biology and so on. In the theory of boundary value problems it serves for estimation of the domain of existence or uniqueness of solutions. We consider the problem of existence of the extremal values for the first eigenvalue for a number of boundary value problems. For example, the boundary value problem of the form $Lu = \lambda V(x)u$ is studied, where L is an elliptic operator in a domain $\Omega \subset \mathbb{R}^n$. The first eigenvalue is estimated from above or from below in terms of the integral $\int_\Omega |V(x)|^\alpha dx$ with some real α.
The applied methods are rather elementary, they use the variational principle, the embedding theorems and some geometrical constructions. For example, for the multidimensional Dirichlet problem the essential characteristic of the domain is the Minkowski dimension of its boundary.

1. On estimates of the first eigenvalue in the Sturm-Liouville problem

Consider the classical Sturm-Liouville problem:

$$y''(x) + \lambda q(x)y(x) = 0, \qquad 0 \le x \le 1,$$
$$y(0) = 0, \qquad y(1) = 0.$$

Suppose that the function q belongs to the class $R_\beta, \beta \neq 0$, of functions given by

$$R_\beta := \left\{ q : [0,1] \to \mathbb{R} : q(x) \ge 0, \int_0^1 q(x)^\beta dx = 1 \right\}. \tag{1.1}$$

Our aim is to estimate the first eigenvalue

$$\lambda_1 = \inf_{y \in C_0^\infty(0,1)} L[q,y]$$

where

$$L[q,y] = \frac{\int_0^1 y'(x)^2 dx}{\int_0^1 q(x)y(x)^2 dx}.$$

In the following we put

$$m_\beta = \inf_{q \in R_\beta} \lambda_1, \qquad M_\beta = \sup_{q \in R_\beta} \lambda_1.$$

We introduce the functional

$$G[y] = \frac{\int_0^1 y'(x)^2 dx}{(\int_0^1 |y(x)|^p dx)^{2/p}}, \qquad p = \frac{2\beta}{\beta - 1},$$

and we use Euler's beta function which is defined for $a > 0$, $b > 0$ as

$$B(a,b) = \int_0^1 x^{a-1}(1-x)^{b-1} dx.$$

Our main result here is the following

Theorem 1.1. *Let*

$$C_\beta = \frac{|\beta - 1|^{1+1/\beta}}{|\beta||2\beta - 1|^{1/\beta}} B^2(\frac{1}{2}, \frac{1}{2} - \frac{1}{2\beta}) \quad \textit{for } \beta > 1 \textit{ and for } \beta < 0,$$

$$C_\beta = \frac{(1-\beta)^{1+1/\beta}}{\beta(1-2\beta)^{1/\beta}} B^2(\frac{1}{2}, \frac{1}{2\beta}) \quad \textit{for } 0 < \beta < 1/2.$$

If $\beta > 1$, *then*

$$m_\beta = C_\beta, \qquad M_\beta = \infty,$$

and there exist functions $q(x) \in R_\beta, u(x) \in H_0^1(0,1)$ *such that*

$$q(x)^{\beta-1} = u(x)^2, \qquad \inf_y L[q,y] = \inf_y G[y] = G[u] = L[q,u] = m_\beta.$$

If $\beta = 1$, *then* $M_1 = \infty$ *and* $m_1 = 4$. *If* $0 < \beta < 1/2$, *then*

$$M_\beta = C_\beta, \qquad m_\beta = 0,$$

and there exist functions $q(x) \in R_\beta, u(x) \in H_0^1(0,1)$ *such that*

$$q(x)^{\beta-1} = u(x)^2, \qquad \inf_y L[q,y] = \inf_y G[y] = G[u] = L[q,u] = M_\beta.$$

If $\beta < 0$, *then*

$$M_\beta = C_\beta, \qquad m_\beta = 0,$$

and there exist functions $q(x) \in R_\beta, u(x) \in H_0^1(0,1)$ *such that*

$$q(x)^{\beta-1} = u(x)^2, \qquad \inf_y L[q,y] = \inf_y G[y] = G[u] = L[q,u] = M_\beta.$$

If $1/2 \leq \beta < 1$, *then* $M_\beta = \infty$ *and* $m_\beta = 0$.

Similar estimates can be obtained for other eigenvalues also. Here the following notion of k-symmetrization introduced by B. SCHWARZ in [Schw] is very useful.

Definition 1.2. The function f_k^+ is the *increasing k-symmetrization* of a function f defined on $[0,1]$ if the functions f and f_k^+ are equimeasurable, f_k^+ is $1/k$-periodic,

$f_k^+(1/2k-x) = f_k^+(1/2k+x)$ for $x \in (0,1/2k)$, and f_k^+ is increasing on the interval $(0,1/2k)$.
The function f_k^- is the *decreasing k-symmetrization* of a function f defined on $[0,1]$ if the functions f and f_k^- are equimeasurable, f_k^- is $1/k$-periodic, $f_k^-(1/2k-x) = f_k^-(1/2k+x)$ for $x \in (0,1/2k)$, and f_k^- is decreasing on the interval $(0,1/2k)$.

Theorem 1.3. *If $\beta > 1$, then*

$$\lambda_k \geq m_\beta k^2 \left(\int_0^1 q(x)^\beta dx\right)^{-1/\beta}, \qquad k = 1, 2, \ldots.$$

Moreover, the corresponding potential function for which the equality holds is $q_k^+(x)$, where q is the function from Theorem 1.1. If $\beta < \frac{1}{2}$, $\beta \neq 0$, then

$$\lambda_k \leq M_\beta k^2 \left(\int_0^1 q(x)^\beta dx\right)^{-1/\beta}, \qquad k = 1, 2, \ldots.$$

Moreover, the corresponding potential function for which the equality holds is $q_k^-(x)$, where q is the function from Theorem 1.1.

We remark that M.G. Krein has obtained a similar result in [Kr] in the case when $\beta = 1$ and $0 < h \leq q(x) \leq H$. More general estimates of λ_k in the multidimensional case have been obtained by M.S. Birman and M.Z. Solomyak in [BS].

2. Other estimates of the first eigenvalue in the Sturm-Liouville problem

Consider the problem

$$(p(x)y'(x))' + \lambda y(x) = 0, \qquad 0 \leq x \leq 1,$$
$$y(0) = 0, \qquad y(1) = 0.$$

Suppose that the function p belongs to the set R_α defined in (1.1) for some $\alpha \neq 0$. The first eigenvalue λ_1 is the minimal value of the functional

$$L[p, y] = \frac{\int_0^1 p(x)y'(x)^2 dx}{\int_0^1 y(x)^2 dx}.$$

We set

$$M_\alpha = \sup_{p \in R_\alpha} \lambda_1, \qquad m_\alpha = \inf_{p \in R_\alpha} \lambda_1,$$

and

$$G[y] = \frac{(\int_0^1 |y'(x)|^r dx)^{2/r}}{\int_0^1 y(x)^2 dx}, \qquad r = \frac{2\alpha}{\alpha - 1}.$$

The extremal values of the functionals L and G are closely connected.

Theorem 2.1. *Let*

$$C(\alpha) = \frac{2\alpha+1}{\alpha}\Big(\frac{\alpha+1}{2\alpha+1}\Big)^{1-1/\alpha} B^2(\frac{1}{2}, \frac{1}{2}+\frac{1}{2\alpha}) \quad \text{for } \alpha < -1 \text{ and for } \alpha > 0,$$

$$C(\alpha) = -\frac{2\alpha+1}{\alpha}\Big(\frac{\alpha+1}{2\alpha+1}\Big)^{1-1/\alpha} B^2(\frac{1}{2}, -\frac{1}{2\alpha}) \quad \text{for } -1/2 < \alpha < 0.$$

If $\alpha > -1/2$, $\alpha \neq 0$, $\alpha \neq 1$, *then* $M_\alpha = C(\alpha)$ *and* $m_\alpha = 0$. *There exist functions* $p \in R_\alpha$, $z \in H_0^1(0,1)$ *such that* $z'(x)^2 = p(x)^{\alpha-1}$ *and*

$$\inf_y L[p,y] = L[p,z] = C(\alpha).$$

If $\alpha = 1$ *then* $M_1 = 12$ *and* $p(x) = 1$, $m_1 = 0$.
If $\alpha < -1$, *then* $m_\alpha = C(\alpha)$ *and* $M_\alpha = \infty$. *There exist functions* $p \in R_\alpha$, $z \in H_0^1(0,1)$ *such that* $z'(x)^2 = p(x)^{\alpha-1}$ *and*

$$\inf_y L[p,y] = \inf_y G[y] = G[z] = L[p,z] = C(\alpha).$$

If $\alpha = -1$, *then* $m_{-1} = 4$ *and* $M_{-1} = \infty$.
If $-1 < \alpha \leq -1/2$, *then* $M_\alpha = \infty$ *and* $m_\alpha = 0$.

As above, we can estimate the other eigenvalues, too.

Theorem 2.2. *If* $\alpha > -1/2$, $\alpha \neq 0$, *then*

$$\lambda_k \geq m_\alpha k^2 \left(\int_0^1 p(x)^\alpha dx\right)^{1/\alpha}, \qquad k = 1, 2, \ldots .$$

Moreover, the corresponding potential for which the equality holds is $p_k^+(x)$, *where* p *is the function from Theorem* 2.1.
If $\alpha < -1$, *then*

$$\lambda_k \leq M_\alpha k^2 \left(\int_0^1 p(x)^\alpha dx\right)^{1/\alpha}, \qquad k = 1, 2, \ldots,$$

and the corresponding potential for which the equality holds is $p_k^-(x)$, *where* p *is the function from Theorem* 2.1.

3. Estimates of the first eigenvalue for a more general Sturm-Liouville problem

Consider the problem

$$(P(x)y'(x))' + \lambda Q(x)y(x) = 0, \qquad 0 < x < 1,$$
$$y(0) = 0, \qquad y(1) = 0,$$

where $P \in R_\alpha$ and $Q \in R_\beta$ with R_α, R_β defined as in (1.1). The first eigenvalue λ_1 of this Sturm-Liouville problem is given by

$$\lambda_1 = \inf_y \frac{\int_0^1 P(x)y'(x)^2dx}{\int_0^1 Q(x)y(x)^2dx}.$$

Put

$$M_{\alpha,\beta} = \sup_{P,Q} \lambda_1, \qquad m_{\alpha,\beta} = \inf_{P,Q} \lambda_1.$$

Then the following result is obtained (see [EK3], [Ka]).

Theorem 3.1. *Let*

$$c_{\alpha,\beta} = \left|\frac{(\beta-1)(\alpha+1)}{\alpha\beta}\right|\left|\frac{2\alpha\beta-\alpha+\beta}{\alpha\beta+\beta}\right|^{1/\alpha}\left|\frac{\alpha\beta-\alpha}{2\alpha\beta-\alpha+\beta}\right|^{1/\beta}$$

for $\alpha \neq 0$, $\alpha \neq -1$, $\beta \neq 0$, $\beta \neq 1$, $2\alpha\beta - \alpha + \beta \neq 0$. *Let*

$$\Omega_1 = \{(\alpha, \beta) \in \mathbb{R}^2 : \alpha < -1,\ \beta > 1\},$$
$$\Omega_2 = \{(\alpha, \beta) \in \mathbb{R}^2 : \alpha > -1/2,\ 2\alpha\beta - \alpha + \beta < 0,\ \alpha \neq 0,\ \beta \neq 0\}.$$

If $(\alpha, \beta) \in \Omega_1$, *then*

$$m_{\alpha,\beta} = c_{\alpha,\beta} B(\frac{1}{2} + \frac{1}{2\alpha}, \frac{1}{2\beta} - \frac{1}{2\alpha})^2, \qquad M_{\alpha,\beta} = \infty.$$

If $(\alpha, \beta) \in \Omega_2$, $\alpha\beta > 0$, *then*

$$M_{\alpha,\beta} = c_{\alpha,\beta} B(\frac{1}{2} + \frac{1}{2\alpha}, \frac{1}{2\beta} - \frac{1}{2\alpha})^2, \qquad m_{\alpha,\beta} = 0.$$

If $(\alpha, \beta) \in \Omega_2$, $\alpha > 0$, $\beta < 0$, *then*

$$M_{\alpha,\beta} = c_{\alpha,\beta} B(\frac{1}{2} + \frac{1}{2\alpha}, \frac{1}{2} - \frac{1}{2\beta})^2, \qquad m_{\alpha,\beta} = 0.$$

If $\beta = 1, \alpha < -1$, *then* $m_{\alpha,1} = 4/(1+\alpha)^2$, $M_{\alpha,1} = \infty$.
If $\alpha = -1, \beta > 1$, *then* $m_{-1,\beta} = 4/(\beta-1)^2$, $M_{-1,\beta} = \infty$.
If $\alpha > -1/2, 1/\alpha - 1/\beta + 2 \leq 0$, *or* $-1 < \alpha \leq -1/2$, *or* $\alpha < -1, \beta < 1$, *or* $\alpha = -1, \beta = 1$, *then* $m_{\alpha,\beta} = 0$ *and* $M_{\alpha,\beta} = \infty$.

As before, we can indicate the extremal values for all eigenvalues of the Sturm-Liouville problem.

Theorem 3.2. *If* $\alpha > -1/2$, $\beta - \alpha + 2\alpha\beta < 0$, *then*

$$\lambda_k \leq M_{\alpha,\beta} \frac{(\int P(x)^\alpha dx)^{1/\alpha}}{(\int Q(x)^\beta dx)^{1/\beta}} \cdot k^2, \qquad k = 1, 2, \ldots.$$

If $\alpha \leq -1$, $\beta \geq 1$, $(\alpha, \beta) \neq (-1, 1)$, *then*

$$\lambda_k \geq m_{\alpha,\beta} \frac{(\int P(x)^\alpha dx)^{1/\alpha}}{(\int Q(x)^\beta dx)^{1/\beta}} \cdot k^2, \qquad k = 1, 2, \ldots.$$

4. On estimates of the first eigenvalue for an operator of higher order

Consider the problem

$$(-1)^{n+1}y^{(2n)}(x) + \lambda q(x)y(x) = 0, \qquad 0 \le x \le 1,$$
$$y(0) = y'(0) = \ldots = y^{(n-1)}(0) = y(1) = y'(1) = \ldots = y^{(n-1)}(1) = 0,$$

assuming that q belongs to the class R_β defined in (1.1), where β is a real number, $\beta \neq 0$. It is well-known that the spectrum is discrete and

$$\lambda_1 = \inf_{y \in C_0^\infty(0,1)} \frac{\int_0^1 y^{(n)}(x)^2 dx}{\int_0^1 q(x)y(x)^2 dx}.$$

Let

$$m_\beta = \inf_{q \in R_\beta} \lambda_1, \qquad M_\beta = \sup_{q \in R_\beta} \lambda_1.$$

Theorem 4.1. *If $\beta \ge 1/2n$, then $M_\beta = \infty$. If $\beta < 1/2n$, then $M_\beta = C(\beta) < \infty$. If $\beta \ge 1$, then $m_\beta \ge 1$. If $\beta < 1$, then $m_\beta = 0$.*

5. Multidimensional problems

Let Ω be a bounded domain in $\mathbb{R}^n$ with a regular boundary. Consider the boundary value problem:

$$(-\Delta)^m u = \lambda P(x)u, \qquad u \in W_{2,0}^m(\Omega),$$

where Δ is the Laplace operator, $m \ge 1$, the function P belongs to the class R_β given in (1.1) where $\beta \neq 0$.

Our aim is to find

$$m_\beta = \inf_{P \in R_\beta} \lambda_1, \qquad M_\beta = \sup_{P \in R_\beta} \lambda_1,$$

where λ_1 is the first eigenvalue of the Dirichlet problem, i.e.,

$$\lambda_1 = \inf_{u \in W_{2,0}^m(\Omega)} \frac{\int_\Omega \sum_{|\alpha|=m} |D^\alpha u|^2 dx}{\int_\Omega P(x)|u(x)|^2 dx}.$$

Theorem 5.1. *Let $n \neq 2m$ and $s = \max(1, n/2m)$. If $\beta < s$, then $m_\beta = 0$. If $\beta \ge s$, then $m_\beta > 0$. If $n = 2m$, then $m_\beta = 0$ for $\beta \le 1$ and $m_\beta > 0$ for $\beta > 1$.*

Theorem 5.2. *Let the boundary of Ω be from the class C^m. If $\beta < 1/2m$, then $M_\beta < \infty$. If $\beta \ge 1/2m$, then $M_\beta = \infty$.*

Let us consider the case $m = 1$ separately.

Theorem 5.3. *Let $m = 1$, $n \geq 2$. If $\beta < n/2$, then $m_\beta = 0$. If $\beta > n/2$, then $m_\beta \geq \lambda$, where $\lambda > 0$ is the least eigenvalue of the following problem:*

$$\Delta u + \lambda u^{(\beta+1)/(\beta-1)} = 0 \ \ in\ K, \qquad u|_{\partial K} = 0, \qquad \int_K |u(x)|^{2\beta/(\beta-1)}dx = 1$$

in the ball K whose measure is equal to the measure of Ω.

The last result is proved using G. TALENTI's constructions [Tal].

The following theorem shows that the regularity of the boundary is essential in this problem.

Let us recall that the Minkowski dimension of a set A is defined as

$$\mu = n - \overline{\lim}_{\rho\to 0} \frac{\ln \operatorname{mes} A_\rho}{\ln \rho},$$

where A_ρ is a ρ-neighbourhood of A.

Theorem 5.4. *Let $m = 1, n \geq 2$. If the Minkowski dimension of $\partial\Omega$ is μ and $\mu < n - 2\beta$, then $M_\beta < \infty$. However, for any μ such that $\mu \geq n - 2\beta$ there exists a domain Ω with the Minkowski dimension of $\partial\Omega$ equal to μ and such that $M_\beta = \infty$.*

Proof. If $\beta < 0$, then

$$\int_\Omega |u(x)|^p dx \leq \left(\int_\Omega P(x)|u(x)|^2 dx\right)^{p/2} \left(\int_\Omega P(x)^\beta dx\right)^{1/(1-\beta)},$$

where $p = 2\beta/(\beta - 1) > 0$. Therefore,

$$M_\beta \leq \frac{\int_\Omega |\nabla u_0|^2 dx}{(\int_\Omega |u_0|^p dx)^{2/p}},$$

where u_0 is a function whose integrals on the right-hand side of the latter inequality are finite. For instance, if $u_0 = d(x)^\gamma$ in a neighbourhood U of the boundary, where $d(x)$ is the distance of x from the boundary and is smooth in $\Omega \setminus U$, then u_0 is admissible for $\gamma > 1/2$, $\gamma > -1/p$.

If $0 < \beta < 1$, then $p < 0$ and

$$\int_\Omega P(x)^\beta dx \leq \left(\int_\Omega P(x) u(x)^2 dx\right)^\beta \left(\int_\Omega u(x)^p dx\right)^{1-\beta}.$$

In this case the function u_0 satisfies the above conditions of convergence if $-1/p > \gamma > 1/2$. Therefore such a γ exists, if $\beta < 1/2$.

If the boundary of Ω is irregular and its Minkowski dimension is $\mu > n-1$, then the function u_0 satisfies the required conditions when $-(n-\mu)/p > \gamma > 1 - (n-\mu)/2$, i.e., when $\mu < n - 2\beta$.

On the other hand, if $\mu > n - 2\beta$, let us take as Ω the domain in the plane of x, y, contained in the square $0 < x < 1$, $0 < y < 1$, and obtained from the square by removing the segments $x = A_n, 1/3 < y < 2/3$, where $A_n = k_0 \sum_{j=1}^{n} j^{-s}$, $s > 1$, $n = 1, 2, \dots$, and k_0 such that $k_0 \sum_{j=1}^{\infty} j^{-s} = 1$. Let us put $a_n = k_0 n^{-s}$, $P(x,y) = h(y) b_n$, for $A_n < x < A_{n+1}$, where $h \in C_0^\infty(1/3, 2/3)$, $h(y) = 0$ for $y < 1/3$ and for $y > 2/3$, and choose the constants b_n so that

$$b_n a_n^2 \to 0 \text{ as } n \to \infty, \qquad \sum_{n=1}^{\infty} b_n^\beta a_n = \infty.$$

Given $\varepsilon > 0$ one can take $P_\varepsilon = 0$ for $x < A_m$ and $x > A_k$, where m is such that $b_m a_m^2 < \varepsilon$ and k is such that

$$\int_{A_m}^{A_k} \int_0^1 P_\varepsilon(x,y)^\beta dx dy \sim 1.$$

Then

$$\int_0^1 \int_0^1 P_\varepsilon(x) u(x)^2 dx dy \leq \varepsilon^2 \int_0^1 \int_0^1 |\nabla u(x)|^2 dx dy,$$

for all $u \in W_1^2(\Omega)$ and therefore $M_\beta = \infty$. It is easy to see that the Minkowski dimension μ of the boundary is equal to $n - 1 + 1/s$. So we can put $b_n = n^{(s-1)/\beta}$ and all the conditions can be satisfied, if $\mu < n - 2\beta$.
If $\mu = n - 1 + 1/s = n - 2\beta$, we can put $b_n = n^{(s-1)/\beta}(\ln n)^{-1/\beta}$. □

Next consider the problem

$$\nabla(P(x)\nabla u(x)) + \lambda u(x) = 0 \text{ in } \Omega,$$
$$u(x) = 0 \text{ on } \partial\Omega.$$

Suppose that the domain Ω has a smooth boundary and that the function P belongs to the class R_α defined in (1.1). The first eigenvalue λ_1 is the minimal value of the functional

$$L[P,u] = \frac{\int_\Omega P(x)|\nabla u(x)|^2 dx}{\int_\Omega u(x)^2 dx}$$

in the class of functions u such that

$$\int_\Omega P(x)|\nabla u(x)|^2 dx < \infty, \qquad \int_\Omega u(x)^2 dx < \infty, \qquad u(x) = 0 \text{ on } \partial\Omega.$$

Theorem 5.5. *Let*

$$M_\alpha = \sup_{P \in R_\alpha} \lambda_1, \qquad m_\alpha = \inf_{P \in R_\alpha} \lambda_1.$$

If $\alpha \leq -n/2$, then $m_\alpha > 0$. For other values of α we have $m_\alpha = 0$.
If $\alpha \geq 1$, then $M_\alpha < \infty$. For other values of α we have $M_\alpha = \infty$.

At last consider the problem

$$\nabla(P(x)\nabla u(x)) + \lambda Q(x)u(x) = 0 \text{ in } \Omega,$$
$$u(x) = 0 \text{ on } \partial\Omega.$$

Suppose that the domain Ω has a smooth boundary and that the functions P, Q belong to the classes R_α, R_β, respectively, introduced in (1.1).

The first eigenvalue λ_1 is the minimal value of the functional

$$L[P, Q, u] = \frac{\int_\Omega P(x)|\nabla u(x)|^2 dx}{\int_\Omega Q(x)u(x)^2 dx}$$

in the class of functions u such that

$$\int_\Omega P(x)|\nabla u(x)|^2 dx < \infty, \quad \int_\Omega Q(x)u(x)^2 dx < \infty, \quad u(x) = 0 \text{ on } \partial\Omega.$$

Theorem 5.6. *Let $n \geq 2$,*

$$M_{\alpha,\beta} = \sup_{P\in R_\alpha, Q\in R_\beta} \lambda_1, \qquad m_{\alpha,\beta} = \inf_{P\in R_\alpha, Q\in R_\beta} \lambda_1.$$

If $\alpha < -n/2$, $n\alpha - n\beta - 2\alpha\beta \leq 0$, then $m_{\alpha,\beta} > 0$. For other values of α, β we have $m_{\alpha,\beta} = 0$.
If $\alpha > 1$, $2\alpha\beta + \beta - \alpha < 0$, then $M_{\alpha,\beta} < \infty$. For other values of α, β we have $M_{\alpha,\beta} = \infty$.

6. First negative eigenvalue

Consider the Schrödinger operator

$$L = -\Delta - V(x)$$

with a real positive potential $V(x)$ in a domain $\Omega \subset \mathbb{R}^n$. Let us estimate its maximal negative eigenvalue which can be defined as

$$\lambda_1 = \sup\left\{\alpha : \int_\Omega V(x)u^2(x)dx + \alpha \int_\Omega u^2(x)dx \leq \int_\Omega |\nabla u(x)|^2 dx\right\}.$$

Ch. Fefferman has obtained in [Fe] the estimates

$$c \cdot \sup_{x,\delta} [Av_{B(x,\delta)}V - C\delta^{-2}] \leq -\lambda_1 \leq C \cdot \sup_{x,\delta} [(Av_{B(x,\delta)}V^p)^{1/p} - c\delta^{-2}],$$

where

$$Av_{B(x,\delta)}f(x) = \delta^{-n} \int_{B(x,\delta)} f(x)dx.$$

These estimates were improved by M. Schechter in [Sch] to the form

$$\sup_{x,\delta}\,[c\cdot Av_{B(x,\delta)}V-\delta^{-2}]\le -\lambda_1\le \sup_{x,\delta}\,[C\cdot (Av_{B(x,\delta)}V^p)^{1/p}-\delta^{-2}],$$

and by R. Kerman and E. Sawyer in [KS] to the form

$$\sup_{Q}\,\{q^{-2}: F(V,Q)\ge C_1\}\le -\lambda_1\le \sup_{Q}\,\{q^{-2}: F(V,Q)\ge C_2\},$$

where

$$F(V,Q)=\frac{\int_Q\int_Q V(x)V(y)|x-y|^{2-n}dxdy}{\int_Q V(x)dx}$$

and Q is a cube with the side length q.

The following result can be obtained from very elementary considerations.

Theorem 6.1. *Let $m\ge 1$, $n>2m$, and let λ_1 be the first negative eigenvalue of the operator L. Then*

$$\sup_Q\left\{\delta^{-2m}:\int_Q V(x)dx\ge C_1\delta^{n-2m}\right\}\le -\lambda_1,$$

$$-\lambda_1\le \sup_Q\left\{b_p\delta^{-2m}:\int_Q V(x)^p|x-x_0|^{2mp-n}dx\ge c_p\right\},$$

where $p\ge n/2m$ and Q is the cube with center at the point x_0 and side length δ.

Suppose that

$$\int_{\mathbb{R}^n} V(x)^p|x|^{2mp-n}dx<\infty,\qquad n>2m,\qquad p\ge n/2m.$$

Let Q be a cube such that

$$\int_{\mathbb{R}^n\setminus Q} V(x)^p|x|^{2mp-n}dx<a_p,$$

where a_p is some constant depending on p only. Using plane sections parallel to the coordinate planes and containing the center of the cube Q, we can divide Q in 2^n smaller cubes $Q_1,\dots,Q_{2^n}$ and choose those of them, for which

$$\int_{Q_j} V(x)^p|x|^{2mp-n}dx\ge a_p.$$

Let K_α be the number of the cubes obtained whose side length is $\le\alpha^{-1/2m}$.

Theorem 6.2. *Let $n>2m$, $p\ge n/2m$, $\int_{\mathbb{R}^n}V(x)^p|x|^{2mp-n}dx<\infty$, and let N_α be the number of the eigenvalues of the operator H that are not greater than $-\alpha$. Then*

$$N_\alpha\le CK_\alpha\quad\textit{and}\quad \sum|\lambda_j|\le C\sum q_j^{-2m},$$

where $q_j\le\alpha^{-1/2m}$.

The following theorem generalizes some results of R. Kerman and E. Sawyer and Ch. Fefferman (see [Fe], [KS]). We set

$$F_m(V,Q) = \frac{\int_Q \int_Q V(x)V(y)|x-y|^{2m-n}dxdy}{\int_Q V(x)dx}.$$

Theorem 6.3. *Let $n > 2m$. There exist positive constants c and C, depending on n only, such that the following statements are valid:*
(A) *Let $\alpha \geq 0$ and $Q_1, \dots, Q_N$ be a set of cubes with side length $\leq \alpha^{-1/2m}$, such that their doubles do not intersect and $F_m(V, Q_j) \geq C$. Then the operator H has at least N eigenvalues, not exceeding $-\alpha$.*
(B) *Conversely, let $\alpha \geq 0$, and let H have at least CN eigenvalues which are $\leq -\alpha$. Then there are non-intersecting cubes $Q_1, \dots, Q_N$ with side lengths not exceeding $\alpha^{-1/2m}$ for which $F_m(V, Q_j) \geq c$, $j = 1, \dots, N$.*

Acknowledgements

This work was partially supported by RFFR, project no. 93-011-16035 and by INTAS, project no. 93-2048.

References

[BS] Birman, M.S., Solomyak, M.Z.: On the leading term of the spectral asymptotics for non-smooth elliptic problems; Funktsional Anal. i Prilozhen. 4 (1974), 1–13.

[EK1] Egorov, Yu.V., Kondratiev, V.A.: On an estimate of the first eigen–value for a self–adjoint elliptic operator; Vestnik Mosk. un–ta, Mathem., Mechanics 3 (1983), 46–52.

[EK2] Egorov, Yu.V., Kondratiev, V.A.: On estimates of the first eigen–value of the Sturm–Liouville problem; Russian Math. Surveys 39:2 (1984), 151–152.

[EK3] Egorov, Yu.V., Kondratiev, V.A.: On an estimate for the first eigen–value of the Sturm–Liouville operator; Vestnik Mosk. un–ta, Mathem., Mechanics 6 (1990), 75–78.

[EK4] Egorov, Yu.V., Kondratiev, V.A.: On an estimate for the principal eigen–value of the Sturm–Liouville operator; Vestnik Mosk. un–ta, Mathem., Mechanics 6 (1991), 5–11.

[EK5] Egorov, Yu.V., Kondratiev, V.A.: On a Lagrange problem; C. R. Acad. Sci. Paris Sér. I 317 (1993), 903–908.

[EK6] Egorov, Yu.V., Kondratiev, V.A.: Spectral Theory of Elliptic Operators; Birkhäuser, Basel 1996.

[Fe] Fefferman, Ch.: The uncertainty principle; Bull. Amer. Math. Soc. 9 (1983), 129–206.

[Ka] KARAA, S.: Extremal eigenvalues end their associated nonlinear equations; Boll. Un. Mat. Ital. B 10 (1996), 625–649.

[KS] KERMAN, R., SAWYER, E.: Weighted norm inequalities for potentials with applications to Schrödinger operators; Bull. Amer. Math. Soc. 312:1 (1985), 112–116.

[Kr] KREIN, M.G.: On certain problems on the maximum and minimum of characteristic values and the Lyapunov zones of stability; Amer. Math. Soc. Transl. Ser. 2 1 (1955), 163–187.

[Sch] SCHECHTER, M.: The spectrum of the Schrödinger operator; Trans. Amer. Math. Soc. 312:1 (1991), 115–128.

[Sch] SCHWARZ, B.: On the extrema of nonhomogeneous string with equimeasurable density; J. Math. Mech. 10 (1961), 401–422.

[Tal] TALENTI, G.: Elliptic equations and rearrangements; Ann. Scuola Norm. Sup. Pisa Cl. Sci. (4) 3 (1976), 697–718.

Université Paul Sabatier, UFR–MIG
118, *route de Narbonne*
31062 *Toulouse*
France
egorov@mip.ups-tlse.fr

Moscow State University
Department of Mathematics
119899 *Moscow*
Russia
kond@nw.math.msu.su

1991 Mathematics Subject Classification: Primary 35P15; Secondary 34B24

Submitted: May 24, 1996

Operator Theory:
Advances and Applications, Vol. 102

Nonsingularity of critical points of some differential and difference operators

A. FLEIGE and B. NAJMAN†

For two different examples of positive definitizable operators in Krein spaces we prove regularity of the critical points 0 and ∞. The first example is the Sturm-Liouville operator $\frac{-1}{\rho(x)}\frac{d^2}{dx^2}$ in $L^2(\mathbb{R},\rho)$ with $\rho(x) = |x|^s \operatorname{sgn} x$. The second example is a difference operator in a Krein space of complex sequences.

1. Introduction

The spectral theory of definitizable operators in a Krein space is a generalization of the spectral theory of selfadjoint operators in a Hilbert space. The main difference is the occurrence of the critical points of the spectral function. If A is a positive operator in a Krein space $\mathcal{K}$ with a nonempty resolvent set, then its only possible critical points are 0 and ∞. It follows from general results, see [9], that if neither of these points is a singular critical point, then A is similar to a selfadjoint operator in the Hilbert space associated with the Krein space $\mathcal{K}$. There has been considerable interest in the nonsingularity of ∞, while the nonsingularity of 0 has attracted less attention. It should be stressed that this question reduces to an algebraic criterion if 0 is a critical point of finite rank (meaning that the spectral projection $E(\Delta)$ has the property that $E(\Delta)\mathcal{K}$ is a Pontrjagin space for some interval Δ containing 0). It follows from [1, Proposition 2.7] that this is the case for a Sturm-Liouville operator with an indefinite weight function, whenever the corresponding differential expression has at least one regular endpoint and the weight function has only finitely many sign changes.

If, however, 0 is a critical point of infinite rank, the question of nonsingularity of 0 is much harder. In the recent paper [2], it has been shown that the operator $\operatorname{sgn} x \, \frac{d^2}{dx^2}$ has this property. In the present note we give two more examples of operators which are positive in a Krein space such that neither 0 nor ∞ are its singular critical points. The first example is a direct generalization of the operator from [2] and the second operator is a difference operator in a Krein space of complex sequences. Both operators could be treated simultaneously by the theory of strings with a nonmonotone mass distribution function, see [10], [8]. Then, of course, a general theorem for such strings would be interesting.

There are two reasons for considering these examples. First, explicit examples are nontrivial and one can hope to obtain better understanding from them. Fur-

†Branko Najman died unexpectedly in August 1996.

ther, in the recent paper [3] it is shown how the results for partial differential operators follow from the corresponding results for ordinary differential operators. We also mention the preprint [4] where further examples are discussed. Note that there exist Sturm-Liouville operators and difference operators such that ∞ is a singular critical point (see [11], [7], [8]).

2. A Sturm-Liouville operator with indefinite weight

Let $\rho(x) = |x|^s \operatorname{sgn} x, s > -1$, and $L^2_\rho := L^2(\mathbb{R}, \rho)$. Let $\mathcal{K} = L^2_\rho$ be the Krein space with the scalar product $[f, g] = \int_{\mathbb{R}} f(x)\overline{g(x)}\rho(x)\, dx$. The operator $(Jy)(x) = (\operatorname{sgn} x)y(x)$ is a fundamental symmetry on $\mathcal{K}$. Let

$$Ay = -\frac{1}{\rho}\, y'', \qquad y \in \mathcal{D}(A) = \left\{ f \in L^2_\rho : f, f' \in AC_{\text{loc}}(\mathbb{R}),\ \frac{1}{\rho} f'' \in L^2_\rho \right\}.$$

Proposition 2.1. *The operator A is a positive definitizable operator in $\mathcal{K}$.*

Proof. Since $[Af, f] = \int_{\mathbb{R}} |f'|^2 dx \geq 0$ for all $f \in D(A)$ with compact support not containing 0, we only have to prove that the resolvent set of A is nonempty. Since $B = JA$ is a closed operator (as it is the maximal Sturm-Liouville operator), the operator A is closed. Therefore, it is sufficient to prove that $B - iJ$ is a bijection of $D(A)$ onto L^2_ρ. Let $g \in L^2_\rho$. Then $(B - iJ)f = g$ is equivalent to the boundary value problem

$$-f''(x) \mp i\, \rho(x)\, f(x) = h(x) := \rho(x)\, g(x), \qquad x \in \mathbb{R}_\pm, \tag{2.1}$$

where $\mathbb{R}_+ = [0, \infty)$ and $\mathbb{R}_- = (-\infty, 0]$. Define the operators $A_\pm$ in $L^2(\mathbb{R}_\pm, \rho)$ on

$$D(A_\pm) = \left\{ f \in L^2_\rho : f, f' \in AC_{\text{loc}}(\mathbb{R}_\pm),\ \frac{1}{\rho} f'' \in L^2_\rho(\mathbb{R}_\pm),\ f(0) = 0 \right\}.$$

Note that the operators $A_\pm$ are selfadjoint since $\frac{1}{\rho}\frac{d^2}{dx^2}$ is in the limit point case at $\pm\infty$. This follows from the fact that the solution $y(x) = x$ is not square integrable, since $\int_{\mathbb{R}} x^2 |\rho(x)|\, dx = \infty$.
Let $g_\pm$ be the restriction of g to $\mathbb{R}_\pm$. From the selfadjointness of $A_\pm$ it follows that we can define the functions $f_\pm := (A_\pm - i)^{-1} g_\pm \in L^2_\rho(\mathbb{R}_\pm)$.
Further, let $\psi_\pm$ be the L^2_ρ solutions of

$$-f''(x) \mp i\, |x|^s\, f(x) = 0, \qquad x \in \mathbb{R}_\pm,$$

(see [5, Theorem 4.3]).

Lemma 2.2. $D = \det \begin{bmatrix} \psi_+(0) & -\psi_-(0) \\ \psi'_+(0) & -\psi'_-(0) \end{bmatrix} \neq 0.$

Proof. Assume $D = 0$. Then

$$\psi_+(0)\psi'_-(0) = \psi'_+(0)\psi_-(0). \tag{2.2}$$

If $\psi_+(0) = 0$, then $\psi'_+(0) \neq 0$ implies $\psi_-(0) = 0$, $\psi'_-(0) \neq 0$. Similarly, if $\psi'_+(0) = 0$, then $\psi'_-(0) = 0$, $\psi_+(0)\psi_-(0) \neq 0$. In both cases there exists a $c \neq 0$ such that $\psi_+(0) = c\psi_-(0)$, $\psi'_+(0) = c\psi'_-(0)$. If $\psi_+(0)\psi'_+(0) \neq 0$, then again such a c exists by (2.2). Now

$$\psi(x) := \begin{cases} \psi_+(x), & x > 0, \\ c\psi_-(x), & x < 0, \end{cases}$$

belongs to L^2_ρ, ψ and ψ' are locally absolutely continuous and $\frac{1}{\rho}\psi'' = -i\psi$ also belongs to L^2_ρ. It follows that i is an eigenvalue of A in contradiction with [10, Lemma 1.1]. This shows $D \neq 0$. □

We proceed with the proof of Proposition 2.1. We search for f in the form

$$f(x) = \begin{cases} f_+(x) + c_+\psi_+(x), & x > 0, \\ f_-(x) + c_-\psi_-(x), & x < 0, \end{cases}$$

with complex numbers c_- and c_+. Then f evidently satisfies the differential equation in $(0, \infty)$ and $(-\infty, 0)$, and the continuity of f and f' at 0 yields

$$0 = f(0+0) - f(0-0) = (f_+(0) - f_-(0)) + c_+\psi_+(0) - c_-\psi_-(0),$$
$$0 = f'(0+0) - f'(0-0) = f'_+(0) - f'_-(0) + c_+\psi'_+(0) - c_-\psi'_-(0).$$

This system of equations has a unique solution by Lemma 2.2. Hence the equation $(A - i)f = Jg$ has a unique solution. Thus $i \in \rho(A)$. □

Proposition 2.3. *The operator A has no eigenvalues. Its spectrum coincides with $\mathbb{R}$. The only critical points of A are 0 and ∞.*

Proof. Denote by U_α, $\alpha \in \mathbb{R} \setminus \{0\}$, the dilation operator: $(U_\alpha f)(x) = f(\alpha x)$. Then U_α is a bounded operator with the bounded inverse $U_{1/\alpha}$. From

$$U_\alpha^{-1} A U_\alpha = (\operatorname{sgn} \alpha)|\alpha|^{s+2} A \tag{2.3}$$

it follows that the spectrum of A is invariant under multiplication by real nonzero numbers. Therefore, $\mathbb{R} \setminus \{0\} \subset \sigma(A)$ and consequently $\sigma(A) = \mathbb{R}$. Since $p(\lambda) = \lambda$ is a definitizing polynomial of A, Theorem 3.1 (4) and Proposition 4.2 in [9] imply that 0 and ∞ are the only critical points of A.
From (2.3) it also follows that the only possible eigenvalue of A is 0. However, all the solutions of the equation $\frac{1}{\rho(x)}y'' = 0$ are of the form $y(x) = ax + b$, and none of these solutions belongs to L^2_ρ. Hence A has no eigenvalues. □

Denote $\mathcal{D} = \mathcal{D}(A)$ and $\mathcal{R} = \mathcal{R}(A) = \{\frac{1}{\rho}y'' : y \in \mathcal{D}\}$. Let $\mathcal{D}_0$ be the space of all functions $y \in \mathcal{D}$ such that $y(0) = y'(0) = 0$ and put $\mathcal{R}_0 = B\mathcal{D}$. The following lemma is evident.

Lemma 2.4. *The spaces $\mathcal{D}_0$ and $\mathcal{R}_0$ are invariant under the fundamental symmetry J.*

Lemma 2.5. *Let $s \in \mathbb{R}$. There exist distinct positive numbers α_j, $j = 1, 2, 3, 4$, such that the matrix*

$$M(\alpha_1, \ldots, \alpha_4; s) = \begin{bmatrix} 1 & 1 & 1 & 1 \\ \alpha_1 & \alpha_2 & \alpha_3 & \alpha_4 \\ \alpha_1^{s+2} & \alpha_2^{s+2} & \alpha_3^{s+2} & \alpha_4^{s+2} \\ \alpha_1^{s+3} & \alpha_2^{s+3} & \alpha_3^{s+3} & \alpha_4^{s+3} \end{bmatrix}$$

is nonsingular.

Proof. Since $\det M(0, 1, N, \frac{1}{N}; s) = N^{s+2} - N^{s+1} - N + \frac{1}{N} + \frac{1}{N^{s+1}} - \frac{1}{N^{s+2}}$, it follows that $\lim_{N\to\infty} |M(0, 1, N, \frac{1}{N}; s)| = \infty$. □

Lemma 2.6.
a) *There exist positive bounded and boundedly invertible operators X and T in the Hilbert space $L^2_{|\rho|}$ such that $X\mathcal{R} \subset \mathcal{R}_0$ and $T\mathcal{D} \subset \mathcal{D}_0$.*
b) *There exist positive bounded and boundedly invertible operators W and V in the Krein space $\mathcal{K}$ such that $W\mathcal{D} \subset \mathcal{D}$ and $V\mathcal{R} \subset \mathcal{R}$.*

Proof. a) We construct only the operator X. The construction of T is similar; see [1, Lemma 3.2]. Let α_j, $j = 1, 2, 3, 4$, be the numbers according to Lemma 2.5 and let ξ_j, $j = 1, 2, 3, 4$, be the solution of the linear system

$$\begin{aligned} \xi_1 + \quad \xi_2 + \quad \xi_3 + \quad \xi_4 &= 1, \\ \alpha_1\xi_1 + \alpha_2\xi_2 + \alpha_3\xi_3 + \alpha_4\xi_4 &= 1, \\ \alpha_1^{s+2}\xi_1 + \alpha_2^{s+2}\xi_2 + \alpha_3^{s+2}\xi_3 + \alpha_4^{s+2}\xi_4 &= -1, \\ \alpha_1^{s+3}\xi_1 + \alpha_2^{s+3}\xi_2 + \alpha_3^{s+3}\xi_3 + \alpha_4^{s+3}\xi_4 &= -1. \end{aligned}$$

Let

$$Z = \sum_{j=1}^{4} \xi_j U_{\alpha_j}, \quad Z_1 = \sum_{j=1}^{4} \xi_j \alpha_j^{s+3} U_{1/\alpha_j}, \quad Y = \sum_{j=1}^{4} \xi_j \alpha_j^{s+2} U_{\alpha_j}.$$

Since $U_\alpha^* = (1/\alpha)^{s+1} U_{1/\alpha}$, we find that

$$Y^* = \sum_{j=1}^{4} \xi_j \alpha_j U_{1/\alpha_j}.$$

Here * denotes the adjoint in the Hilbert space $L^2_{|\rho|}$. As linear combinations of bounded operators, the operators Z, Z_1, Y and Y^* are bounded in $L^2(\mathbb{R}, |\rho|)$. Note that $YB = BZ$ and $Y^*B = BZ_1$ on $\mathcal{D}$. Therefore,

$$(Y^*Y + I)B = B(Z_1 Z + I) \text{ on } \mathcal{D}.$$

By the choice of $\xi_j, j = 1,2,3,4$, it follows that

$$(Zu)^{(k)}(0) = u^{(k)}(0) \quad \text{and} \quad (Z_1u)^{(k)}(0) = -u^{(k)}(0), \;\; k = 1,2,$$

for all $u \in \mathcal{D}$. Therefore $(Z_1Z + I)\mathcal{D} \subset \mathcal{D}_0$. Put $X = Y^*Y + I$. Then

$$X\mathcal{R} = XB\mathcal{D} = B(Z_1Z + I)\mathcal{D} \subset B\mathcal{D} = \mathcal{R}_0.$$

Therefore X has all required properties.
b) Put $W = JX, V = JX$ and apply Lemma 2.4. □

Theorem 2.7.
a) *The points* 0 *and* ∞ *are regular critical points of* A.
b) *The operator* A *is similar to a selfadjoint operator in the Hilbert space* $L^2(\mathbb{R}, |\rho|)$.

Proof. a) follows from the previously proved results exactly as in the proof of [2, Theorem 6] .
b) follows from a) exactly as in the proof of [2, Corollary 7]. □

3. A difference operator

Let $x, m > 1$. In the Hilbert space $(l^2(m^k), (\cdot\,,\cdot)_+)$, defined by

$$\begin{aligned} l^2(m^k) &:= \left\{ (f_k)_{k\in\mathbb{Z}} : \sum_{k\in\mathbb{Z}} |f_k|^2 m^k < \infty \right\}, \\ (f,g)_+ &:= \sum_{k\in\mathbb{Z}} f_k \overline{g_k} m^k, \qquad f = (f_k),\, g = (g_k) \in l^2(m^k), \end{aligned}$$

we introduce the "maximal" operator $G_{\max}$ by

$$\begin{aligned} \mathcal{D}(G_{\max}) &:= \left\{ (f_k) \in l^2(m^k) : \sum_{k\in\mathbb{Z}} \frac{1}{m^k} \left| \frac{f_k - f_{k-1}}{x^k - x^{k-1}} - \frac{f_{k+1} - f_k}{x^{k+1} - x^k} \right|^2 < \infty \right\}, \\ G_{\max} f &:= \left(\frac{1}{m^k} \left(\frac{f_k - f_{k-1}}{x^k - x^{k-1}} - \frac{f_{k+1} - f_k}{x^{k+1} - x^k} \right) \right)_{k\in\mathbb{Z}}, \qquad f = (f_k) \in l^2(m^k). \end{aligned}$$

For $f \in \mathcal{D}(G_{\max})$, $h := G_{\max} f (\in l^2(m^k))$ and $l \in \mathbb{Z}$, the limit

$$f'(0) := \lim_{n\to-\infty} \frac{f_{n+1} - f_n}{x^{n+1} - x^n} = \lim_{n\to-\infty} \sum_{n+1\le j\le l} h_j m^j + \frac{f_{l+1} - f_l}{x^{l+1} - x^l}$$

exists since $|h_j m^j| \le m^{\frac{j}{2}}$ for all sufficiently small $j \in \mathbb{Z}$. Therefore, for $k \in \mathbb{Z}$ also the limit

$$(3.1) \qquad f(0) \;:=\; \lim_{n\to-\infty} f_n = f_k - \lim_{n\to-\infty} \sum_{n\le l\le k-1} f_{l+1} - f_l$$

$$= \quad f_k - \lim_{n\to-\infty} \sum_{n\le l\le k-1} \left(f'(0) - \sum_{j\le l} h_j m^j \right) (x^{l+1} - x^l)$$

$$= \quad f_k - f'(0)x^k + \sum_{l\le k-1} \left(\sum_{j\le l} h_j m^j \right) (x^{l+1} - x^l)$$

exists. This relation allows us to apply the theory of strings from [6, Chapter 5] (with the function $m(x) := \sum_{x^k\le x} m^k$). Since $\sum_{k\in\mathbb{Z}} x^{2k} m^k = \infty$, it follows from [6, Section 5.2] that the operator G, defined by

$$\mathcal{D}(G) := \{f \in \mathcal{D}(G_{\max}) : f'(0) = 0\}, \qquad Gf := G_{\max} f,$$

is selfadjoint in$(l^2(m^k), (\cdot\,,\,\cdot)_+)$. Moreover, by [6, Section 5.4] for every $\lambda \in \mathbb{C}\setminus\mathbb{R}$, there exists a sequence $\psi^\lambda = (\psi_k^\lambda)_{k\in\mathbb{Z}} \in \mathcal{D}(G_{\max})$ with

$$G_{\max}\psi^\lambda = \lambda\psi^\lambda, \qquad (\psi^\lambda)'(0) = -1. \tag{3.2}$$

For $f = (f_k) \in \mathcal{D}(G_{\max})$ and $i < j$ we have

$$\sum_{k=i}^{j} \left(\frac{f_k - f_{k-1}}{x^k - x^{k-1}} - \frac{f_{k+1} - f_k}{x^{k+1} - x^k} \right) \overline{f_k}$$

$$= \frac{f_i - f_{i-1}}{x^i - x^{i-1}} \overline{f_i} - \frac{f_{j+1} - f_j}{x^{j+1} - x^j} \overline{f_j} + \sum_{k=i+1}^{j} \frac{|f_k - f_{k-1}|^2}{x^k - x^{k-1}}.$$

Now $\sum_{j=1}^{\infty} |f_j|^2 m^j < \infty$ implies $f_j \to 0$, $j \to \infty$. Letting $i \to -\infty, j \to \infty$ and noting that

$$\left| \frac{f_{j+1} - f_j}{x^{j+1} - x^j} \right| \le \frac{2}{x-1} \cdot \sup_{k\in\mathbb{N}} |f_k| < \infty, \qquad j \in \mathbb{N},$$

we get the formula

$$(G_{\max} f, f)_+ = f'(0)\overline{f(0)} + \sum_{k\in\mathbb{Z}} \frac{|f_{k+1} - f_k|^2}{x^{k+1} - x^k}. \tag{3.3}$$

Define $\mathcal{K} := l^2(m^k) \times l^2(m^k)$, and for $f = \begin{pmatrix} f^+ \\ f^- \end{pmatrix}$, $g = \begin{pmatrix} g^+ \\ g^- \end{pmatrix} \in \mathcal{K}$,

$$[f, g] := (f^+, g^+)_+ - (f^-, g^-)_+, \quad (f, g) := (f^+, g^+)_+ + (f^-, g^-)_+.$$

Then $(\mathcal{K}, [\cdot\,,\,\cdot])$ is a Krein space, the operator $Jf := \begin{pmatrix} f^+ \\ -f^- \end{pmatrix}$ is a fundamental symmetry in $(\mathcal{K}, [\cdot\,,\,\cdot])$ and $(\cdot\,,\,\cdot)$ is the corresponding positive definite scalar product. In $\mathcal{K}$ we consider the operator A, given by

$$\mathcal{D}(A) := \left\{ \begin{pmatrix} f^+ \\ f^- \end{pmatrix} \in (\mathcal{D}(G_{\max}))^2 : f^+(0) = f^-(0),\ (f^+)'(0) = -(f^-)'(0) \right\},$$

$$A\begin{pmatrix} f^+ \\ f^- \end{pmatrix} := \begin{pmatrix} G_{\max} f^+ \\ -G_{\max} f^- \end{pmatrix}.$$

Lemma 3.1. *The operator A is symmetric and positive in $(\mathcal{K}, [\,\cdot\,,\,\cdot\,])$.*

Proof. For $f = \begin{pmatrix} f^+ \\ f^- \end{pmatrix} \in \mathcal{D}(A)$ it follows from (3.3) that

$$\begin{aligned} [Af, f] &= (G_{\max} f^+, f^+) + (G_{\max} f^-, f^-) \\ &= \sum_{k\in\mathbb{Z}} \frac{|f^+_{k+1} - f^+_k|^2}{x^{k+1} - x^k} + \sum_{k\in\mathbb{Z}} \frac{|f^-_{k+1} - f^-_k|^2}{x^{k+1} - x^k} \geq 0. \end{aligned}$$

It is easy to see that the linear functional $\delta(f) := f(0)$, defined on the dense subspace $\mathcal{D}(G)$ of $l^2(m^k)$, is not continuous with respect to $(\,\cdot\,,\,\cdot\,)_+$. Therefore, the kernel of δ, i.e. $\{f \in \mathcal{D}(G_{\max}) : f'(0) = f(0) = 0\}$, is dense in $l^2(m^k)$, hence $\mathcal{D}(A)$ is dense in $\mathcal{K}$. □

Lemma 3.2. *For $\lambda \in \mathbb{C} \setminus \mathbb{R}$ we have $\psi^\lambda(0) \neq -\psi^{-\lambda}(0)$.*

Proof. Assume $\psi^\lambda(0) = -\psi^{-\lambda}(0)$. It follows from (3.2) that

$$\Psi^\lambda := \begin{pmatrix} \psi^\lambda \\ -\psi^{-\lambda} \end{pmatrix} \in \mathcal{D}(A) \quad \text{and} \quad A\Psi^\lambda = \begin{pmatrix} G_{\max}\psi^\lambda \\ G_{\max}\psi^{-\lambda} \end{pmatrix} = \lambda \Psi^\lambda.$$

This is a contradiction since by Lemma 3.1 and [10, Lemma 1.1] A cannot have a nonreal eigenvalue. □

Proposition 3.3. *The operator A is definitizable in $(\mathcal{K}, [\,\cdot\,,\,\cdot\,])$.*

Proof. Let $\lambda \in \mathbb{C} \setminus \mathbb{R}$ $(\subset \rho(G))$ and $g = \begin{pmatrix} g^+ \\ g^- \end{pmatrix} \in \mathcal{K}$. With

$$h^+ := (G - \lambda)^{-1} g^+, \qquad h^- := -(G + \lambda)^{-1} g^- \in \mathcal{D}(G),$$

we consider the sequences

$$f^+ = h^+ + c^+ \psi^\lambda, \qquad f^- := h^- + c^- \psi^{-\lambda} \in \mathcal{D}(G_{\max}),$$

where $(c^+, c^-) \in \mathbb{C}^2$ is the solution of the system of equations

$$\begin{aligned} 0 &= h^+(0) - h^-(0) + c^+\psi^\lambda(0) - c^-\psi^{-\lambda}(0) \quad (= f^+(0) - f^-(0)), \\ 0 &= -c^+ - c^- \quad (= (h^+)'(0) + (h^-)'(0) + c^+(\psi^\lambda)'(0) + c^-(\psi^{-\lambda})'(0) \\ &= (f^+)'(0) + (f^-)'(0)\,). \end{aligned}$$

Note that indeed this system is solvable since by Lemma 3.2,

$$\det\begin{pmatrix} \psi^{\lambda}(0) & -\psi^{-\lambda}(0) \\ -1 & -1 \end{pmatrix} = -\psi^{\lambda}(0) - \psi^{-\lambda}(0) \neq 0.$$

Then

$$f := \begin{pmatrix} f^+ \\ f^- \end{pmatrix} \in \mathcal{D}(A) \quad \text{and} \quad (A-\lambda)f = \begin{pmatrix} G_{\max} - \lambda)f^+ \\ -(G_{\max} + \lambda)f^- \end{pmatrix} = \begin{pmatrix} g^+ \\ g^- \end{pmatrix} = g.$$

Applying this result for $\lambda = i$ and $\lambda = -i$ and [10, Lemma 1.1] it follows that $\mathcal{R}(A-i) = \mathcal{R}(A+i) = \mathcal{K}$, $\mathcal{N}(A-i) = \mathcal{N}(A+i) = \{0\}$. From Lemma 3.1 it follows that A is selfadjoint and $\pm i \in \rho(A)$. □

From Lemma 3.1 it follows that $\sigma(A) \subset \mathbb{R}$. In order to find out additional spectral properties of A for $l \in \mathbb{Z}$ we consider the shift operator U_l in $\mathcal{K}$ given by

$$U_l f := \begin{pmatrix} (f^+_{k+l})_{k\in\mathbb{Z}} \\ (f^-_{k+l})_{k\in\mathbb{Z}} \end{pmatrix}, \qquad f = \begin{pmatrix} f^+ \\ f^- \end{pmatrix} \in \mathcal{K}.$$

Let U_l^* denote its adjoint in the Hilbert space $(\mathcal{K}, (\cdot\,,\cdot))$.

Lemma 3.4. *For all $l \in \mathbb{Z}$ the operator U_l is bounded and boundedly invertible in $\mathcal{K}$ with $U_l^{-1} = U_{-l}$ and $U_l^* = m^{-l}U_{-l}$. For $f = \begin{pmatrix} f^+ \\ f^- \end{pmatrix} \in \mathcal{D}(A)$ it holds*

$$U_l f \in \mathcal{D}(A), \qquad AU_l f = (x\cdot m)^l U_l A f, \tag{3.4}$$

$$(U_l f)^{\pm}(0) = f^{\pm}(0), \qquad \left((U_l f)^{\pm}\right)'(0) = x^l (f^{\pm})'(0). \tag{3.5}$$

Proof. For $f = \begin{pmatrix} f^+ \\ f^- \end{pmatrix} \in \mathcal{K}$ we have

$$(U_l f, U_l f) = m^{-l}\left(\sum_{k\in\mathbb{Z}} |f^+_{k+l}|^2 m^{k+l} + \sum_{k\in\mathbb{Z}} |f^-_{k+l}|^2 m^{k+l}\right) = m^{-l}(f,f).$$

If $f \in \mathcal{D}(A)$, then for $g = \begin{pmatrix} g^+ \\ g^- \end{pmatrix} := U_l f$ it holds

$$\begin{aligned} &\frac{1}{m^k}\left(\frac{g_k^{\pm} - g_{k-1}^{\pm}}{x^k - x^{k-1}} - \frac{g_{k+1}^{\pm} - g_k^{\pm}}{x^{k+1} - x^k}\right) \\ &= (x\cdot m)^l \frac{1}{m^{k+l}}\left(\frac{f^{\pm}_{k+l} - f^{\pm}_{k+l-1}}{x^{k+l} - x^{k+l-1}} - \frac{f^{\pm}_{k+l+1} - f^{\pm}_{k+l}}{x^{k+l+1} - x^{k+l}}\right), \\ &g_k^{\pm} = f^{\pm}_{k+l} \to f^{\pm}(0) \qquad k \to -\infty, \\ &\frac{g_{k+1}^{\pm} - g_k^{\pm}}{x^{k+1} - x^k} = x^l \frac{f^{\pm}_{k+l+1} - f^{\pm}_{k+l}}{x^{k+l+1} - x^{k+l}} \to x^l (f^{\pm})'(0), \qquad k \to -\infty. \end{aligned}$$

This completes the proof. □

Since $\mathcal{D}(A) \neq \mathcal{K}$, the operator A is unbounded. By [9, II.2, Corollary 2], $\sigma(A)$ is unbounded. From Lemma 3.4 it follows that $U_l(\mathcal{D}(A)) = \mathcal{D}(A)$ and $U_l^{-1}AU_l = (x \cdot m)^l A$. Therefore, we have

$$\sigma(A) = \sigma(U_l^{-1}AU_l) = (x \cdot m)^l \cdot \sigma(A) \tag{3.6}$$

and hence $0 \in \sigma(A)$. However, 0 is not an eigenvalue of A, because by (3.1) $A\begin{pmatrix} f^+ \\ f^- \end{pmatrix} = 0$ implies

$$(f_k^\pm)_{k\in\mathbb{Z}} = (f^\pm(0) + (f^\pm)'(0)x^k)_{k\in\mathbb{Z}} \in l^2(m^k),$$

which is only satisfied if $f^\pm(0) = (f^\pm)'(0) = 0$.

Proposition 3.5. *The spectrum of A is symmetric with respect to 0 (i.e., $\sigma(A) = -\sigma(A)$) and $0, \infty$ are accumulation points of $\sigma(A)$. The only critical points of A are 0 and ∞.*

Proof. Because of (3.6) it remains to show that $\sigma(A) = -\sigma(A)$. The operator

$$P : \mathcal{K} \to \mathcal{K}, \qquad P\begin{pmatrix} f^+ \\ f^- \end{pmatrix} := \begin{pmatrix} f^- \\ f^+ \end{pmatrix}$$

is bounded with $P^{-1} = P$. For $f = \begin{pmatrix} f^+ \\ f^- \end{pmatrix} \in \mathcal{D}(A)$ we have

$$Pf \in \mathcal{D}(A), \qquad APf = \begin{pmatrix} G_{\max} f^- \\ -G_{\max} f^+ \end{pmatrix} = -PAf.$$

This implies $P(\mathcal{D}(A)) = \mathcal{D}(A), P^{-1}AP = -A$. □

Lemma 3.6. *The spaces*

$$\mathcal{D}_0 := \left\{ \begin{pmatrix} f^+ \\ f^- \end{pmatrix} \in \mathcal{D}(A) : f^\pm(0) = (f^\pm)'(0) = 0 \right\}, \qquad \mathcal{R}_0 := A(\mathcal{D}_0)$$

are invariant under the fundamental symmetry J.

Proof. For $f = \begin{pmatrix} f^+ \\ f^- \end{pmatrix} \in \mathcal{D}_0$ we have $Jf \in \mathcal{D}_0$ and $AJf = \begin{pmatrix} G_{\max} f^+ \\ G_{\max} f^- \end{pmatrix} = JAf$. □

Lemma 3.7.
a) *There exists a positive, bounded and boundedly invertible operator X in the Hilbert space $(\mathcal{K}, (\cdot\,,\cdot))$ such that $X(\mathcal{D}(A)) \subset \mathcal{D}_0$ and $X(\mathcal{R}(A)) \subset \mathcal{R}_0$.*
b) *There exists a positive, bounded and boundedly invertible operator W in the Krein space $(\mathcal{K}, [\cdot\,,\cdot])$ such that $W(\mathcal{D}(A)) \subset \mathcal{D}(A)$ and $W(\mathcal{R}(A)) \subset \mathcal{R}(A)$.*

Proof. a) Let $(\xi_1, \xi_2, \xi_3, \xi_4) \in \mathbb{R}^4$ be the solution of the system of equations

$$
\begin{array}{rrrrl}
(xm)^{-1}\xi_1 + & (xm)^{-2}\xi_2 + & (xm)^{-3}\xi_3 + & (xm)^{-4}\xi_4 = & 1, \\
x\xi_1 + & x^2\xi_2 + & x^3\xi_3 + & x^4\xi_4 = & -1, \\
m^{-1}\xi_1 + & m^{-2}\xi_2 + & m^{-3}\xi_3 + & m^{-4}\xi_4 = & 1, \\
\xi_1 + & \xi_2 + & \xi_3 + & \xi_4 = & -1.
\end{array}
\tag{3.7}
$$

Note that this system is solvable since the corresponding Vandermonde determinant is nonzero. Let

$$Y := \sum_{l=1}^{4} \xi_l U_l, \quad Z_1 := \sum_{l=1}^{4} \xi_l x^l U_{-l}, \quad Z_2 := \sum_{l=1}^{4} \xi_l (xm)^{-l} U_l.$$

By Lemma 3.4 it holds $Y^* = \sum_{l=1}^{4} \xi_l m^{-l} U_{-l}$,

$$Y(\mathcal{D}(A)) \subset \mathcal{D}(A), \qquad Y^*(\mathcal{D}(A)) \subset \mathcal{D}(A), \qquad Z_i(\mathcal{D}(A)) \subset \mathcal{D}(A), \quad i = 1, 2,$$

and

$$(Y^*Y + I)Af = Y^*AZ_2 f + Af = A(Z_1 Z_2 f + f)$$

for $f = \begin{pmatrix} f^+ \\ f^- \end{pmatrix} \in \mathcal{D}(A)$. Then from (3.5) and (3.7) it follows that

$$
\begin{aligned}
(Y^*Yf)^{\pm}(0) &= \left(\sum_{l=1}^{4} \xi_l m^{-l}\right) \cdot \left(\sum_{l=1}^{4} \xi_l\right) \cdot f^{\pm}(0) = -f^{\pm}(0), \\
\left((Y^*Yf)^{\pm}\right)'(0) &= \left(\sum_{l=1}^{4} \xi_l (xm)^{-l}\right) \cdot \left(\sum_{l=1}^{4} \xi_l x^l\right) \cdot (f^{\pm})'(0) = -(f^{\pm})'(0), \\
(Z_1 Z_2 f)^{\pm}(0) &= \left(\sum_{l=1}^{4} \xi_l x^l\right) \cdot \left(\sum_{l=1}^{4} \xi_l (xm)^{-l}\right) \cdot f^{\pm}(0) = -f^{\pm}(0), \\
\left((Z_1 Z_2 f)^{\pm}\right)'(0) &= \left(\sum_{l=1}^{4} \xi_l\right) \cdot \left(\sum_{l=1}^{4} \xi_l m^{-l}\right) \cdot (f^{\pm})'(0) = -(f^{\pm})'(0).
\end{aligned}
$$

Therefore, $Y^*Yf + f \in \mathcal{D}_0$ and $Z_1 Z_2 f + f \in \mathcal{D}_0$. Put $X := Y^*Y + I$. Then $X(\mathcal{D}(A)) \subset \mathcal{D}_0$, and

$$X(\mathcal{R}(A)) = XA(\mathcal{D}(A)) = A(Z_1 Z_2 + I)(\mathcal{D}(A)) \subset A(\mathcal{D}_0) = \mathcal{R}_0.$$

Obviously, X is positive, bounded and boundedly invertible.
b) Put $W := JX$. □

Theorem 3.8.
a) *The points* 0 *and* ∞ *are regular critical points of* A.
b) *The operator* A *is similar to a selfadjoint operator in the Hilbert space* $(\mathcal{K}, (\cdot, \cdot))$.

Proof. The proof of Theorem 3.8 is the same as the proof of Theorem 2.7. □

References

[1] ĆURGUS, B., LANGER, H.: A Krein space approach to symmetric ordinary differential operators with an indefinite weight function; J. Differential Equations 79 (1989), 31–61.

[2] ĆURGUS, B., NAJMAN, B.: The operator $(\operatorname{sgn} x)\frac{d^2}{dx^2}$ is similar to a selfadjoint operator in $L^2(\mathbb{R})$; Proc. Amer. Math. Soc. 123 (1995), 1125–1128.

[3] ĆURGUS, B., NAJMAN, B.: Positive differential operators in Krein space $L^2(\mathbb{R}^n)$; Preprint.

[4] ĆURGUS, B., NAJMAN, B. : Examples of positive operators in Krein space with 0 a regular critical point of infinite type; Preprint.

[5] DAHO, K., LANGER, H.: Sturm-Liouville operators with an indefinite weight function; Proc. Royal Soc. Edinburgh Sect. A 78 (1977), 161–191.

[6] DYM, H., MCKEAN, H.P.: Gaussian processes, function theory, and the inverse spectral problem; Academic Press, New York, San Francisco, London 1976.

[7] FLEIGE, A.: A counterexample to completeness properties for indefinite Sturm-Liouville problems; Math. Nachr., to appear.

[8] FLEIGE, A.: Spectral theory of indefinite Krein–Feller differential operators; Mathematical Research 98, Akademie Verlag, Berlin 1996.

[9] LANGER, H.: Spectral function of definitizable operators in Krein spaces; Functional Analysis, Proceedings, Dubrovnik 1981, Lecture Notes in Math. 948 (1982), 1–46.

[10] LANGER, H.: Zur Spektraltheorie verallgemeinerter gewöhnlicher Differentialoperatoren zweiter Ordnung mit einer nichtmonotonen Gewichtsfunktion; Report # 14, Department of Mathematics, Univ. of Jyväskylä, 1972.

[11] VOLKMER, H.: Sturm–Liouville problems with indefinite weights and Everitt's inequality; Proc. Roy. Soc. Edinburgh Sect. A 126 (1996), 1097–1112.

Weisbachstr. 18
44139 Dortmund
Germany
fleige@feat.mathematik.uni-essen.de

Department of Mathematics
University of Zagreb
Bijenička 30
41000 Zagreb
Croatia

1991 Mathematics Subject Classification: Primary 47B50; Secondary 47B39, 39A70, 47E05, 47A10

Submitted: March 22, 1996

Operator Theory:
Advances and Applications, Vol. 102

A nonlinear spectral problem with periodic coefficients occurring in magnetohydrodynamic stability theory

A. LIFSCHITZ

Three-dimensional quasi-helical plasma equilibria with flow are considered. A method for studying their instabilities and waves is described. The method uses the Fourier transformation to reduce the original spectral problem for a partial differential operator with linear coefficients in the physical space to a family of spectral problems for ordinary differential operators with periodic coefficients in the Fourier space. A qualitative description of the spectrum (including bounds for nonimaginary eigenvalues) is obtained. An efficient numerical procedure for finding the spectrum is presented. A complicated interplay between the flow and the magnetic field is illustrated.

1. Introductory remarks

Ideal magnetohydrodynamics (MHD) studies slow, large-scale motions of perfect magnetized fluids (plasmas). For numerous physical and engineering applications stretching from astrophysics, to geophysics, to controlled thermonuclear fusion it is vital to analyze possible MHD instabilities and waves. The usual way of carrying out such an analysis can be summarized as follows. First, certain steady solutions of the nonlinear MHD equations representing plasma equilibria are found. One can distinguish static plasma equilibria without flow (for which the velocity field is zero while the magnetic field is nonzero), and more general plasma equilibria with flow (for which both the velocity and magnetic fields are nonzero). Second, the MHD equations are linearized in the vicinity of a certain equilibrium and their solutions are studied either directly or via spectral methods.

The stability theory is an integral part of MHD. This theory studies the impact of initially small perturbations on a given steady plasma equilibrium (with or without flow). An equilibrium is called stable and can occur in nature if perturbations do not have a profound effect on its properties. An equilibrium is called unstable if under the influence of perturbations it either evolves into a different equilibrium, or looses its steady character altogether. Over a period of years many classical stability problems were solved, however, several important stability problems are still open. There are many stability definitions and we have to be careful to choose the one which is the most appropriate for the physical problem under consideration. For the sake ot the present study we use a rather crude definition of stability. Namely, we call an equilibrium stable if the spectrum of the linearized operator describing the evolution of small perturbations depending on time as $\exp(\sigma t)$ does not contain points σ with $\Re(\sigma) > 0$, otherwise we call the corresponding equilibrium unstable. It is well-known that the spectrum of the linearized MHD equations is symmetric with respect to the imaginary axis (see below), so that the spectra of stable equilibria are located on the imaginary axis.

The wave theory studies the behavior of small perturbations in the vicinity of stable plasma equilibria. The main issues include the analysis of the distribution of the spectrum on the imaginary axis, the structure of the corresponding eigenfunctions, the eigenfunction expansion formulas, and the asymptotic behavior of perturbations in time.

For many years most of the investigations of plasma instabilities and waves were carried out under the assumption that the background equilibria are static, while equilibria with flow received little attention (cf, e.g, [4], [6], [9]). Generally speaking, the above assumption is satisfied in the laboratory context but violated in the astrophysical and geophysical contexts, [1], [4]. In order to address a variety of problems occurring in astrophysics and geophysics it is very important to generalize the results concerning instabilities and waves for equilibria without flow to the case of equilibria with flow. However, such a generalization is rather nontrivial. The reason is that for equilibria without flow the corresponding analysis boils down to studying a self-adjoint operator [2], [9], while for equilibria with flow it reduces to studying a quadratic operator pencil [5]. Although some properties of such pencils are known (cf., e.g., [7], [12], [13]), by no means they are understood with the same degree of completeness as the corresponding self-adjoint operators.

In the present paper we consider a special class of plasma equilibria with flow which we call quasi-helical equilibria, and analyze the corresponding instabilities and waves. To achieve our goals we extend the technique proposed in a recent paper [10] for studying the analogous problem for two-dimensional quasi-circular equilibria. Our results can be applied in order to describe instabilities and waves in various astrophysical situations.

2. Three-dimensional quasi-helical plasma equilibria with flow

In Cartesian coordinates x_1, x_2, x_3 the nonlinear MHD equations describing the evolution of an ideal incompressible plasma of constant density have the form

$$\begin{gathered}\partial_t \mathbf{V}_f + \mathbf{V}_f \cdot \nabla \mathbf{V}_f - \mathbf{V}_a \cdot \nabla \mathbf{V}_a + \nabla P = 0,\\ \partial_t \mathbf{V}_a + \mathbf{V}_f \cdot \nabla \mathbf{V}_a - \mathbf{V}_a \cdot \nabla \mathbf{V}_f = 0,\\ \nabla \cdot \mathbf{V}_f = 0, \qquad \nabla \cdot \mathbf{V}_a = 0,\end{gathered} \tag{2.1}$$

where $\mathbf{V}_f$ is the fluid velocity, $\mathbf{V}_a$ is the Alfvén velocity representing the magnetic field, and P is the total kinematic pressure.

These equations supplied with appropriate initial and boundary conditions describe an extremely rich and complicated variety of plasma motions. We are interested in a special class of steady solutions of these equations which we call quasi-helical equilibria. For such equilibria the corresponding $\mathbf{V}_f, \mathbf{V}_a, P$ have the form

$$\begin{aligned} &\mathbf{V}_f = \mathbf{V}_{f\perp} + \mathbf{V}_{f3}, \qquad \mathbf{V}_{f\perp} = c_f \Omega \mathcal{L}_\perp \mathbf{x}_\perp, \qquad \mathbf{V}_{f3} = \nu_f \mathbf{e}_3, \\ &\mathbf{V}_a = \mathbf{V}_{a\perp} + \mathbf{V}_{a3}, \qquad \mathbf{V}_{a\perp} = c_a \Omega \mathcal{L}_\perp \mathbf{x}_\perp, \qquad \mathbf{V}_{a3} = \nu_a \mathbf{e}_3, \\ &P = \tfrac{1}{2}(c_f^2 - c_a^2)|\mathbf{x}_\perp|^2 + P_0, \end{aligned} \tag{2.2}$$

where the subscripts $\perp, 3$ indicate the projections of matrices and vectors onto the horizontal (x_1, x_2) plane and the vertical x_3 axis. Depending on the context we understand $\mathbf{V}_\perp$ either as a three-component vector, $\mathbf{V}_\perp = (V_1, V_2, 0)$, or as a two-component one, $\mathbf{V}_\perp = (V_1, V_2)$. We treat matrices in a similar fashion. Here Ω is the characteristic angular velocity, $\mathcal{L}_\perp$ is a nondimensional 2×2 matrix such that $\mathrm{tr}\mathcal{L}_\perp = 0$, $\det\mathcal{L}_\perp = 1$, c_f, c_a are constants characterizing the relative magnitude of the horizontal fluid and Alfvén velocities, ν_f, ν_a are constants characterizing the vertical fluid and Alfvén velocities, and P_0 is the pressure at the origin. Since the MHD equations are invariant with respect to rotations in the horizontal plane, we may assume without any loss of generality the matrix $\mathcal{L}_\perp$ is antidiagonal and write it in the form

$$\mathcal{L}_\perp = \begin{pmatrix} 0 & -a_1/a_2 \\ a_2/a_1 & 0 \end{pmatrix}, \tag{2.3}$$

where a_1, a_2, are certain characteristic lengths in the x_1 and x_2 directions, and $a_1 \geq a_2$. We also assume that the coefficients c_f, c_a are normalized in such a way that $c_f^2 + c_a^2 = 1$, $c_a > 0$, and represent them in the form $c_f = \cos\chi, c_a = \sin\chi$ with $0 \leq \chi \leq \pi$.

When $a_1 = a_2$ equilibria (2.2) have helical streamlines and lines of force projecting onto circles in the (x_1, x_2) plane. When $a_1 \neq a_2$ streamlines and lines of force are distorted helices projecting onto ellipses in the horizontal plane. For this reason we call general equilibria (2.2) quasi-helical. The departure of the general equilibria from helical equilibria is characterized by a nondimensional ellipticity parameter $\delta = (a_1^2 - a_2^2)/(a_1^2 + a_2^2)$. Although quasi-helical equilibria are very interesting in many respects, they do have certain deficiencies related to the unphysical behavior of the variables $\mathbf{V}_f, \mathbf{V}_a, P$ at infinity including a possibility for P to become negative there. For the purpose of the present study we disregard these deficiencies.

3. Basic equations

Instabilities and waves associated with quasi-helical equilibria are governed by the linearized MHD equations of the form

$$\begin{aligned}
&\sigma \mathbf{v}_f + (c_f \Omega \mathcal{L}_\perp \mathbf{x}_\perp \cdot \nabla_\perp + \nu_f \partial_{x_3})\mathbf{v}_f + c_f \Omega \mathcal{L}_\perp \mathbf{v}_{f\perp} \\
&-(c_a \Omega \mathcal{L}_\perp \mathbf{x}_\perp \cdot \nabla_\perp + \nu_a \partial_{x_3})\mathbf{v}_a - c_a \Omega \mathcal{L}_\perp \mathbf{v}_{a\perp} + \nabla p = 0, \\
&\sigma \mathbf{v}_a + (c_f \Omega \mathcal{L}_\perp \mathbf{x}_\perp \cdot \nabla_\perp + \nu_f \partial_{x_3})\mathbf{v}_a + c_a \Omega \mathcal{L}_\perp \mathbf{v}_{f\perp} \\
&-(c_a \Omega \mathcal{L}_\perp \mathbf{x}_\perp \cdot \nabla_\perp + \nu_a \partial_{x_3})\mathbf{v}_f - c_f \Omega \mathcal{L}_\perp \mathbf{v}_{a\perp} = 0, \\
&\nabla \cdot \mathbf{v}_f = 0, \qquad \nabla \cdot \mathbf{v}_a = 0,
\end{aligned} \tag{3.1}$$

where $\mathbf{v}_f, \mathbf{v}_a, p$ are the perturbed fluid and Alfvén velocities, and pressure, respectively, and the regularity conditions at infinity. Here we assume that the perturbed quantities depend on time as $\exp(\sigma t)$. We first consider the corresponding spectral problem in the space of smooth vector functions with compact support and then take its suitable extension in the space of square-integrable vector functions. The spectrum of this problem is denoted by $\Sigma(\delta, \Omega, \chi, \nu_f, \nu_a)$. Below (cf. problems (3.3), (3.6), etc.,) we understand the regularity conditions at infinity in a similar way.

To simplify equations (3.1) we introduce the characteristic scale along the x_3 axis $a_3 = [(a_1^2 + a_2^2)/2]^{1/2}$, define the nondimensional stretched variables by

$$x_i' = \frac{x_i}{a_i}, \qquad v_{\alpha i}' = \frac{v_{\alpha i}}{\Omega a_i}, \qquad \sigma' = \frac{\sigma}{\Omega}, \qquad p' = \frac{p}{\Omega^2 a_3^2}, \qquad \nu_\alpha' = \frac{\nu_\alpha}{\Omega a_3}, \tag{3.2}$$

where $i = 1, 2, 3, \alpha = f, a$, and rewrite equations (3.1) in the form

$$\begin{aligned}
&\sigma' \mathbf{v}_f' + (c_f \mathcal{J}_\perp \mathbf{x}_\perp' \cdot \nabla_\perp' + \nu_f' \partial_{x_3'})\mathbf{v}_f' + c_f \mathcal{J}_\perp \mathbf{v}_{f\perp}' \\
&-(c_a \mathcal{J}_\perp \mathbf{x}_\perp' \cdot \nabla_\perp' + \nu_a' \partial_{x_3'})\mathbf{v}_a' - c_a \mathcal{J}_\perp \mathbf{v}_{a\perp}' + \mathcal{G}^{-1} \nabla' p' = 0, \\
&\sigma' \mathbf{v}_a' + (c_f \mathcal{J}_\perp \mathbf{x}_\perp' \cdot \nabla_\perp' + \nu_f' \partial_{x_3'})\mathbf{v}_a' + c_a \mathcal{J}_\perp \mathbf{v}_{f\perp}' \\
&-(c_a \mathcal{J}_\perp \mathbf{x}_\perp' \cdot \nabla_\perp' + \nu_a' \partial_{x_3'})\mathbf{v}_f' - c_f \mathcal{J}_\perp \mathbf{v}_{a\perp}' = 0, \\
&\nabla' \cdot \mathbf{v}_f' = 0, \qquad \nabla' \cdot \mathbf{v}_a' = 0,
\end{aligned} \tag{3.3}$$

where $\mathcal{J}_\perp = \begin{pmatrix} 0 & -1 \\ 1 & 0 \end{pmatrix}$ is the rotation matrix in the horizontal plane and $\mathcal{G}$ is the matrix of metric coefficients, $\mathcal{G} = \mathrm{diag}(1 + \delta, 1 - \delta, 1)$. As before, we augment these equations with the regularity conditions. We denote the spectrum of the corresponding problem by $\Sigma'(\delta, \chi, \nu_f', \nu_a')$. It is clear that $\Sigma(\delta, \Omega, \chi, \nu_f, \nu_a) = \Omega\Sigma'(\delta, \chi, \nu_f', \nu_a')$. Below we study the nondimensional spectral problem; we omit primes for the sake of brevity.

In order to replace dynamical equations (3.3) by a single equation we introduce the incompressible Lagrangian displacement $\boldsymbol{\xi}$, $\nabla \cdot \boldsymbol{\xi} = 0$, related to the fluid velocity by

$$\mathbf{v}_f = (\sigma + c_f \mathcal{J}_\perp \mathbf{x}_\perp \cdot \nabla_\perp + \nu_f \partial_{x_3})\boldsymbol{\xi} - c_f \mathcal{J}_\perp \boldsymbol{\xi}_\perp. \tag{3.4}$$

We integrate the second equation (3.3) and express the Alfvén velocity in terms of $\boldsymbol{\xi}$ as:

$$\mathbf{v}_a = (c_a \mathcal{J}_\perp \mathbf{x}_\perp \cdot \nabla_\perp + \nu_a \partial_{x_3})\boldsymbol{\xi} - c_a \mathcal{J}_\perp \boldsymbol{\xi}_\perp. \tag{3.5}$$

We substitute expressions (3.4), (3.5) into the first equation (3.3) and obtain the following equations for $\boldsymbol{\xi}$

$$\begin{gathered}
[(\sigma + c_f \mathcal{J}_\perp \mathbf{x}_\perp \cdot \nabla_\perp + \nu_f \partial_{x_3})^2 - (c_a \mathcal{J}_\perp \mathbf{x}_\perp \cdot \nabla_\perp + \nu_a \partial_{x_3})^2]\boldsymbol{\xi} \\
+(c_f^2 - c_a^2)\boldsymbol{\xi}_\perp + \mathcal{G}^{-1}\nabla p = 0, \\
\nabla \cdot \boldsymbol{\xi} = 0.
\end{gathered} \tag{3.6}$$

As before, we augment these equations with the regularity conditions. The reduction of the first-order linear MHD equations for $\mathbf{v}_f, \mathbf{v}_a, p$ to a second-order equation for $\boldsymbol{\xi}, p$ is possible for the most general equilibria with flow (cf. [5]).

Our objective is to solve the spectral problem for the corresponding quadratic pencil. The spectrum of this problem coincides with $\Sigma(\delta, \chi, \nu_f, \nu_a)$. Considering the adjoint and complex conjugate problems, we can easily prove that this spectrum is symmetric with respect to the imaginary and real axes:

$$\Sigma(\delta, \chi, \nu_f, \nu_a) = -\bar{\Sigma}(\delta, \chi, \nu_f, \nu_a), \qquad \Sigma(\delta, \chi, \nu_f, \nu_a) = \bar{\Sigma}(\delta, \chi, \nu_f, \nu_a). \tag{3.7}$$

In order to reduce the corresponding spectral problem to a simpler form we consider the Fourier transformation of equations (3.6) and obtain the following equations in the Fourier space with coordinates k_1, k_2, k_3

$$\begin{gathered}
[(\sigma + c_f \mathcal{J}_\perp \mathbf{k}_\perp \cdot \nabla_\perp + ik_3\nu_f)^2 - (c_a \mathcal{J}_\perp \mathbf{k}_\perp \cdot \nabla_\perp + ik_3\nu_a)^2]\tilde{\boldsymbol{\xi}} \\
+(c_f^2 - c_a^2)\tilde{\boldsymbol{\xi}}_\perp + i\tilde{p}\mathcal{G}^{-1}\mathbf{k} = 0, \\
\mathbf{k} \cdot \tilde{\boldsymbol{\xi}} = 0,
\end{gathered} \tag{3.8}$$

where $\tilde{\boldsymbol{\xi}}, \tilde{p}$ denote the Fourier images of $\boldsymbol{\xi}$ and p. We augment these equations with the regularity conditions. Since equation (3.8) does not contain differentiation with respect to k_3 the third coordinate k_3 can be treated as a parameter, rather than an independent variable on a par with coordinates k_1 and k_2. In particular, the spectrum of the corresponding problem parametrically depends on k_3. We denote this spectrum by $\Sigma(\delta, \chi, \nu_f, \nu_a, k_3)$. It is clear that

$$\Sigma(\delta, \chi, \nu_f, \nu_a) = \overline{\bigcup_{-\infty<k_3<\infty} \Sigma(\delta, \chi, \nu_f, \nu_a, k_3)}, \tag{3.9}$$

and

$$\begin{aligned}
\Sigma(\delta, \chi, \nu_f, \nu_a, k_3) &= -\overline{\Sigma}(\delta, \chi, \nu_f, \nu_a, k_3), \\
\Sigma(\delta, \chi, \nu_f, \nu_a, k_3) &= \overline{\Sigma}(\delta, \chi, \nu_f, \nu_a, -k_3).
\end{aligned} \tag{3.10}$$

Equations (3.8) suggest that without loss of generality we may assume that $\nu_f = 0$, since the spectrum of the problem with $\nu_f \neq 0$ can be obtained from the spectrum of the problem with $\nu_f = 0$ via a shift along the imaginary axis:

$$\Sigma(\delta, \chi, \nu_f, \nu_a, k_3) = \Sigma(\delta, \chi, 0, \nu_a, k_3) - ik_3\nu_f. \tag{3.11}$$

Physically equation (3.11) describes the Doppler shift of the frequency. Below we assume that $\nu_f = 0$ and use the simplified notation $\Sigma(\delta, \chi, \nu_a, k_3)$ instead of $\Sigma(\delta, \chi, 0, \nu_a, k_3)$.

In the Fourier space we can eliminate the pressure in favor of the Lagrangian displacement by virtue of the incompressibility condition and rewrite the spectral problem in terms of the Lagrangian displacement alone. Applying the operators $(\mathcal{J}_\perp \mathbf{k}_\perp \cdot \nabla_\perp)$ and $(\mathcal{J}_\perp \mathbf{k}_\perp \cdot \nabla_\perp)^2$ to the incompressibility condition we obtain

$$\begin{aligned} &(\mathcal{J}_\perp \mathbf{k}_\perp \cdot \nabla_\perp)\tilde{\boldsymbol{\xi}} \cdot \mathbf{k} + \tilde{\boldsymbol{\xi}} \cdot \mathcal{J}_\perp \mathbf{k} = 0, \\ &(\mathcal{J}_\perp \mathbf{k}_\perp \cdot \nabla_\perp)^2 \tilde{\boldsymbol{\xi}} \cdot \mathbf{k} + 2(\mathcal{J}_\perp \mathbf{k}_\perp \cdot \nabla_\perp)\tilde{\boldsymbol{\xi}} \cdot \mathcal{J}_\perp \mathbf{k} - \tilde{\boldsymbol{\xi}}_\perp \cdot \mathbf{k}_\perp = 0, \end{aligned} \tag{3.12}$$

Taking the scalar product of the first equation (3.8) with $\mathbf{k}$ and using the incompressibility condition together with equations (3.12), we obtain the following expression for $i\tilde{p}$:

$$\begin{aligned} i\tilde{p} = \tfrac{2}{\mathcal{G}^{-1}\mathbf{k}\cdot\mathbf{k}} \Big\{ &(c_f^2 - c_a^2)[(\mathcal{J}_\perp \mathbf{k}_\perp \cdot \nabla_\perp)\tilde{\boldsymbol{\xi}} \cdot \mathcal{J}_\perp \mathbf{k} - \tilde{\boldsymbol{\xi}}_\perp \cdot \mathbf{k}_\perp] \\ &+ (\sigma c_f - i k_3 \nu_a c_a)\tilde{\boldsymbol{\xi}} \cdot \mathcal{J}_\perp \mathbf{k} \Big\}. \end{aligned} \tag{3.13}$$

Substituting this expression in equations (3.8) we obtain the equations for $\tilde{\boldsymbol{\xi}}$ alone:

$$\begin{aligned} &\sigma^2 \tilde{\boldsymbol{\xi}} + 2\sigma c_f (\mathcal{J}_\perp \mathbf{k}_\perp \cdot \nabla_\perp + \mathcal{Q})\tilde{\boldsymbol{\xi}} \\ &+ (c_f^2 - c_a^2)(\mathcal{J}_\perp \mathbf{k}_\perp \cdot \nabla_\perp + \mathcal{J}_\perp + 2\mathcal{Q})(\mathcal{J}_\perp \mathbf{k}_\perp \cdot \nabla_\perp - \mathcal{J}_\perp)\tilde{\boldsymbol{\xi}} \\ &- 2 i k_3 \nu_a c_a (\mathcal{J}_\perp \mathbf{k}_\perp \cdot \nabla_\perp + \mathcal{Q})\tilde{\boldsymbol{\xi}} + k_3^2 \nu_a^2 \tilde{\boldsymbol{\xi}} = 0, \\ &\tilde{\boldsymbol{\xi}} \cdot \mathbf{k} = 0. \end{aligned} \tag{3.14}$$

Here $\mathcal{Q}$ is the projector of the form

$$\mathcal{Q} = \frac{\mathcal{G}^{-1}\mathbf{k} \otimes \mathcal{J}_\perp \mathbf{k}_\perp}{\mathcal{G}^{-1}\mathbf{k} \cdot \mathbf{k}}, \tag{3.15}$$

where $\otimes$ denotes the standard tensor product of $\mathcal{G}^{-1}\mathbf{k}$ and $\mathcal{J}_\perp \mathbf{k}_\perp$. Finally, we project the first equation (3.14) onto the horizontal plane and obtain the following family of equations for $\mathbf{s} \equiv \tilde{\boldsymbol{\xi}}_\perp$ parametrically depending on k_3:

$$\begin{aligned} &\sigma^2 \mathbf{s} + 2\sigma c_f (\mathcal{J}_\perp \mathbf{k}_\perp \cdot \nabla_\perp + \mathcal{Q}_\perp)\mathbf{s} \\ &+ (c_f^2 - c_a^2)(\mathcal{J}_\perp \mathbf{k}_\perp \cdot \nabla_\perp + \mathcal{J}_\perp + 2\mathcal{Q}_\perp)(\mathcal{J}_\perp \mathbf{k}_\perp \cdot \nabla_\perp - \mathcal{J}_\perp)\mathbf{s} \\ &- 2 i k_3 \nu_a c_a (\mathcal{J}_\perp \mathbf{k}_\perp \cdot \nabla_\perp + \mathcal{Q}_\perp)\mathbf{s} + k_3^2 \nu_a^2 \mathbf{s} = 0. \end{aligned} \tag{3.16}$$

The vertical component of the Lagrangian displacement $\tilde{\xi}_3$ can be expressed in terms of $\mathbf{s}$ by virtue of the incompressibility condition as $\tilde{\xi}_3 = -\mathbf{s} \cdot \mathbf{k}_\perp / k_3$. Here we tacitly assume that $k_3 \neq 0$, the opposite case can easily be treated separately. The spectrum of problem (3.16) coincides with $\Sigma(\delta, \chi, \nu_a, k_3)$.

In order to rewrite equation (3.16) as a family of ordinary differential equations we introduce scaled polar coordinates (ρ, ψ) in the (k_1, k_2) plane such that $k_1 = |k_3|\rho\cos\psi$, $k_2 = |k_3|\rho\sin\psi$, and write equation (3.16) in the form

$$\begin{aligned} &\sigma^2\mathbf{s} + 2\sigma c_f(\tfrac{d}{d\psi} + \mathcal{N})\mathbf{s} + (c_f^2 - c_a^2)(\tfrac{d}{d\psi} + 2\mathcal{N})\tfrac{d}{d\psi}\mathbf{s} \\ &-2ik_3\nu_a c_a(\tfrac{d}{d\psi} + \mathcal{N})\mathbf{s} + k_3^2\nu_a^2\mathbf{s} = 0, \end{aligned} \tag{3.17}$$

where $\mathbf{s} = (s_\rho, s_\psi)$, $s_\rho = \mathbf{s}\cdot\mathbf{k}_\perp/k_\perp$, $s_\psi = \mathbf{s}\cdot\mathcal{J}_\perp\mathbf{k}_\perp/k_\perp$, and

$$\mathcal{N} = \begin{pmatrix} 0 & -\frac{1-\delta^2}{\rho^2(1-\delta\cos 2\psi)+1-\delta^2} \\ 1 & \frac{\delta\rho^2\sin 2\psi}{\rho^2(1-\delta\cos 2\psi)+1-\delta^2} \end{pmatrix}. \tag{3.18}$$

Since equation (3.17) involves differentiation *only* with respect to ψ, it parametrically depends on *both* k_3 and ρ. This equation is augmented with the periodicity conditions in ψ and regularity conditions in ρ. The spectrum of the corresponding problem is denoted by $\Sigma(\delta, \chi, \nu_a, k_3, \rho)$. By virtue of the parametric dependence of equation (3.17) of ρ we have

$$\Sigma(\delta, \chi, \nu_a, k_3) = \overline{\bigcup_{0\le\rho<\infty} \Sigma(\delta, \chi, \nu_a, k_3, \rho)}. \tag{3.19}$$

In addition, we have

$$\begin{aligned} \Sigma(\delta, \chi, \nu_a, k_3, \rho) &= -\overline{\Sigma}(\delta, \chi, \nu_a, k_3, \rho), \\ \Sigma(\delta, \chi, \nu_a, k_3, \rho) &= \overline{\Sigma}(\delta, \chi, \nu_a, -k_3, \rho). \end{aligned} \tag{3.20}$$

Thus, the Fourier transformation reduces the original constrained spectral problem for a partial differential operator with linear coefficients to an unconstrained spectral problem for a family of ordinary differential operators with periodic coefficients. In general, the latter spectral problem is rather difficult to solve. However, in some special cases its reasonably complete solution can be obtained. In particular, for helical equilibria ($\delta = 0$) this problem has an explicit solution which is described in the next section. Also, for equilibria without flow ($c_f = 0$) the corresponding problem reduces to a linear spectral problem with respect to σ^2 for a self-adjoint operator and can be studied in detail. In this case we can prove that the spectrum is either purely real or purely imaginary which greatly simplifies its analysis. Other special cases include equilibria without magnetic field ($c_a = 0, \nu_a = 0$) and equilibria for which the horizontal fluid and Alfvén velocities exactly balance each other ($c_f^2 = c_a^2$).

4. Instabilities and waves for helical flows

In this section we study spectral problem (3.17) for helical flows with $\delta = 0$. Helical flows can be studied via other techniques as well (cf., e.g., [4], [8], [11], [13], [14]).

First, we notice that for $\delta = 0$ the matrix $\mathcal{N}$ reduces to the antidiagonal matrix $\mathcal{N}_0$ of the form

$$\mathcal{N}_0 = \begin{pmatrix} 0 & -\tau^2 \\ 1 & 0 \end{pmatrix}, \tag{4.1}$$

where $\tau = 1/\sqrt{1+\rho^2}$, $0 \le \tau \le 1$, so that equation (3.17) becomes an equations with constant coefficients. Its solutions can be chosen in the form $\mathbf{s}(\psi) = \mathbf{s}_n \exp(in\psi)$, where $\mathbf{s}_n$ is a constant vector. For a fixed n the characteristic equation for σ has the form

$$\begin{aligned} &[\sigma^2 + 2i\sigma c_f n - (c_f^2 - c_a^2)n^2 + 2k_3\nu_a c_a n + k_3^2\nu_a^2]^2 \\ &\quad + 4\tau^2[\sigma c_f + i(c_f^2 - c_a^2)n - ik_3\nu_a c_a]^2 = 0. \end{aligned} \tag{4.2}$$

Solving this equation is clearly equivalent to solving two quadratic equations, so that all four solutions can be found explicitly. These solutions can be written as

$$\sigma_{\alpha\beta}(n,\tau) = -ic_f(n+\beta\tau) + \alpha\{(c_a^2 - c_f^2)\tau^2 - [c_a(n+\beta\tau) + k_3\nu_a]^2\}^{1/2}, \tag{4.3}$$

where $\alpha = \pm, \beta = \pm$. These expressions clearly show that the spectrum Σ has symmetry properties (3.20).

The spectrum of (3.17) coincides with the union of the points $\sigma_{\alpha\beta}(n,\tau)$. When $c_f^2 \ge c_a^2$ *all* of them are purely imaginary and are not associated with instabilities. However, for $c_f^2 < c_a^2$ it is no longer the case. It is clear that $\Re(\sigma_{\alpha\beta}) \ne 0$ when

$$\left[\sqrt{c_a^2 - c_f^2}\,\tau + c_a(n+\beta\tau) + k_3\nu_a\right]\left[\sqrt{c_a^2 - c_f^2}\,\tau - c_a(n+\beta\tau) - k_3\nu_a\right] > 0, \tag{4.4}$$

for some τ such that $0 \le \tau \le 1$. This product is positive on the interval $[\tau_{l\beta}, \tau_{r\beta}]$ where

$$\begin{aligned} \tau_{l\beta} &= \min\left\{-\frac{c_a n + k_3\nu_a}{\sqrt{c_a^2 - c_f^2} + \beta c_a}, \frac{c_a n + k_3\nu_a}{\sqrt{c_a^2 - c_f^2} - \beta c_a}\right\}, \\ \tau_{r\beta} &= \max\left\{-\frac{c_a n + k_3\nu_a}{\sqrt{c_a^2 - c_f^2} + \beta c_a}, \frac{c_a n + k_3\nu_a}{\sqrt{c_a^2 - c_f^2} - \beta c_a}\right\}. \end{aligned} \tag{4.5}$$

Thus, nonimaginary eigenvalues exist for such n that the intervals $[0,1]$ and $[\tau_{l\beta}, \tau_{r\beta}]$ intersect. It is easy to verify that this intersection occurs when

$$0 < -\beta\frac{c_a n + k_3\nu_a}{\sqrt{c_a^2 - c_f^2} + c_a} < 1, \tag{4.6}$$

or, equivalently, when

$$\begin{aligned} &-\frac{\sqrt{c_a^2-c_f^2}}{c_a}-1-\frac{k_3\nu_a}{c_a}<n<-\frac{k_3\nu_a}{c_a}, && \beta=+,\\ &-\frac{k_3\nu_a}{c_a}<n<\frac{\sqrt{c_a^2-c_f^2}}{c_a}+1-\frac{k_3\nu_a}{c_a}, && \beta=-. \end{aligned} \tag{4.7}$$

These inequalities have no less than two and no more than four integer solutions which represent unstable modes. Accordingly, nonimaginary spectrum contains no less than two and no more than four branches. For a fixed unstable pair n, β, and $\alpha = \pm 1$, the corresponding branch is the image of the interval $[\tau_{l\beta}, \tau_{r\beta}] \cap [0, 1]$ under the mapping (4.3). It is easy to prove that the corresponding image is an arc of a circle of radius R centered at the point $i\sigma_0$ on the imaginary axis, where

$$R=\sqrt{c_a^2-c_f^2}\,\frac{|c_a n+k_3\nu_a|}{c_f}, \qquad \sigma_0=\frac{(c_a^2-c_f^2)n+k_3\nu_a c_a}{c_f}. \tag{4.8}$$

When $\tau_{r\beta} \le 1$ the corresponding arc coincides with the circle, when $\tau_{r\beta} > 1$ the arc forms a proper part of the circle. In the latter case we can use an obvious relation $\sigma_{\alpha+}(n, 1) = \sigma_{\alpha-}(n + 2, 1)$ to prove that two arcs corresponding to n and $n + 2$, respectively, meet at the nonimaginary points $\sigma_\pm = \sigma_{\pm+}(n, 1) = \sigma_{\pm-}(n + 2, 1)$. Various possibilities are illustrated in Figures 1–3 showing the spectrum of problem (3.17) for representative values of $\chi, k_3\nu_a$.

We see that all helical equilibria without flow are unstable, while equilibria with flow can be stabilized by sufficiently strong horizontal mass flows such that $c_f^2 \ge c_a^2$ (cf., also, [4], [8], [13], [14]).

5. The location of the spectrum for general quasi-helical equilibria

It is shown above that for helical equilibria with $\delta = 0$ the stability diagram in the parameter space with coordinates $\chi, k_3\nu_a$ is rather simple. Namely, all equilibria with $c_f^2 \ge c_a^2$ are stable, while equilibria with $c_f^2 < c_a^2$ are unstable. For quasi-helical equilibria with $\delta > 0$ the stability diagram in the parameter space with coordinates $\delta, \chi, k_3\nu_a$ is much more complicated. In particular, it is shown in [10] that all distorted helical equilibria without magnetic field ($c_a = 0, \nu_a = 0$) are unstable. The main objective of the quantitative analysis of the spectra of problems (3.17) is to construct the stability diagram in the $(\delta, \chi, k_3\nu_a)$ space and to discriminate between stable and unstable quasi-helical equilibria. In this section we establish some qualitative results concerning the location of the spectrum in the complex plane which can eventually facilitate the construction of the stability diagram.

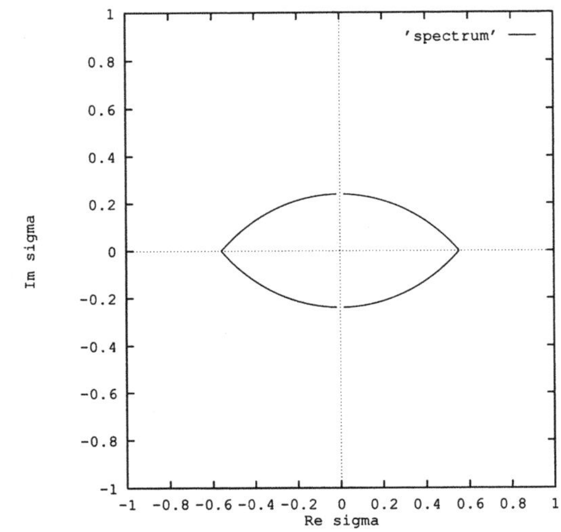

Figure 1: The nonimaginary part of the spectra $\Sigma(0, \chi, \nu_a, k_3, \rho)$ for $\chi = 0.3\pi, \nu_a = 0.0, k_3 = \pm 1.0, 0 \leq \rho < \infty$. The arcs in the complex plane correspond to $n = \pm 1$.

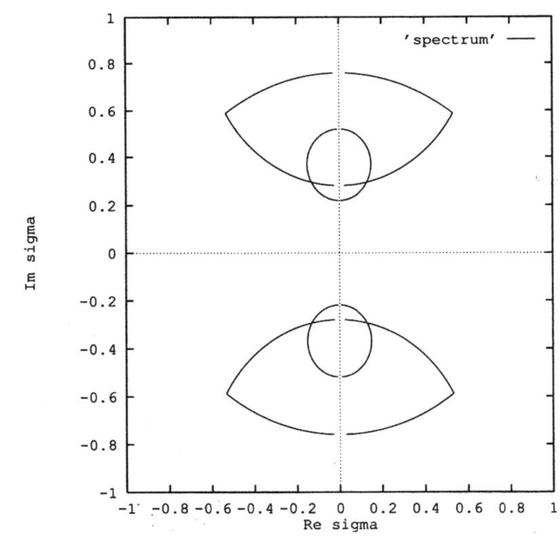

Figure 2: The same spectra as in Figure 1 but for $\chi = 0.3\pi, \nu_a = 0.65, k_3 = \pm 1.0, 0 \leq \rho < \infty$. The arcs in the complex plane correspond to $n = 0, \pm 1, \pm 2$.

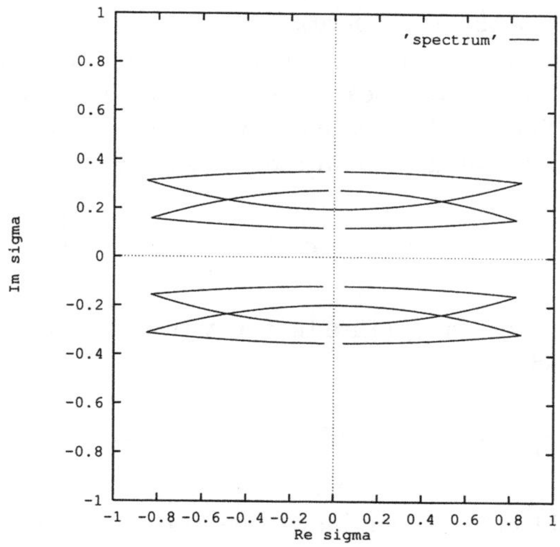

Figure 3: The same spectra as in Figure 1 but for $\chi = 0.45\pi, \nu_a = 1.5, k_3 = \pm 1.0, 0 \leq \rho < \infty$. The arcs in the complex plane correspond to $n = 0, \pm 1, \pm 2, \pm 3$.

(A) We notice that for $\rho = 0$ the matrix $\mathcal{N}$ turns into the constant matrix $\mathcal{J}_\perp$ (regardless of the value of δ). Accordingly, we can use formulas (4.3) with $\tau = 1$ in order to describe a sequence of points $\sigma_{\alpha\beta}(n, 1)$ belonging to the spectrum of the problem (3.17). It is clear that for $c_f^2 \geq c_a^2$ all these points are purely imaginary and no large-scale instabilities are present. Needless to say that this fact does not imply that equilibria with $c_f^2 \geq c_a^2$ are stable. For $c_f^2 < c_a^2$ the situation is more complicated. Inspection of expressions (4.3) shows that $\sigma_{\alpha\beta}$ have nonzero real parts provided that n satisfies one of the following inequalities:

$$(5.1) \qquad -\frac{\sqrt{c_a^2 - c_f^2} - \beta c_a - k_3 \nu_a}{c_a} < n < \frac{\sqrt{c_a^2 - c_f^2} - \beta c_a - k_3 \nu_a}{c_a}.$$

Accordingly, all quasi-helical equilibria with $c_f^2 < c_a^2$ and

$$\min\left\{\left[\frac{k_3 \nu_a}{c_a}\right], 1 - \left[\frac{k_3 \nu_a}{c_a}\right]\right\} < \frac{\sqrt{c_a^2 - c_f^2}}{c_a},$$

where $\left[\frac{k_3 \nu_a}{c_a}\right]$ is the fractional part of $\frac{k_3 \nu_a}{c_a}$, are unstable. Once again, it does not mean that other equilibria with $c_f^2 < c_a^2$ are stable.

(B) We notice that operator (3.17) acts invariantly on the subspaces of functions periodic and antiperiodic with period π, so that its spectrum can be studied separately in each of these subspaces.

(C) We find some bounds for the real parts of eigenvalues of the problem (3.17). To obtain such a bound, we linearize the corresponding quadratic pencil. We use a version of the linearization procedure which is particularly well suited for the problem in question. Namely, we introduce two new variables

$$\boldsymbol{\eta}_\pm = \left[\sigma + (c_f \pm c_a)\frac{d}{d\psi} \pm ik_3\nu_a\right]\mathbf{s}, \tag{5.2}$$

where $\boldsymbol{\eta}_\pm = \mathbf{v}_{f\perp} \pm \mathbf{v}_{a\perp}$, and notice that by virtue of equation (3.17) the following relations are satisfied

$$\left[\sigma + (c_f \mp c_a)\left(\frac{d}{d\psi} + \mathcal{N}\right) \mp ik_3\nu_a\right]\boldsymbol{\eta}_\pm + (c_f \pm c_a)\mathcal{N}\boldsymbol{\eta}_\mp = 0. \tag{5.3}$$

Accordingly, problem (3.17) can be written in the linear form

$$\begin{pmatrix} (c_f - c_a)\left(\frac{d}{d\psi} + \mathcal{N}\right) - ik_3\nu_a & (c_f + c_a)\mathcal{N} \\ (c_f - c_a)\mathcal{N} & (c_f + c_a)\left(\frac{d}{d\psi} + \mathcal{N}\right) + ik_3\nu_a \end{pmatrix}\begin{pmatrix}\boldsymbol{\eta}_+ \\ \boldsymbol{\eta}_-\end{pmatrix} + \sigma\begin{pmatrix}\boldsymbol{\eta}_+ \\ \boldsymbol{\eta}_-\end{pmatrix} = 0. \tag{5.4}$$

The symmetric part of the corresponding differential expression has the form

$$\begin{pmatrix} (c_f - c_a)\mathcal{N}_s & c_f\mathcal{N}_s + c_a\mathcal{N}_a \\ c_f\mathcal{N}_s - c_a\mathcal{N}_a & (c_f + c_a)\mathcal{N}_s \end{pmatrix}, \tag{5.5}$$

where $\mathcal{N}_s, \mathcal{N}_a$ are symmetric and antisymmetric parts of $\mathcal{N}$. Since it is bounded, the real part of any eigenvalue σ is bounded too. The corresponding bound (which can be obtained via a standard procedure) has the form

$$|\Re(\sigma)| \le (|c_f - c_a| + |c_f + c_a| + |c_f|) \parallel \mathcal{N}_s \parallel + |c_a| \parallel \mathcal{N}_a \parallel . \tag{5.6}$$

This estimate shows that all the eigenvalues of problems (5.4) and (3.17) are located in a vertical strip. Although this estimate is easy to obtain, it has an important physical meaning since it implies that the growth rates of the instabilities of quasi-helical flows are bounded uniformly in ρ, k_3, so that explosive growth of perturbations *à la* Hadamard is not possible for the equilibria under consideration.

(D) To advance further, we symmetrize pencil (3.17). To this end we notice that

$$\frac{d\mathcal{M}}{d\psi} - 2\mathcal{M}\mathcal{N} + 2\mathcal{J}_\perp = 0, \tag{5.7}$$

where $\mathcal{M}$ is a symmetric matrix of the form

$$\begin{pmatrix} 1/\tau^2+\delta\cos 2\psi & -\delta\sin 2\psi \\ -\delta\sin 2\psi & 1-\delta\cos 2\psi \end{pmatrix}. \tag{5.8}$$

Applying this matrix to equation (3.17), we rewrite it in the following symmetric form

$$\begin{aligned} &\sigma^2\mathcal{M}\mathbf{s}+\sigma c_f\left(\mathcal{M}\tfrac{d}{d\psi}+\tfrac{d}{d\psi}\mathcal{M}+2\mathcal{J}_\perp\right)\mathbf{s} \\ &+\Big[(c_f^2-c_a^2)\left(\tfrac{d}{d\psi}\mathcal{M}\tfrac{d}{d\psi}+2\mathcal{J}_\perp\tfrac{d}{d\psi}\right) \\ &-ik_3\nu_a c_a\left(\mathcal{M}\tfrac{d}{d\psi}+\tfrac{d}{d\psi}\mathcal{M}+2\mathcal{J}_\perp\right)+k_3^2\nu_a^2\mathcal{M}\Big]\mathbf{s}=0. \end{aligned} \tag{5.9}$$

Introducing two differential operators

$$\hat{D}=\frac{d}{d\psi}+\mathcal{M}^{-1}\mathcal{J}_\perp, \qquad \hat{D}^*=-\frac{d}{d\psi}-\mathcal{J}_\perp\mathcal{M}^{-1}, \tag{5.10}$$

we rewrite (5.9) as

$$\begin{aligned} &\sigma^2\mathcal{M}\mathbf{s}+\sigma c_f\left(\mathcal{M}\hat{D}-\hat{D}^*\mathcal{M}\right)\mathbf{s}+\Big[(c_f^2-c_a^2)\left(-\hat{D}^*\mathcal{M}\hat{D}+\mu\mathcal{M}\right) \\ &-ik_3\nu_a c_a\left(\mathcal{M}\hat{D}-\hat{D}^*\mathcal{M}\right)+k_3\nu_a^2\mathcal{M}\Big]\mathbf{s}=0, \end{aligned} \tag{5.11}$$

where $\mu=1/\det\mathcal{M}$.

Symbolically pencil (5.11) can be written as

$$\sigma^2\hat{A}\mathbf{s}+\sigma\hat{B}\mathbf{s}+\hat{C}\mathbf{s}=0, \tag{5.12}$$

where $\hat{A}, i\hat{B}, \hat{C}$ are Hermitian operators. Equation (5.11) clearly implies that the spectrum has symmetry properties (3.20). Pencil (5.12) is similar in some respects to the pencil used in [3] in order to study the stability of Josephson junctions.

(E) It is shown above that the spectra of helical equilibria with $c_f^2 > c_a^2$ differ in many respects from the spectra of equilibria with $c_f^2 < c_a^2$. It turns out that important distinction persist for quasi-helical equilibria as well. Namely, for equilibria with $c_f^2 < c_a^2$ nonimaginary eigenvalues are located in a circle of finite radius centered at the origin while for equilibria with $c_f^2 > c_a^2$ they can be located anywhere in the strip defined by condition (5.6). To prove the above statement we assume that σ is an eigenvalue with nonzero real part and $\mathbf{s}_\sigma$ is the corresponding eigenvector normalized with respect to the scalar product defined by the matrix $\mathcal{M}$, multiply equation (5.9) by $\mathbf{s}_\sigma$ and obtain the following quadratic equation for σ:

$$\begin{aligned} &\sigma^2+2i\sigma c_f\Im\langle\hat{D}\mathbf{s}_\sigma,\mathbf{s}_\sigma\rangle+(c_f^2-c_a^2)(-\langle\hat{D}\mathbf{s}_\sigma,\hat{D}\mathbf{s}_\sigma\rangle+\langle\mu\mathbf{s}_\sigma,\mathbf{s}_\sigma\rangle) \\ &+2k_3\nu_a c_a\Im\langle\hat{D}\mathbf{s}_\sigma,\mathbf{s}_\sigma\rangle+k_3^2\nu_a^2=0, \end{aligned} \tag{5.13}$$

where $\langle \cdot , \cdot \rangle$ denotes the scalar product defined by $\mathcal{M}$. Solutions of this equation can be written as

$$\begin{aligned}\sigma_{\pm} &= -ic_f \Im\langle \hat{D}\mathbf{s}_\sigma, \mathbf{s}_\sigma\rangle \\ &\pm \Big\{ -c_f^2(\Im\langle \hat{D}\mathbf{s}_\sigma, \mathbf{s}_\sigma\rangle)^2 + (c_f^2 - c_a^2)(\langle \hat{D}\mathbf{s}_\sigma, \hat{D}\mathbf{s}_\sigma\rangle - \langle \mu\mathbf{s}_\sigma, \mathbf{s}_\sigma\rangle) \\ &\quad -2k_3\nu_a c_a \Im\langle \hat{D}\mathbf{s}_\sigma, \mathbf{s}_\sigma\rangle - k_3^2\nu_a^2 \Big\}^{1/2}. \end{aligned} \tag{5.14}$$

These solutions have nonzero real parts provided that the corresponding discriminant is positive. For such solutions $|\sigma|^2$ can be written as

$$|\sigma|^2 = (c_f^2 - c_a^2)(\langle \hat{D}\mathbf{s}_\sigma, \hat{D}\mathbf{s}_\sigma\rangle - \langle \mu\mathbf{s}_\sigma, \mathbf{s}_\sigma\rangle) - 2k_3\nu_a c_a \Im\langle \hat{D}\mathbf{s}_\sigma, \mathbf{s}_\sigma\rangle - k_3^2\nu_a^2. \tag{5.15}$$

It is clear that $|\sigma|^2$ can be arbitrarily large when $c_f^2 > c_a^2$; when $c_f^2 < c_a^2$ it can be estimated from above as follows

$$\begin{aligned} |\sigma|^2 &\leq (c_a^2 - c_f^2)(-\langle \hat{D}\mathbf{s}_\sigma, \hat{D}\mathbf{s}_\sigma\rangle + \mu_{\max}) \\ &\quad +2|k_3\nu_a c_a||\Im\langle \hat{D}\mathbf{s}_\sigma, \mathbf{s}_\sigma\rangle| - k_3^2\nu_a^2 \\ &\leq (c_a^2 - c_f^2)(-\langle \hat{D}\mathbf{s}_\sigma, \hat{D}\mathbf{s}_\sigma\rangle + \mu_{\max}) \\ &\quad +|k_3\nu_a c_a|(\kappa\langle \hat{D}\mathbf{s}_\sigma, \hat{D}\mathbf{s}_\sigma\rangle| + 1/\kappa) - k_3^2\nu_a^2 \\ &= [-(c_a^2 - c_f^2) + \kappa|k_3\nu_a c_a|]\langle \hat{D}\mathbf{s}_\sigma, \hat{D}\mathbf{s}_\sigma\rangle \\ &\quad +(c_a^2 - c_f^2)\mu_{\max} + |k_3\nu_a c_a|/\kappa - k_3^2\nu_a^2, \end{aligned} \tag{5.16}$$

where $\mu_{\max} = \max_{0\leq\psi\leq 2\pi}\mu(\psi)$, and κ is an arbitrary constant. By choosing $\kappa = \frac{c_a^2 - c_f^2}{|k_3\nu_a c_a|}$ we finally obtain the estimate we sought

$$|\sigma|^2 \leq (c_a^2 - c_f^2)\mu_{\max} + \frac{c_f^2 k_3^2\nu_a^2}{c_a^2 - c_f^2}. \tag{5.17}$$

Among other things this estimate shows that all quasi-helical equilibria with $c_f^2 = c_a^2$ and $\nu_a = 0$ are stable.

It is difficult to obtain further information about the spectrum of problems (3.17), (5.9) and we have to use asymptotic and numerical methods to do so. Weakly distorted quasi-equilibria with $\delta \ll 1$ can be treated via theory of parametric resonance which provides a number of interesting insights, however, for the sake of brevity, in the present paper we concentrate on general quasi-equilibria with $\delta \sim 1$ and study them numerically.

6. A numerical study of the spectrum

In this section we describe a method for studying the spectrum of the symmetric quadratic pencil (5.12) and give a few representative examples illustrating its basic properties.

First, we choose the standard basis in the space of periodic vector functions on $[0, 2\pi]$ of the form

$$\begin{aligned}
&\mathbf{E}_{0,1} = (\tfrac{1}{\sqrt{2\pi}}, 0)^T, && \mathbf{E}_{0,2} = (0, \tfrac{1}{\sqrt{2\pi}})^T, && \\
&\mathbf{E}_{m,1} = \cos(m\psi)(\tfrac{1}{\sqrt{\pi}}, 0)^T, && \mathbf{E}_{m,2} = \cos(m\psi)(0, \tfrac{1}{\sqrt{\pi}})^T, && m \geq 1, \\
&\mathbf{E}_{m,3} = \sin(m\psi)(\tfrac{1}{\sqrt{\pi}}, 0)^T, && \mathbf{E}_{m,4} = \sin(m\psi)(0, \tfrac{1}{\sqrt{\pi}})^T, && m \geq 1.
\end{aligned} \tag{6.1}$$

In this basis the problem for the operators $\hat{A}, \hat{B}, \hat{C}$ reduces to an equivalent problem for infinite matrices $\mathcal{A}, \mathcal{B}, \mathcal{C}$ (when $\delta = 0$ the corresponding matrices are block-diagonal). Since the operators $\hat{A}, \hat{B}, \hat{C}$ act invariantly on the subspaces of functions periodic and antiperiodic with period π, the matrices $\mathcal{A}, \mathcal{B}, \mathcal{C}$ act invariantly on the subspaces spanned by $\mathbf{E}_{m,i}$ with even and odd m. Thus, we can treat even and odd m independently.

We start with a simpler case and consider odd m. We represent the operators $\hat{A}, \hat{B}, \hat{C}$ as block matrices with blocks of order 4×4. A straightforward computation shows the blocks $\mathcal{A}_{mn}$ corresponding to $\hat{A}$ have the form

$$\mathcal{A}_{mn} = (A_{mn,ij})$$

$$= \begin{pmatrix} \delta_{mn}/\tau^2 + \zeta_{1mn} & 0 & 0 & -\zeta_{2mn} \\ 0 & \delta_{mn} - \zeta_{1mn} & -\zeta_{2mn} & 0 \\ 0 & -\zeta_{2nm} & \delta_{mn}/\tau^2 + \zeta_{3mn} & 0 \\ -\zeta_{2nm} & 0 & 0 & \delta_{mn} - \zeta_{3mn} \end{pmatrix}, \tag{6.2}$$

where δ_{mn} is the Kronecker delta (not to be confused with the ellipticity parameter δ defined above), and

$$\begin{aligned}
\zeta_{1mn} &= \frac{\delta}{2}\left(\delta_{(m+n-2)0} + \delta_{(m-n-2)0} + \delta_{(m-n+2)0}\right), \\
\zeta_{2mn} &= \frac{\delta}{2}\left(\delta_{(m+n-2)0} - \delta_{(m-n-2)0} + \delta_{(m-n+2)0}\right), \\
\zeta_{3mn} &= \frac{\delta}{2}\left(-\delta_{(m+n-2)0} + \delta_{(m-n-2)0} + \delta_{(m-n+2)0}\right).
\end{aligned} \tag{6.3}$$

It is convenient to further split each of the 4×4 blocks $\mathcal{A}_{mn}$ into 2×2 blocks $\mathcal{A}_{mn}^{\alpha\beta}$, $\alpha, \beta = 1, 2$ and write it as

$$\mathcal{A}_{mn} = \begin{pmatrix} \mathcal{A}_{mn}^{11} & \mathcal{A}_{mn}^{12} \\ \mathcal{A}_{mn}^{21} & \mathcal{A}_{mn}^{22} \end{pmatrix}. \tag{6.4}$$

Using this notation we can represent the blocks $\mathcal{B}_{mn}, \mathcal{C}_{mn}$ corresponding to $\hat{B}, \hat{C}$ in the form

$$\text{(6.5)}\quad \mathcal{B}_{mn} = \left(B_{mn}^{ij}\right) = \begin{pmatrix} \mathcal{B}_{mn}^{11} & \mathcal{B}_{mn}^{12} \\ \mathcal{B}_{mn}^{21} & \mathcal{B}_{mn}^{22} \end{pmatrix}$$

$$= c_f \begin{pmatrix} -m\mathcal{A}_{mn}^{21} + n\mathcal{A}_{mn}^{12} - 2\delta_{mn}\mathcal{J}_\perp & -m\mathcal{A}_{mn}^{22} - n\mathcal{A}_{mn}^{11} \\ m\mathcal{A}_{mn}^{11} + n\mathcal{A}_{mn}^{22} & m\mathcal{A}_{mn}^{12} - n\mathcal{A}_{mn}^{21} - 2\delta_{mn}\mathcal{J}_\perp \end{pmatrix},$$

$$(6.6)\quad \mathcal{C}_{mn} = (C_{mn}^{ij}) = \begin{pmatrix} \mathcal{C}_{mn}^{11} & \mathcal{C}_{mn}^{12} \\ \mathcal{C}_{mn}^{21} & \mathcal{C}_{mn}^{22} \end{pmatrix}$$

$$= (c_f^2 - c_a^2) \begin{pmatrix} -mn\mathcal{A}_{mn}^{22} & mn\mathcal{A}_{mn}^{21} + 2m\delta_{mn}\mathcal{J}_\perp \\ mn\mathcal{A}_{mn}^{12} - 2m\delta_{mn}\mathcal{J}_\perp & -mn\mathcal{A}_{mn}^{11} \end{pmatrix}$$

$$-ik_3\nu_a c_a \begin{pmatrix} \mathcal{B}_{mn}^{11} & \mathcal{B}_{mn}^{12} \\ \mathcal{B}_{mn}^{21} & \mathcal{B}_{mn}^{22} \end{pmatrix} + k_3^2\nu_a^2 \begin{pmatrix} \mathcal{A}_{mn}^{11} & \mathcal{A}_{mn}^{12} \\ \mathcal{A}_{mn}^{21} & \mathcal{A}_{mn}^{22} \end{pmatrix}.$$

Formulas (6.1), (6.5), (6.6) clearly show that the matrices $\mathcal{A}, i\mathcal{B}, \mathcal{C}$ are symmetric three-diagonal block matrices. To compute the spectrum corresponding to the odd subspace numerically, we choose a certain large M and project the matrices $\mathcal{A}, \mathcal{B}, \mathcal{C}$ onto the finite-dimensional subspace spanned by $E_{mi}, m = 1, 3, \ldots, 2M-1$. Denoting the corresponding projections by $\mathcal{A}_M, \mathcal{B}_M, \mathcal{C}_M$, respectively, we obtain the following finite-dimensional spectral problem

$$(6.7)\qquad \sigma^2\mathcal{A}_M\mathbf{s}_M + \sigma\mathcal{B}_M\mathbf{s}_M + \mathcal{C}_M\mathbf{s}_M = 0.$$

Its spectrum $\Sigma_{M,\text{odd}}(\delta, \chi, \nu_a, k_3, \rho)$ consists of $4M$ points and can easily be computed by virtue of a standard eigenvalue solver such as EISPACK.

For even m the strategy is essentially the same. For $m > 0, n > 0$ the blocks $\mathcal{A}_{mn}, \mathcal{B}_{mn}, \mathcal{C}_{mn}$ are given by expressions (6.2), (6.5), (6.6). The only complication is due to the fact that there are only two basis functions $\mathbf{E}_{01}, \mathbf{E}_{02}$ for $m = 0$ rather than four for all other m. For this reason we have to consider additional 2×2, 4×2 and 2×4 blocks. These blocks have the form

$$\mathcal{A}_{00} = \begin{pmatrix} 1/\tau^2 & 0 \\ 0 & 1 \end{pmatrix},$$

$$(6.8)\quad \mathcal{A}_{m0} = \frac{1}{\sqrt{2}} \begin{pmatrix} \zeta_{1m0} & 0 \\ 0 & -\zeta_{1m0} \\ 0 & -\zeta_{20m} \\ -\zeta_{20m} & 0 \end{pmatrix} = \begin{pmatrix} \mathcal{A}_{m0}^{11} \\ \mathcal{A}_{m0}^{21} \end{pmatrix},$$

$$\mathcal{A}_{0n} = \frac{1}{\sqrt{2}} \begin{pmatrix} \zeta_{1n0} & 0 & 0 & -\zeta_{20n} \\ 0 & -\zeta_{1n0} & -\zeta_{20n} & 0 \end{pmatrix} = \begin{pmatrix} \mathcal{A}_{0n}^{11} & \mathcal{A}_{0n}^{12} \end{pmatrix},$$

$$\mathcal{B}_{00} = c_f \begin{pmatrix} 0 & 2 \\ -2 & 0 \end{pmatrix},$$

$$(6.9)\quad \mathcal{B}_{m0} = c_f \begin{pmatrix} -m\mathcal{A}_{m0}^{21} \\ m\mathcal{A}_{m0}^{11} \end{pmatrix},$$

$$\mathcal{B}_{0n} = c_f \begin{pmatrix} n\mathcal{A}_{0n}^{12} & -n\mathcal{A}_{0n}^{11} \end{pmatrix},$$

$$
\begin{aligned}
\mathcal{C}_{00} &= \begin{pmatrix} k_3^2\nu_a^2/\tau^2 & -2ik_3\nu_a c_a \\ 2ik_3\nu_a c_a & k_3^2\nu_a^2 \end{pmatrix}, \\
\mathcal{C}_{m0} &= -ik_3\nu_a c_a \begin{pmatrix} -m\mathcal{A}_{m0}^{21} \\ m\mathcal{A}_{m0}^{11} \end{pmatrix} + k_3^2\nu_a^2 \begin{pmatrix} \mathcal{A}_{m0}^{11} \\ \mathcal{A}_{m0}^{21} \end{pmatrix}, \qquad (6.10) \\
\mathcal{C}_{0n} &= -ik_3\nu_a c_a \,(n\mathcal{A}_{0n}^{12} \quad -n\mathcal{A}_{0n}^{11}) + k_3^2\nu_a^2 \,(\mathcal{A}_{0n}^{11} \quad \mathcal{A}_{0n}^{12}).
\end{aligned}
$$

We choose a certain large M and use formulas (6.2), (6.5), (6.6), (6.8), (6.9), (6.10) to construct the finite-dimensional projections $\mathcal{A}_M, \mathcal{B}_M, \mathcal{C}_M$ of the matrices $\mathcal{A}, i\mathcal{B}, \mathcal{C}$ on the space spanned by $E_{mi}, m = 0, 2, \ldots, 2M$. As a result we obtain a finite-dimensional pencil similar to (6.7). Its spectrum is denoted by $\Sigma_{M,\mathrm{even}}(\delta, \chi, \nu_a, k_3, \rho)$, it consists of $4M + 2$ points and can easily be computed numerically.

It can be shown that when $M \to \infty$ the union of the spectra $\Sigma_{M,\mathrm{odd}}(\delta, \chi, \nu_a, k_3, \rho)$ and $\Sigma_{M,\mathrm{even}}(\delta, \chi, \nu_a, k_3, \rho)$ approximates the spectrum $\Sigma(\delta, \chi, \nu_a, k_3, \rho)$.

To illustrate the validity of our numerical procedure, in Figure 4 we show the union of $\Sigma_{M,\mathrm{odd}}$ and $\Sigma_{M,\mathrm{even}}$ for $M = 10$ and 100 values of ρ uniformly distributed on the interval $[0, 5]$. We use the same parameters as in Figure 2 and obtain an excellent agreement between the spectra computed analytically and numerically. In Figures 5–7 we show $\Sigma_{M,\mathrm{odd}}$ and $\Sigma_{M,\mathrm{even}}$ for $M = 10$, $\rho = 0, 0.05, 0.1, \ldots, 5.0$, and representative values of $\delta, \chi, k_3\nu_a$, for which the analytical treatment is not possible. Figure 5 illustrates the fact that for equilibria without flow the spectrum is located on the real and imaginary axes. Figure 6 shows the "backbone" structure of the spectrum for equilibria without magnetic field ($c_a = 0, \nu_a = 0$) discovered in [10]. The comparison of these figures clearly shows that the nonimaginary part of the spectrum is located near the origin when $c_f^2 < c_a^2$, while it can extend to infinity (along the imaginary axis) when $c_f^2 > c_a^2$. Finally, Figure 7 shows the spectrum of a generic equilibrium with $\delta, \chi, k_3\nu_a \sim 1$.

7. Concluding remarks

In the present paper it is shown how to reduce a rather complicated spectral problem for a partial differential operator describing instabilities and waves in quasi-helical MHD equilibria with flow to a family of (relatively simple) problems for ordinary differential operators with periodic coefficients. The corresponding spectral problems are analyzed via a combination of analytical and numerical methods. It is proved that all helical equilibria with dominant poloidal magnetic field are unstable. At the same time, general equilibria can be both stable and unstable depending on the ellipticity and the relative magnitudes of the horizontal and vertical fluid and Alfvén velocities.

A number of issues associated with problems (3.16), (3.17), (5.12) remains open. In particular, one needs to prove that eigenfunctions and associated eigenvectors

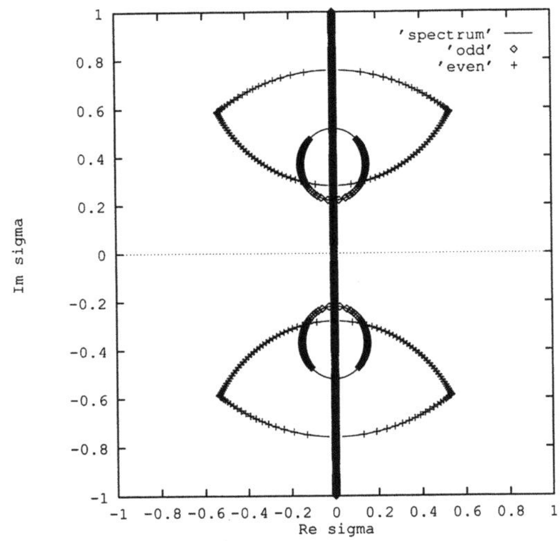

Figure 4: The spectra $\Sigma_{M,\mathrm{odd}}(0, \chi, \nu_a, k_3, \rho)$ (diamonds), $\Sigma_{M,\mathrm{even}}(0, \chi, \nu_a, -k_3, \rho)$ (crosses), and $\Sigma(0, \chi, \nu_a, \pm k_3, \rho)$ (solid lines) for $\delta = 0, \chi = 0.3\pi, \nu_a = 0.65, k_3 = \pm 1.0, 0 \leq \rho \leq 5$.

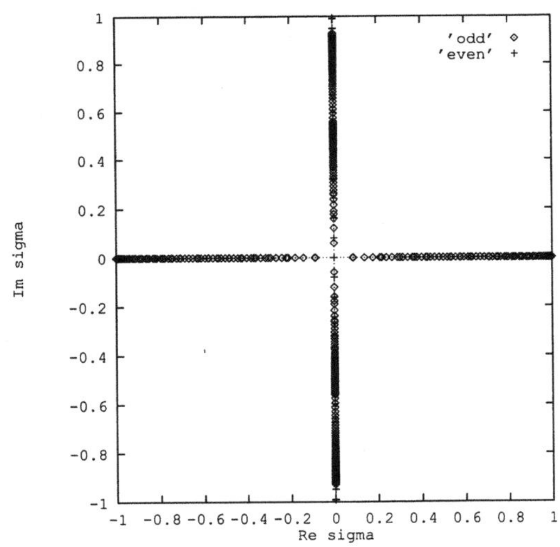

Figure 5: The spectra $\Sigma_{M,\mathrm{odd}}(0, \chi, \nu_a, k_3, \rho)$ (diamonds) and $\Sigma_{M,\mathrm{even}}(0, \chi, \nu_a, -k_3, \rho)$ (crosses) for $\delta = 0.5, \chi = 0.5\pi, \nu_a = 2.0, k_3 = \pm 1.0, 0 \leq \rho \leq 5$.

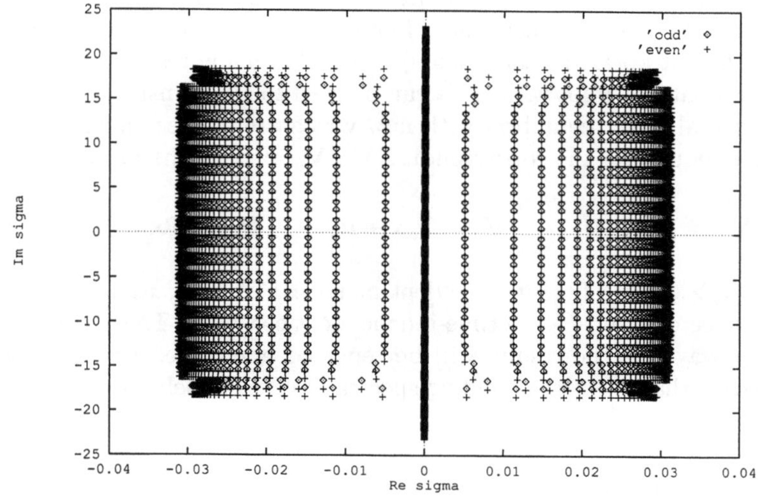

Figure 6: The same spectra as in Figure 5 but for $\delta = 0.5, \chi = 0.0, \nu_a = 2.0, k_3 = \pm 1$.

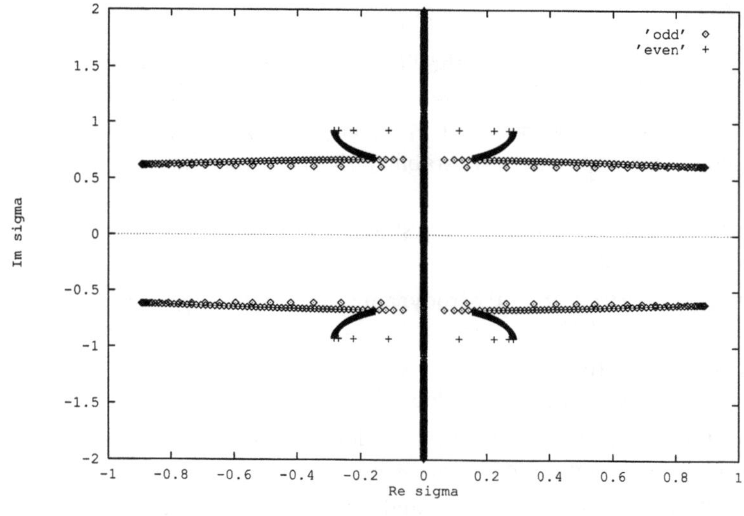

Figure 7: The same spectra as in Figure 5 but for $\delta = 0.7, \chi = 0.4\pi, \nu_a = 2.0, k_3 = \pm 1$.

for problems (3.17), (5.12) with fixed ρ form a complete system. Once this is done, the eigenfunction expansion formula for problem (3.16) has to be obtained.

The reduction of a spectral problem for a partial differential operator with linear coefficients to a problem for a family of ordinary differential operators (not necessarily with periodic coefficients) is useful in many respects. This reduction is of general nature. In particular, it can be used to study instabilities and waves for more general linear equilibria with flow which are of great physical interest. For the latter equilibria the corresponding $\mathbf{V}_f, \mathbf{V}_a, P$ have the form

$$\text{(7.1)}\quad \mathbf{V}_f = \mathcal{V}_f\mathbf{x} + \boldsymbol{\nu}_f, \qquad \mathbf{V}_a = \mathcal{V}_a\mathbf{x} + \boldsymbol{\nu}_a, \qquad P = \frac{1}{2}\mathcal{P}\mathbf{x}\cdot\mathbf{x} + \boldsymbol{\pi}\cdot\mathbf{x} + P_0,$$

where $\mathcal{V}_f, \mathcal{V}_a, \mathcal{P}$ are time-independent matrices, $\mathcal{P}^* = \mathcal{P}$, $\boldsymbol{\nu}_f, \boldsymbol{\nu}_a, \boldsymbol{\pi}$ are time-independent vectors, and P_0 is a time-independent constant. The corresponding study is under way and the results will be reported elsewhere. One can say, and not be in error, that the corresponding spectra have extremely unusual and intriguing properties.

Acknowledgements

We are grateful to the organizers of IWOTA 95 Prof. Dr. Reinhard Mennicken and Dr. Christiane Tretter for their kind invitation to deliver a lecture in Regensburg. The atmosphere of the workshop was very pleasant and stimulating. This work was supported in part by a grant from the National Science Foundation.

References

[1] ACHESON, D.J.: On the instability of toroidal magnetic fields and differential rotation in stars; Phil. Trans. Roy. Soc. London Ser. A 289 (1978), 459–500.

[2] BERNSTEIN, I.B., FRIEMAN, E.A., KRUSKAL, M.D., KULSRUD, R.M.: An energy principle for hydromagnetic stability theory; Proc. Roy. Soc. London Ser. A 224 (1958), 17–40.

[3] BURKOV, S.E., LIFSCHITZ. A.: Stability of moving soliton lattices; Wave Motion 5 (1983), 197–213.

[4] CHANDRASEKHAR, S.: Hydrodynamic and Hydromagnetic Stability; Clarendon Press, Oxford 1961.

[5] FRIEMAN, E., ROTENBERG, M.: On hydromagnetic stability of stationary equilibria; Rev. Modern Phys. 32 (1960), 898–902.

[6] GOEDBLOED, J.P.: Lecture Notes on Ideal Magnetohydrodynamics; FOM Institute for Plasma Physics, Rijnhuizen 1983.

[7] HAMEIRI, E.: The equilibrium and stability of rotating plasmas; Phys. Fluids 26 (1983), 230–237.

[8] HOWARD, L.N., GUPTA, A.S.: On hydrodynamic and hydromagnetic stability of swirling flows; J. Fluid Mech. 14 (1962), 463–476.

[9] LIFSCHITZ, A.: Magnetohydrodynamics and Spectral Theory; Kluwer Academic Publishers, Dordrecht 1989.

[10] LIFSCHITZ, A.: Exact description of the spectrum of elliptical vortices in hydrodynamics and magnetohydrodynamics; Phys. Fluids 7 (1995), 1626–1636.

[11] ROBERTS, P.H., Twisted magnetic fields; Astrophys. J. 124 (1956), 430–442.

[12] SPIES, G.O.: Magnetohydrodynamic spectrum of instabilities due to plasma rotation; Phys. Fluids 21 (1978), 580–587.

[13] TATARONIS, J.A., MOND, M.: Magnetohydrodynamic stability of plasmas with aligned mass flow; Phys. Fluids 30 (1987), 84–90.

[14] TREHAN, S.K.: The effect of fluid motions on the stability of twisted magnetic fields; Astrophys. J. 129 (1959), 475–482.

Department of Mathematics
University of Illinois
851 S. Morgan Str.
Chicago IL 60607 USA
alexli@uic.edu

1991 Mathematics Subject Classification: Primary 76W05; Secondary 47E05

Submitted: June 13, 1996

Operator Theory:
Advances and Applications, Vol. 102

An evolutionary problem of a flow of a nonlinear viscous fluid in a deformable viscoelastic tube

W.G. Litvinov

We consider the problem of a nonsteady flow of a nonlinear viscous fluid in an oscillating tube. In this problem, the oscillations of the tube define the shape of the domain in which the fluid flows, this domain is changing in time, and the flow of the fluid influences the oscillations of the tube. So one has to solve a coupled system of equations of forced oscillations of the tube and of a flow of a fluid in a varying domain, and this domain is to be found. Such a problem is formulated and studied, and for small data, the existence of a solution is proven.

1. Introduction

Various processes of engineering and biomedicine are connected with nonsteady flows of viscous or nonlinear viscous fluids in deformable tubes. For such flows, deformations (oscillations) of the tube influence the flow of the fluid, which, in turn, influences the oscillations of the tube. So, one has to solve a coupled system of equations of forced oscillations of the tube, and of a nonsteady flow of the fluid in a domain which changes in time. Moreover, the shape of the domain at each moment of time is unknown because it depends on a function of displacements (oscillations) of the tube to be found.

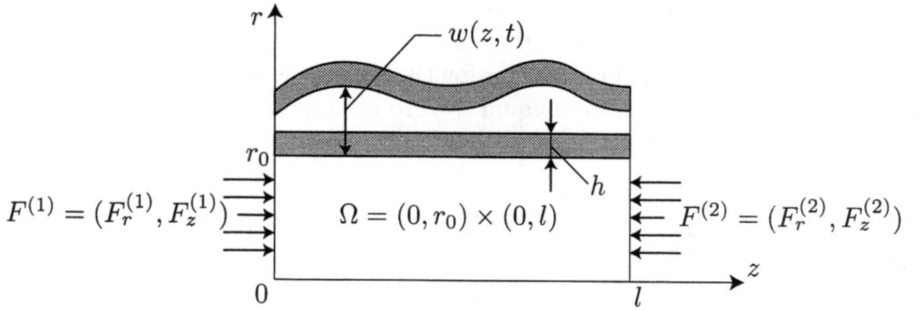

Fig. 1

A scheme of a flow of a fluid in a deformable tube is shown in Figure 1. We use the cylindrical coordinates r, α, z and suppose that the flow of the fluid is axially symmetric. The longitudinal section of the tube at the initial time $t = 0$ is the rectangle $\Omega = (0, r_0) \times (0, l)$, where r_0 is the radius of the tube at the undeformed state, and l is its length, the cross-sections $z = 0$ and $z = l$ being the inflow and outflow of the fluid. We prescribe functions of surface forces $F^{(i)} = (F_r^{(i)}, F_z^{(i)})$, $i = 1, 2$, at the inflow and outflow. The longitudinal section of the deformed tube at instant t is defined by a function of displacements (oscillations) w that depends on points z of the tube and on t.

This problem of a nonsteady flow of a nonlinear viscous fluid in a deformable tube was formulated and studied in [3], where under some assumptions the existence of a local solution (i.e., a solution for a small interval of time) was proved. Below, we continue the investigation of this problem, and under the assumption that the radial component of the velocity of the fluid on the boundary of the tube is equal to 0 (this assumption is acceptable when the rate of oscillations of the tube is small compared with the rate of the fluid), we prove the existence of a nonlocal solution for small surface forces.

2. Problem of forced oscillations of a tube

We assume that the tube is viscoelastic and the displacements of its points $w(z,t)$ are small compared with its radius r_0. Then, the forced oscillations of the tube are described by the following equation, see [3, 7]:

$$\alpha \frac{\partial^2 w}{\partial t^2} + \gamma L \frac{\partial w}{\partial t} + Lw = q(w) \qquad \text{in } G = (0,l) \times (0,T), \tag{2.1}$$

where

$$Lw = \mathcal{D} \frac{\partial^4 w}{\partial z^4} + \beta w. \tag{2.2}$$

Here, w is the function of displacements (oscillations), α, γ, $\mathcal{D}$, β are positive constants, T is a finite number, $q(w)$ is a function of the load that is defined by means of the solution of the problem for the fluid. We suppose that the bending moments and cutting forces are equal to zero at the edges of the tube. Then, the boundary conditions are the following:

$$\begin{aligned} &\frac{\partial^2 w}{\partial z^2}(0,t) = \frac{\partial^2 w}{\partial z^2}(l,t) = 0, \qquad && t \in (0,T), \\ &\frac{\partial^3 w}{\partial z^3}(0,t) = \frac{\partial^3 w}{\partial z^3}(l,t) = 0, \qquad && t \in (0,T). \end{aligned} \tag{2.3}$$

We prescribe the following initial conditions

$$w(z,0) = 0, \qquad \frac{\partial w}{\partial t}(z,0) = 0, \qquad z \in (0,l). \tag{2.4}$$

3. Problem for the fluid and the function of the load of the tube

We denote by $\Omega_w(t)$ the longitudinal section of the tube at time t for a function of displacements w. We have, see Figure 1,

$$\Omega_w(t) = \{\, (r, z) \mid z \in (0, l),\ r \in (0, r_0 + w(z, t)) \,\}, \tag{3.1}$$

and (2.4) gives

$$\Omega_w(0) = \Omega = \{\, (r, z) \in (0, r_0) \times (0, l) \,\}. \tag{3.2}$$

We denote

$$Q_w = \{\, (r, z, t) \mid t \in (0, T),\ (r, z) \in \Omega_w(t) \,\}, \tag{3.3}$$

and let Γ_{w1}, Γ_{w2}, and Γ_{w3} be the parts of the boundary of Q_w corresponding to the inflow, outflow, and the wall of the tube, i.e.,

$$\begin{aligned} \Gamma_{w1} &= \{\, (r, z, t) \mid t \in (0, T),\ z = 0,\ r \in (0, r_0 + w(0, t)) \,\}, \\ \Gamma_{w2} &= \{\, (r, z, t) \mid t \in (0, T),\ z = l,\ r \in (0, r_0 + w(l, t)) \,\}, \\ \Gamma_{w3} &= (r, z, t) \mid t \in (0, T),\ z \in (0, l),\ r = r_0 + w(z, t)) \,\}. \end{aligned} \tag{3.4}$$

We suppose the flow to be axially symmetric and the fluid to be nonlinear viscous. Then, in the cylindrical coordinates r, α, z, the velocity vector v is given by $v = (v_r, v_z)$, $v_\alpha = 0$, and the components of the rate of the deformation tensor $e(v) = \{e_{ij}(v)\}_{i,j=1}^3$ have the form

$$\begin{aligned} &e_{11}(v) = \frac{\partial v_r}{\partial r}, \qquad e_{22}(v) = \frac{v_r}{r}, \qquad e_{33}(v) = \frac{\partial v_z}{\partial z}, \\ &e_{12}(v) = e_{21}(v) = e_{23}(v) = e_{32}(v) = 0, \\ &e_{13}(v) = e_{31}(v) = \frac{1}{2}\left(\frac{\partial v_z}{\partial r} + \frac{\partial v_r}{\partial z}\right). \end{aligned} \tag{3.5}$$

The constitutive equation of a nonlinear viscous fluid is defined by, see [2, 1],

$$\sigma_{ij}(p, v) = -p\delta_{ij} + 2\phi(I(v))e_{ij}(v), \qquad i, j = 1, 2, 3. \tag{3.6}$$

Here, $\sigma_{ij}(p, v)$ are the components of the stress tensor $\sigma(p, v)$ depending on the functions of pressure p and velocity v, δ_{ij} is the Kronecker delta, ϕ is a viscosity function depending on the second invariant of the rate of the deformation tensor $I(v)$,

$$I(v) = \sum_{i,j=1}^{3} \left(e_{ij}(v)\right)^2. \tag{3.7}$$

According to (3.5), we have

$$I(v) = \left(\frac{\partial v_r}{\partial r}\right)^2 + \left(\frac{v_r}{r}\right)^2 + \left(\frac{\partial v_z}{\partial z}\right)^2 + \frac{1}{2}\left(\frac{\partial v_z}{\partial r} + \frac{\partial v_r}{\partial z}\right)^2. \tag{3.8}$$

For a given function of displacements w, let us consider the problem for the fluid with the constitutive equation (3.6). We suppose that the velocities of the fluid are not high and that its viscosity is high. Then, in the equations of motion, the nonlinear terms in the inertia forces can be neglected, and these equations take the form

$$\rho\frac{\partial v_r}{\partial t}+\frac{\partial p}{\partial r}-2\left\{\frac{\partial\left[\phi(I(v))\frac{\partial v_r}{\partial r}\right]}{\partial r}+\frac{1}{2}\frac{\partial\left[\phi(I(v))\left(\frac{\partial v_z}{\partial r}+\frac{\partial v_r}{\partial z}\right)\right]}{\partial z}+\frac{\phi(I(v))\left(\frac{\partial v_r}{\partial r}-\frac{v_r}{r}\right)}{r}\right\}=0 \quad \text{in } Q_w, \tag{3.9}$$

$$\rho\frac{\partial v_z}{\partial t}+\frac{\partial p}{\partial z}-2\left\{\frac{\partial\left[\phi(I(v))\frac{\partial v_z}{\partial z}\right]}{\partial z}+\frac{1}{2}\frac{\partial\left[\phi(I(v))\left(\frac{\partial v_z}{\partial r}+\frac{\partial v_r}{\partial z}\right)\right]}{\partial r}+\frac{1}{2}\frac{\phi(I(v))\left(\frac{\partial v_r}{\partial z}+\frac{\partial v_z}{\partial r}\right)}{r}\right\}=0 \quad \text{in } Q_w. \tag{3.10}$$

If ϕ is a constant, then (3.9) and (3.10) are the Stokes equations in the cylindrical coordinates. The equation of incompressibility is given by

$$\operatorname{div}_c v=\frac{\partial v_r}{\partial r}+\frac{\partial v_z}{\partial z}+\frac{v_r}{r}=0 \quad \text{in } Q_w. \tag{3.11}$$

Here and below, $\operatorname{div}_c v$ denotes the operator of divergence in the cylindrical coordinates for an axially symmetric flow.

We consider mixed boundary conditions and prescribe surface forces on Γ_{w1} and Γ_{w2} and velocities on Γ_{w3}. The conditions on Γ_{w1} and Γ_{w2} are the following:

$$\begin{aligned}\sigma_{33}(p,v)\big|_{\Gamma_{w1}}&=F_z^{(1)}(w), &\quad \sigma_{31}(p,v)\big|_{\Gamma_{w1}}&=F_r^{(1)}(w),\\ \sigma_{33}(p,v)\big|_{\Gamma_{w2}}&=F_z^{(2)}(w), &\quad \sigma_{31}(p,v)\big|_{\Gamma_{w2}}&=F_r^{(2)}(w),\end{aligned} \tag{3.12}$$

where $\sigma_{ij}(p,v)$ are defined by (3.5) and (3.6). We suppose that the functions $w\to F^{(i)}(w)=(F_r^{(i)}(w),F_z^{(i)}(w))$, $i=1,2$, are given. Assuming that the fluid adheres to the wall of the tube and the rate of oscillations of the tube $\frac{\partial w}{\partial t}$ is small, we take the following boundary conditions on Γ_{w3}:

$$v_z\big|_{\Gamma_{w3}}=0, \qquad v_r\big|_{\Gamma_{w3}}=0. \tag{3.13}$$

The case when $v_z\big|_{\Gamma_{w3}}=0$ and $v_r(r_0+w(z,t),z,t)=\frac{\partial w}{\partial t}(z,t)$ is considered in [3], but the results of [3] are weaker than those of this paper. We assign an initial condition

(3.14) $$v(r, z, 0) = v_0(r, z), \qquad (r, z) \in \Omega = \Omega_w(0).$$

Let v, p be solutions of the problem (3.9)–(3.14). Then, the function of the load of the tube $q(w)$ – the right hand side of (2.1) – is given by

(3.15) $$q(w) = \mathcal{P}\Big\{\Big[p - 2\phi(I(v))\,\frac{\partial v_r}{\partial r}\Big]\Big|_{\Gamma_{w3}}\Big\}.$$

Here, $\mathcal{P}$ is the operator of translation from Γ_{w3} onto Γ_0,

(3.16) $$\Gamma_0 = \{\,(r, z, t) \mid r = r_0,\ z \in (0, l),\ t \in (0, T)\,\}.$$

Thus, Γ_0 is identified with $G = (0, l) \times (0, T)$. If f is a function given on Γ_{w3}, then

(3.17) $$(\mathcal{P}f)(z, t) = f(r_0 + w(z, t), z, t), \qquad (z, t) \in \overline{G}.$$

We introduce the operator $\mathcal{P}$ because the problem (2.1) for the tube is associated with Γ_0.

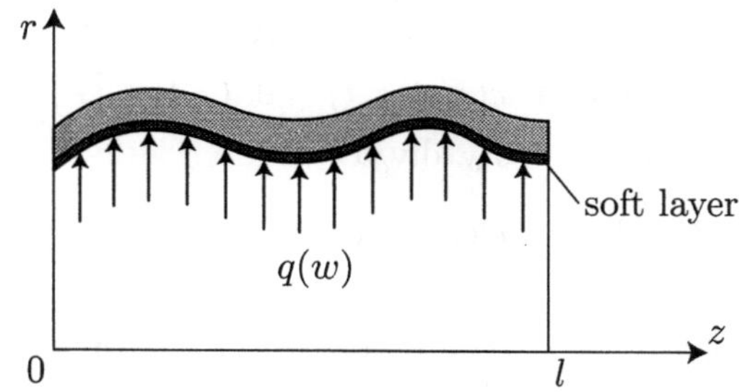

Fig. 2

We suppose that the tube has a soft thin viscoelastic inner layer that softens the load of the tube, making it spread and smooth, see Figure 2. Vessels of people and animals have such a layer, and in many cases, the tubes that are used in engineering do also have a soft inner layer. We simulate the soft layer by a smoothing operator (an averaging operator), so that instead of (3.15) we determine the load of the tube as follows:

(3.18) $$q(w) = \mathcal{P}\Big\{\Big[\mathcal{R}p - 2\phi(I(\mathcal{R}v))\,\frac{\partial(\mathcal{R}v_r)}{\partial r}\Big]\Big|_{\Gamma_{w3}}\Big\},$$

where $\mathcal{R}$ is the operator of smoothing over the space and time variables.

Thus, the problem of nonsteady flow of the nonlinear viscous fluid in the deformable tube consists in solving the problem (2.1)–(2.4) with $q(w)$ defined by (3.18), where the pair (v, p) is the solution of the problem for the fluid (3.9)–(3.14) that depends on w.

4. Generalized solution of the problem for the fluid

Let us introduce a set of admissible displacements W as follows:

$$
\begin{aligned}
W &= \{ w \mid w \in C^{1,\beta}(\overline{G}),\ G = (0,l) \times (0,T),\ |w(z,t)| \leq b,\ (z,t) \in \overline{G}, \\
&\qquad b \text{ is a constant},\ 0 < b < r_0,\ w(z,0) = 0,\ \tfrac{\partial w}{\partial t}(z,0) = 0,\ z \in [0,l] \}.
\end{aligned} \tag{4.1}
$$

The set W is equipped with the topology generated by that of $C^{1,\beta}(\overline{G})$, the space of functions continuously differentiable in $\overline{G}$ whose derivatives satisfy the Hölder condition with index $\beta \in (0,1)$, the norm in $C^{1,\beta}(\overline{G})$ being defined by

$$
\begin{aligned}
\|w\|_{C^{1,\beta}(\overline{G})} = \|w\|_{C^1(\overline{G})} &+ \sup_{(z,t),\,(z',t')\in G} \frac{\left|\frac{\partial w}{\partial z}(z,t) - \frac{\partial w}{\partial z}(z',t')\right|}{\left[(z-z')^2 + (t-t')^2\right]^{\beta/2}} \\
&+ \sup_{(z,t),\,(z',t')\in G} \frac{\left|\frac{\partial w}{\partial t}(z,t) - \frac{\partial w}{\partial t}(z',t')\right|}{\left[(z-z')^2 + (t-t')^2\right]^{\beta/2}}.
\end{aligned}
$$

For a function $w \in W$, we define the following sets

$$
\begin{aligned}
J_w &= \{ v \mid v = (v_r, v_z) \in C^\infty(\overline{Q}_w)^2,\ v_r(0,z,t) = 0, \\
&\qquad v(r_0 + w(z,t), z, t) = 0,\ (z,t) \in \overline{G} \}, \\
J_{ws} &= \{ v \mid v \in J_w,\ \operatorname{div}_c v = 0 \},
\end{aligned} \tag{4.2}
$$

and we let H_w and H_{ws} be the closures of J_w and J_{ws} with respect to the norm

$$
\|v\|_{H_w} = \left(\int_{Q_w} \sum_{i,j=1}^{3} \big(e_{ij}(v)\big)^2 r\, dr\, dz\, dt \right)^{1/2}. \tag{4.3}
$$

Let also H_{w0} and M_w be the spaces of vector-valued functions $v = (v_r, v_z)$ and scalar functions μ, respectively, which are square integrable in Q_w with respect to the measure $r\,dr\,dz\,dt$.

By $N(\Gamma_{w1})$ and $N(\Gamma_{w2})$ we denote the spaces of functions square integrable in Γ_{w1} and Γ_{w2} with respect to the measure $r\,dr\,dt$.

Let us determine sets $\mathcal{I}_1$ and $\mathcal{I}$ as follows:

$$
\begin{aligned}
\mathcal{I}_1 &= \{ u \mid u = (u_r, u_z) \in C^\infty(\overline{\Omega}),\ \Omega = (0, r_0) \times (0, l), \\
&\qquad u_r(0,z) = 0,\ u(r_0, z) = 0,\ z \in [0,l] \}, \\
\mathcal{I} &= \{ u \mid u \in \mathcal{I}_1,\ \operatorname{div}_c u = 0 \},
\end{aligned} \tag{4.4}
$$

and let V_1 and V be the closures of $\mathcal{I}_1$ and $\mathcal{I}$ with respect to the norm

$$
\|u\|_V = \left(\int_\Omega \sum_{i,j=1}^{3} \big(e_{ij}(u)\big)^2 r\, dr\, dz \right)^{1/2}. \tag{4.5}
$$

Concerning the viscosity function, we suppose that ϕ is a function twice continuously differentiable in $\mathbb{R}_+ = [0,\infty)$, and there exist positive constants $a_1,\dots,a_4$ such that, for an arbitrary $y \in \mathbb{R}_+$, the following inequalities hold:

$$a_1 \le \phi(y) \le a_2, \qquad \phi(y) + 2y\,\frac{d\phi}{dy}(y) \ge a_3, \qquad y\left|\frac{d\phi}{dy}(y)\right| \le a_4. \tag{4.6}$$

For the physical meaning of the inequalities in (4.6), see [3].

We suppose that, for each $w \in W$, functions of surface forces at the inflow and outflow, $F^{(1)}(w)$ and $F^{(2)}(w)$, are given and (see (3.12))

$$F^{(i)}(w) = (F_r^{(i)}(w),\ F_z^{(i)}(w)) \in N(\Gamma_{wi})^2, \qquad i = 1,2. \tag{4.7}$$

We also assume that

$$v_0 \in V, \tag{4.8}$$

where v_0 is the function of the initial velocities of (3.14).

We define a function of the load of the fluid, $\mathcal{F}_w \in H_w^*$, as follows:

$$(\mathcal{F}(w), h) = \sum_{i=1}^{2} \int_{\Gamma_{wi}} \left[F_r^{(i)}(w)h_r + F_z^{(i)}(w)h_z\right] r\,dr\,dt, \qquad h \in H_w. \tag{4.9}$$

Further we introduce operators

$$\begin{aligned} &L_w\colon H_w \to H_w^*,\\ &(L_w(v), h) = 2\int_{Q_w} \phi(I(v))e_{ij}(v)e_{ij}(h)r\,dr\,dz\,dt, \qquad v,h \in H_w, \end{aligned} \tag{4.10}$$

$$\begin{aligned} &B_w \in \mathcal{L}(H_w, M_w^*),\\ &(B_w v, \mu) = \int_{Q_w} (\mathrm{div}_c v)\mu r\,dr\,dz\,dt, \qquad v \in H_w,\ \mu \in M_w, \end{aligned} \tag{4.11}$$

$$\begin{aligned} &A_w \in \mathcal{L}(H_{w0}, H_{w0}^*) \quad \text{and} \quad A_w \in \mathcal{L}(H_w^*, H_w^*),\\ &(A_w v, h) = \rho\int_{Q_w} (v_r h_r + v_z h_z)r\,dr\,dz\,dt, \qquad v,h \in H_{w0}. \end{aligned} \tag{4.12}$$

In the expression (4.10), the summation over repeated indices is implied. We consider the problem: Find a pair of functions (v,p) satisfying

$$v \in H_w, \qquad \frac{\partial v}{\partial t} \in H_w^*, \qquad p \in M_w, \tag{4.13}$$

$$\left(A_w\frac{\partial v}{\partial t}, h\right) + (L_w(v), h) - (B_w^* p, h) = (\mathcal{F}(w), h), \qquad h \in H_w, \tag{4.14}$$

$$(B_w v, \mu) = 0, \qquad \mu \in M_w, \tag{4.15}$$

$$u(\cdot\,,\cdot\,,0) = v_0. \tag{4.16}$$

Here, B_w^* is the operator adjoint of B_w. By using Green's formula, one can verify that, if (v,p) is a smooth solution of the problem (3.9)–(3.14), then (v,p) is a solution of the problem (4.13)–(4.16). Conversely,, if (v,p) is a solution of (4.13)–(4.16), then (v,p) is a solution of (3.9)–(3.14) in the distribution sense, i.e., it is a generalized solution of this problem.

We determine the set W_1 by

$$W_1 = \{ w \mid w \in W,\ \|w\|_{C^{1,\beta}(\overline{G})} \leq \tilde{c} \}, \tag{4.17}$$

where $\tilde{c}$ is a positive constant and W is defined by (4.1). The set W_1 is equipped with the topology generated by that of $C^{1,\beta}(\overline{G})$.

It is clear that the solution (v,p) of the problem (4.13)–(4.16) depends on $w \in W$, and so we denote it by $(v(w), p(w))$.

The next theorem follows from the results of [3].

Theorem 4.1. *Suppose the conditions* (4.6)–(4.8) *hold and* $w \in W$. *Then, there exist unique solutions* $v = v(w)$, $p = p(w)$ *of the problem* (4.13)–(4.16) *and positive constants* μ_1, $\mu(\tilde{c})$ *such that the following estimates hold:*

$$\|v(w)\|_{H_w} \leq \mu_1(\|\mathcal{F}(w)\|_{H_w^*} + \|v_0\|_V), \qquad w \in W, \tag{4.18}$$

$$\|p(w)\|_{H_{w0}} \leq \mu(\tilde{c})(\|\mathcal{F}(w)\|_{H_w^*} + \|v_0\|_V), \qquad w \in W, \tag{4.19}$$

where the constant $\mu(\tilde{c})$ *depends on* $\tilde{c}$ *from* (4.17).

5. On the w-dependence of the velocity and the pressure functions

We introduce the notations

$$Q = Q_w \quad \text{if } w = 0, \tag{5.1}$$

$$\Gamma_i = \Gamma_{wi} \quad \text{if } w = 0,\ i = 1,2,3, \tag{5.2}$$

where Q_w and Γ_{wi} are defined by (3.3) and (3.4).

Let b_1 be a constant such that

$$0 < b_1 < r_0 - b, \tag{5.3}$$

where b is the constant from (4.1).

For each $w \in W$, we determine a mapping $\mathcal{P}_w : \overline{Q} \to \overline{Q}_w$ by

$$\begin{aligned} &\overline{Q} \ni (r,z,t) \to \mathcal{P}_w(r,z,t) \\ &= \begin{cases} (r,z,t), & r \in [0,b_1),\ (z,t) \in \overline{G}, \\ ((\mathcal{P}_w)_1(r,z,t), (\mathcal{P}_w)_2(r,z,t), (\mathcal{P}_w)_3(r,z,t)), & r \in [b_1, r_0],\ (z,t) \in \overline{G}, \end{cases} \end{aligned} \tag{5.4}$$

where

$$(5.5)\quad (\mathcal{P}_w)_1(r,z,t) = e_0 + e_1 r + e_2 r^2, \quad (\mathcal{P}_w)_2(r,z,t) = z, \quad (\mathcal{P}_w)_3(r,z,t) = t,$$

with

$$e_0 = w(z,t)\left(1 + \frac{2b_1 r_0 - r_0^2}{(r_0 - b_1)^2}\right), \qquad e_1 = 1 - \frac{2b_1 w(z,t)}{(r_0 - b_1)^2}, \qquad e_2 = \frac{w(z,t)}{(r_0 - b_1)^2}.$$

It is not difficult to verify that, for an arbitrary $w \in W$, the mapping $\mathcal{P}_w$ defined by (5.4), (5.5) is a diffeomorphism of class C^1 from $\overline{Q}$ onto $\overline{Q}_w$.

We denote by H the space H_w if $w = 0$ and introduce an element $\mathcal{F}_1(w) \in H^*$ by

$$(5.6)\qquad (\mathcal{F}_1(w), h) = (\mathcal{F}(w), h \circ \mathcal{P}_w^{-1}), \qquad h \in H,$$

where $\mathcal{P}_w^{-1}$ is the inverse mapping of $\mathcal{P}_w$ and $\mathcal{F}(w)$ is defined by (4.9). We suppose that

$$(5.7)\qquad \begin{cases} \text{the function } w \to \mathcal{F}_1(w) \text{ is a continuous mapping of } W \text{ equipped} \\ \text{with the topology generated by that of } C^1(\overline{G}) \text{ into } H^*. \end{cases}$$

By using the diffeomorphism $\mathcal{P}_w$, we transform the problem (4.13)–(4.16) in the noncylindrical domain Q_w into the problem in the cylindrical domain Q. Denoting the solutions of the problem in Q by $\tilde{v}(w)$, $\tilde{p}(w)$, we have $\tilde{v}(w) = v(w) \circ \mathcal{P}_w$, $\tilde{p}(w) = p(w) \circ \mathcal{P}_w$, where $v(w) = v$, $p(w) = p$ are the solutions of (4.13)–(4.16).

The following theorem follows from the results of [3].

Theorem 5.1. *Suppose the conditions* (4.6)–(4.8) *and* (5.7) *are satisfied. Let* $\{w_n\}_{n=1}^{\infty} \subset W_1$, *where* W_1 *is defined by* (4.17) *and* $w_n \to w_0$ *in* $C^1(\overline{G})$, *let* $\tilde{v}(w_n) = v(w_n) \circ \mathcal{P}_{w_n}$, $\tilde{p}(w_n) = p(w_n) \circ \mathcal{P}_{w_n}$, *where* $v(w_n) = v$, $p(w_n) = p$ *are the solutions of the problem* (4.13)–(4.16) *for* $w = w_n$, $n = 0, 1, 2\ldots.$ *Then,* $\tilde{v}(w_n) \to \tilde{v}(w_0)$ *weakly in* H *and* $\tilde{p}(w_n) \to \tilde{p}(w_0)$ *weakly in* M, *where* M *is the space* M_w *if* $w = 0$.

6. Existence theorem

We suppose that the smoothing operator $\mathcal{R}$ from (3.18) satisfies the condition

$$(6.1)\qquad \mathcal{R} \in \mathcal{L}(M_w, C^\infty(\overline{Q}_w)), \qquad w \in W.$$

Theorem 6.1. *Suppose the conditions* (4.6)–(4.8), (5.7), *and* (6.1) *hold. Then, there exists a constant* $\theta > 0$ *such that, if*

$$(6.2)\qquad \sup_{w \in W_1} \|\mathcal{F}_1(w)\|_{H^*} + \|v_0\|_V = \theta_1 \le \theta,$$

where $\mathcal{F}_1(w)$ *is defined by* (4.9) *and* (5.6), *there exists a function* $w \in W_1$ *which is a solution of the problem* (2.1)–(2.4) *with* $q(w)$ *defined by* (3.18).

Proof. For a given function $f \in C^1([0,T];H^1(0,l))$, let us consider the following problem: Find a function $u(f)$ satisfying

$$(6.3)\quad \begin{aligned} &\alpha \frac{\partial^2 u(f)}{\partial t^2} + \gamma L \frac{\partial u(f)}{\partial t} + Lu(f) = f && \text{in } G = (0,l) \times (0,T), \\ &\frac{\partial^2 u(f)}{\partial z^2}(0,t) = \frac{\partial^2 u(f)}{\partial z^2}(l,t) = 0, && t \in (0,T), \\ &\frac{\partial^3 u(f)}{\partial z^3}(0,t) = \frac{\partial^3 u(f)}{\partial z^3}(l,t) = 0, && t \in (0,T), \\ &u(f)(z,0) = \frac{\partial u(f)}{\partial t}(z,0) = 0, && z \in (0,l), \end{aligned}$$

where L is the operator defined by (2.2), and α, γ are positive constants. It follows from the results of [5, 6] that there exists a unique function $u(f) \in C^{1,\nu}(\overline{G})$, $\nu \in (0,1)$, satisfying (6.3) and such that

$$(6.4)\qquad \|u(f)\|_{C^{1,\nu}(\overline{G})} \leq \eta \|f\|_{C^1([0,T];H^1(0,l))},$$

where η is independent of f.
Consider the problem: Find a function w satisfying

$$(6.5)\qquad w \in W_1, \qquad w = u(q(w)),$$

where $q(w)$ is defined by (3.18) and $u(q(w))$ is the solution of the problem (6.3) for $f = q(w)$. Clearly, if w is a solution of the problem (6.5), then w is a solution of the problem (2.1)–(2.4) with $q(w)$ defined by (3.18).
We take in (6.4) an arbitrary $\nu \in (0,1)$, $\nu > \beta$, where β is the index from (4.17). It follows from (3.18), (4.18), (4.19), (6.1), and (6.4) that there exists a constant $\theta > 0$ such that, under the condition (6.2), the function $w \longrightarrow u(q(w))$ maps W_1 into W_1. By virtue of (6.1), (6.4), and Theorem 5.1, the function $w \longrightarrow u(q(w))$ is a compact mapping of W_1 into W. Now, the existence of a solution of the problem (6.5) follows from the Schauder principle, see, e.g., [4]. □

References

[1] ASTARITA, G., MARRUCCI, G.: Principles of Non-Newtonian Fluid Mechanics; Mc. Graw-Hill, New York 1974.

[2] LITVINOV, W.G.: Motion of Nonlinear Viscous Fluid; Nauka, Moscow 1982 (in Russian).

[3] LITVINOV, W.G.: A problem on nonsteady flow of a nonlinear viscous fluid in a deformable pipe; Methods Funct. Anal. Topology 2 (1996), 85–113.

[4] NIRENBERG, L.: Topics in Nonlinear Functional Analysis; New York University, New York 1974.

[5] ORLOV, V.P.: On the stability of the zero solution of a one-dimensional mathematical model of viscoelasticity; Differerential Integral Equations 4 (1991), 89–101.

[6] SHKLIAR, A.YA.: Cauchy Problem for Differential Equations with Commuting Normal Operational Coefficients; Ph.D. Thesis, Institute of Mathematics of Ukrainian Academy of Sciences, Kiev 1990 (in Russian).

[7] TIMOSHENKO, S., WOINOWSKY-KRIEGER, S.: Theory of Plates and Shells; Mc. Graw-Hill, New York 1959.

Institute of Mechanics
National Academy of Sciences of Ukraine
3 Nesterov St.
252057 Kiev
Ukraine
wlit@resistance.freenet.kiev.ua

1991 Mathematics Subject Classification: Primary 35Q35, 35Q72

Submitted: April 29, 1996

Operator Theory:
Advances and Applications, Vol. 102

Quantum compound Poisson processes and white noise analysis

E.W. LYTVYNOV

We consider a family B of self-adjoint commuting operators $B_\zeta = \int \zeta(t)\, dB_t$ where B_t is a quantum compound Poisson process in a Fock space. By using the projection spectral theorem, we construct the Fourier transform in generalized joint eigenvectors of the family B which is unitary between the Fock space and the L^2-space of compound Poisson white noise, (L^2_{CP}). This construction gives the possibility of introducing spaces of test and generalized functions the dual pairing of which is determined by the inner product of (L^2_{CP}).

1. Introduction

This note is aimed to be an introduction to analysis of test and generalized functions the dual pairing of which is determined by the inner product of the L^2-space $(L^2_{\mathrm{CP}}) \equiv L^2(S'(T), d\mu_{\mathrm{CP}})$, where $S'(T)$ is the Schwartz space of tempered distributions on $T \subset \mathbb{R}^d$ and μ_{CP} is the measure of compound Poisson white noise (CPWN), whose Fourier transform is given by

$$(1.1) \quad \int_{S'(T)} e^{i\langle \omega, \zeta\rangle}\, d\mu_{\mathrm{CP}}(\omega) = \exp\left[\int_\alpha \int_T (e^{is\zeta(t)} - 1)\, d\rho(s)\, d\nu(t)\right], \quad \zeta \in S(T),$$

where ν is a non-atomic measure on T, α is a bounded interval of $\mathbb{R}$, and ρ a finite measure on α.

Our particular interest in the measure μ_{CP} is inspired by the recent work [8], where a compound Poisson field was used in the investigation of some models of statistical mechanics of continuous systems.

Generally speaking, there are two approaches to studying functions defined on the space $S'(T)$ with a non-Gaussian measure μ. In the first approach [10, 11, 2, 4, 5, 16, 17], see also [20] and references therein, one constructs an orthogonal decomposition of $L^2(d\mu)$ in homogeneous chaoses and obtains in such a way a unitary mapping between $L^2(d\mu)$ and a Fock space. Then, one constructs a rigging of the Fock space, which under the unitary mapping goes over into a rigging of $L^2(d\mu)$. Unfortunately, this technique cannot be applied to a large class of probability measures. In the second approach [1, 15], one constructs a system of Appell polynomials in $L^2(d\mu)$ and its dual (biorthogonal) system, and then constructs spaces of test and generalized functions by using these systems. But apart from the case of Gaussian white noise, this does not lead to a unitary mapping between a Fock space and $L^2(d\mu)$.

In this note, we will follow, in general, the first approach, though, because of the fact that the compound Poisson process does not possess the chaotic representation property [22, 6, 20], some new points will appear.

In the forthcoming paper [19], see also [18], it will be shown that the spaces of test and generalized functions constructed in this paper (with $\kappa = 1$) coincide with those constructed by applying the second approach to the (L^2_{CP}) space.

2. Basic standard triples

Let us remind some special construction of a standard triple from [17]. Let T be a subset of $\mathbb{R}^d$, $d \in \mathbb{N}$ (which may coincide with the whole $\mathbb{R}^d$), and let ν be a Borel, regular, non-atomic measure on T which is finite on bounded subsets of T and, in the case of unbounded T, satisfies some additional conditions at infinity (see [17] for details). For example, one can take ν to be the Lebesgue measure dt. Then, for each $p \geq 1$, define $\hat{S}_p(\mathbb{R})$ to be the real Hilbert space spanned by the orthonormal basis $\big(e_j(2j+2)^{-p}\big)_{j=0}^{\infty}$, where $e_j = e_j(t)$ are the Hermite functions on $\mathbb{R}$, and let $\hat{S}_p(\mathbb{R}^d) \equiv \hat{S}_p^{\otimes d}(\mathbb{R})$. By putting $\nu(\mathbb{R}^d \setminus T) = 0$, we can consider ν as a measure on $\mathbb{R}^d$. The above mentioned conditions on ν ensure the existence of $k \geq 0$ such that the space $\hat{S}_{1+k}(\mathbb{R}^d)$ is embedded into $L^2(\mathbb{R}^d, d\nu)$, and moreover, the embedding $\hat{S}_{1+k}(\mathbb{R}^d) \hookrightarrow L^2(\mathbb{R}^d, d\nu)$ is of Hilbert–Schmidt type (in case of the Lebesgue measure this holds for $k = 0$). Put $S_p(T)$ to be the factor Hilbert space $S_p(T) \equiv \hat{S}_{p+k}(\mathbb{R}^d)/\ker O_{p+k}$, where $O_{p+k} : \hat{S}_{p+k}(\mathbb{R}^d) \hookrightarrow L^2(\mathbb{R}^d, d\nu)$ is the embedding operator. Let $S_{-p}(T)$ denote the dual of $S_p(T)$ with respect to the zero space $L^2(T, d\nu)$. Thus, one constructs the standard triple

$$S'(T) = \operatorname*{ind\,lim}_{p\to\infty} S_{-p}(T) \supset L^2(T, d\nu) \equiv L^2(T) \supset \operatorname*{proj\,lim}_{p\to\infty} S_p(T) = S(T). \tag{2.1}$$

Here, $S(T)$ is the Schwartz test space and $S'(T)$ the Schwartz space of tempered distributions on T (the dual of $S(T)$ with respect to $L^2(T)$). Note that, in case of a bounded T, $S(T)$ is, in fact, the space $\mathcal{D}(T)$ of infinitely differentiable functions on T.

Also, given a compactum $\alpha \subset \mathbb{R}$, we consider a Borel, finite, regular measure ρ on α whose support consists of more than one point, and either $0 \notin \alpha$ or $\rho(\{0\}) = 0$, i.e., 0 is not an atom for ρ. Absolutely analogously to (2.1), we construct the following standard triple:

$$\mathcal{D}'(\alpha) = \operatorname*{ind\,lim}_{p\to\infty} \mathcal{D}_{-p}(\alpha) \supset L^2(\alpha, d\rho) \equiv L^2(\alpha) \supset \operatorname*{proj\,lim}_{p\to\infty} \mathcal{D}_p(\alpha) = \mathcal{D}(\alpha). \tag{2.2}$$

Denoting $L^2(\alpha \times T) \equiv L^2(\alpha \times T, d\rho\, d\nu) = L^2(\alpha) \otimes L^2(T)$, $S_p(\alpha \times T) \equiv \mathcal{D}_p(\alpha) \otimes$

$S_p(T)$ and "taking the tensor product of the riggings (2.1) and (2.2)", we obtain

$$\begin{aligned} S'(\alpha\times T) &\equiv \mathcal{D}'(\alpha)\otimes S'(T)=\operatorname*{ind\,lim}_{p\to\infty} S_{-p}(\alpha\times T)\supset L^2(\alpha\times T)\\ &= S_0(\alpha\times T)\supset \operatorname*{proj\,lim}_{p\to\infty} S_p(\alpha\times T)=\mathcal{D}(\alpha)\otimes S(T)\\ &\equiv S(\alpha\times T). \end{aligned}\tag{2.3}$$

Let us note that $S(\alpha\times T)$ is an algebra under pointwise multiplication of functions and $|\xi\eta|_p\le C_p|\xi|_p|\eta|_p$, where $\xi,\eta\in S_p(\alpha\times T)$. Here and below, $|\cdot|_p$ stands for the norm of the spaces $S_p(T)$, $\mathcal{D}_p(T)$, and $S_p(\alpha\times T)$, as well as of their tensor powers.

Then, for every $n\in\mathbb{Z}_+$, one constructs the complex standard triple

$$\begin{aligned} S_c'^{\hat\otimes n}(\alpha\times T) &= \operatorname*{ind\,lim}_{p\to\infty} S_{-p,c}'^{\hat\otimes n}(\alpha\times T)\supset \hat L^2(\alpha^n\times T^n)=\left(L^2(\alpha\times T)\right)_c^{\hat\otimes n}\\ &\supset \operatorname*{proj\,lim}_{p\to\infty} S_{p,c}^{\hat\otimes n}(\alpha\times T)\supset S_c^{\hat\otimes n}(\alpha\times T), \end{aligned}\tag{2.4}$$

where $\hat\otimes$ denotes the symmetric tensor product, the index c stands for the complexification of a real space, and $S_{p,c}^{\hat\otimes 0}=\mathbb{C}$. Notice that, taking away all the indices c in (2.4), one gets the corresponding real standard triple.

Next, for any $p\in\mathbb{R}$, $q\in\mathbb{Z}$, and $\kappa\in\mathbb{R}$, define a weighted (Boson) Fock space $\Gamma_{q,\kappa}(S_p(\alpha\times T))$ to be the complex Hilbert space with norm $\|\cdot\|_{p,q,\kappa}$ consisting of sequences $\left(f^{(n)}\right)_{n=0}^\infty$ such that $f^{(n)}\in S_{p,c}^{\hat\otimes n}(\alpha\times T)$ and

$$\left\|\left(f^{(n)}\right)_{n=0}^\infty\right\|_{p,q,\kappa}^2=\sum_{n=0}^\infty \left|f^{(n)}\right|_p^2 2^{nq}(n!)^{1+\kappa},\qquad 0!\equiv 1.$$

Particularly, by setting $q=\kappa=0$, one gets the usual Fock space $\Gamma(S_p(\alpha\times T))$. Then, for each $\kappa\ge 0$, one gets the following standard triple (cf. [3, 14, 13]):

$$\begin{aligned} \Gamma_{-\kappa}(S'(\alpha\times T)) &= \operatorname*{ind\,lim}_{p,q\to\infty}\Gamma_{-q,-\kappa}(S_{-p}(\alpha\times T))\supset\Gamma(L^2(\alpha\times T))\\ &\supset \operatorname*{proj\,lim}_{p,q\to\infty}\Gamma_{q,\kappa}(S_p(\alpha\times T))\supset\Gamma_\kappa(S(\alpha\times T)). \end{aligned}\tag{2.5}$$

We only note that each $\Gamma_\kappa(S(\alpha\times T))$ is a nuclear space, and for each $F=\left(F^{(n)}\right)_{n=0}^\infty\in\Gamma_{-\kappa}(S'(\alpha\times T))$ and each $f=\left(f^{(n)}\right)_{n=0}^\infty\in\Gamma_\kappa(S(\alpha\times T))$ the dual pairing between F and f is given by $\langle\!\langle F,f\rangle\!\rangle=\sum_{n=0}^\infty\langle\overline{F^{(n)}},f^{(n)}\rangle n!$, where $\overline{F^{(n)}}$ denotes the complex conjugate of $F^{(n)}$ and $\langle\cdot,\cdot\rangle$ stands for the dual pairing between $S_c'^{\hat\otimes n}(\alpha\times T)$ and $S_c^{\hat\otimes n}(\alpha\times T)$, which is supposed to be linear in both dots.

We will also use the spaces $\Gamma_{\mathrm{fin}}(L^2(\alpha\times T))$ and $\Gamma_{\mathrm{fin}}(S(\alpha\times T))$ that are defined as topological sums of the spaces $\hat L^2(\alpha^n\times T^n)$ and $S_c^{\hat\otimes n}(\alpha\times T)$. The space $\Gamma_{\mathrm{fin}}(S(\alpha\times T))$ is a nuclear space, and its dual space $\Gamma_{\mathrm{fin}}^*(S(\alpha\times T))$ consists of all sequences of the form $\left(F^{(n)}\right)_{n=0}^\infty$, $F^{(n)}\in S_c'^{\hat\otimes n}(\alpha\times T)$. Thus, we have also constructed the nuclear rigging

$$\Gamma_{\mathrm{fin}}^*(S(\alpha\times T))\supset\Gamma(L^2(\alpha\times T))\supset\Gamma_{\mathrm{fin}}(S(\alpha\times T)).$$

3. Compound Poisson white noise on T – a spectral approach

Let $\mu_{\rm P}$ be the measure of (usual) Poisson white noise (PWN) on $\alpha \times T$ with intensity $\rho\nu$:

$$\int_{S'(\alpha\times T)} e^{i\langle x,\xi\rangle}\, d\mu_{\rm P}(x) = \exp\left[\int_\alpha\int_T (e^{i\xi(s,t)}-1)\, d\rho(s)\, d\nu(t)\right], \tag{3.1}$$

where $\xi \in S(\alpha \times T)$. Let ζ be an arbitrary element of $S(T)$, and put $\xi = s \otimes \zeta$, i.e., $\xi(s,t) = s\zeta(t)$, which belongs to $S(\alpha \times T)$ since $s \in \mathcal{D}(\alpha)$ (this is the point where we use the condition of the boundedness of the set α). By (3.1),

$$\int_{S'(\alpha\times T)} e^{i\langle x,s\otimes\zeta\rangle}\, d\mu_{\rm P}(x) = \exp\left[\int_\alpha\int_T (e^{is\zeta(t)}-1)\, d\rho(s)\, d\nu(t)\right]. \tag{3.2}$$

Define the mapping $S'(\alpha \times T) \ni x \to Kx = \langle x, s\rangle \in S'(T)$. Note that, for each $x \in S'(\alpha\times T)$, there is $p \geq 0$ such that $x \in S_{-p}(\alpha \times T)$, and so

$$|\langle Kx,\zeta\rangle| \leq |x|_{-p}|s\otimes\zeta|_p = |x|_{-p}|s|_p|\zeta|_p,$$

whence $|Kx|_{-p} \leq |x|_{-p}|s|_p$. Thus, $\langle x, s\rangle$ indeed belongs to $S'(T)$. The mapping K is evidently measurable and "on." Hence, $\mu_{\rm P}$ generates, via K, some measure $\mu_{\rm CP}$ on the measure space $(S'(T), \mathcal{B}(S'(T)))$: $\mu_{\rm CP}(\beta) \equiv \mu_{\rm P}(K^{-1}\beta)$, $\beta \in \mathcal{B}(S'(T))$, $\mathcal{B}(S'(T))$ the Borel σ-algebra on $S'(t)$. Taking to notice (3.2), we conclude that the equality (1.1) holds for the measure $\mu_{\rm CP}$ just constructed, so that it is the measure of CPWN on T with Lévy measure $\rho\nu$.

It is well-known, e.g., [12], that $\mu_{\rm P}$ is concentrated on the set of the series of delta functions

$$x = \sum \delta_{(s_i,t_i)}, \qquad (s_i,t_i) \in \alpha\times T, \qquad t_i \neq t_j \quad \text{if} \quad i \neq j. \tag{3.3}$$

If T is bounded, each series is finite, and if T is unbounded, the series (3.3) are infinite and such that, on every bounded subset of $\alpha \times T$, there are only a finite number of atoms (s_i, t_i). On the other hand, $\mu_{\rm CP}$ is concentrated on the set of the series

$$\omega = \sum s_i\delta_{t_i}, \qquad (s_i,t_i) \in \alpha\times T, \quad t_i \neq t_j \ \text{if} \ i \neq j, \tag{3.4}$$

Since 0 is not an atom for ρ, the mapping K establishes a *one-to-one* correspondence between the sets of the series (3.3) and (3.4), which gives us the possibility of introducing the unitary mapping (cf. [6])

$$(L^2_{\rm P}) \ni f = f(x) \to \mathcal{K}f = (\mathcal{K}f)(\omega) \equiv f(K^{-1}\omega) \in (L^2_{\rm CP}). \tag{3.5}$$

Since $(L^2_{\rm P})$ is isometrically isomorphic to the Fock space $\Gamma(L^2(\alpha \times T))$, e.g., [21, 10, 11, 16], so is the space $(L^2_{\rm CP})$.

We will now show that the latter unitary mapping can also be constructed by using the projection spectral theorem [3] in a way parallel to the case of the usual Poisson, which seems to be useful for further investigation.

For any $h \in H = L^2(\alpha \times T) \cap L^1(\alpha \times T) \cap L^\infty(\alpha \times T)$, we define, in the Fock space $\Gamma(L^2(\alpha \times T))$, a linear Hermitian operator $\mathcal{A}_h$ with domain $\Gamma_{\text{fin}}(L^2(\alpha \times T))$ by

$$\mathcal{A}_h \equiv a^+(h) + a^-(h) + a^0(h) + \int_\alpha \int_T h(s,t)\, d\rho(s)\, d\nu(t),$$

where $a^+(h)$ is the creation operator: $a^+(h)f^{(n)} = h\hat{\otimes} f^{(n)}$, $a^-(h)$ is the annihilation operator: $a^-(h)f^{(n)} = n\langle f^{(n)}, h\rangle$, and $a^0(h)$ is the preservation operator:

$$a^0(h)f^{(n)}(s_1,t_1,\dots,s_n,t_n) = n((f^{(n)}(s_1,t_1,\dots,s_n,t_n)h(s_n,t_n))^{\widehat{}},$$

where $n \in \mathbb{Z}_+$, $\widehat{}$ denotes the symmetrization of a function, and $f^{(n)} \in \hat{L}^2(\alpha^n \times T^n)$. It is worth noting that $a^0(h)$ is actually the differential second quantization $d\Gamma(h\,\cdot)$ of the operator $h\,\cdot$ of multiplication by the function h in the space $L^2(\alpha \times T)$.

For each $\Delta \in \Lambda$, where Λ is the set of Borel bounded subsets of $\alpha \times T$, we put $A_\Delta = \mathcal{A}_{\chi_\Delta}$, where χ_Δ is the indicator of Δ defined on $\alpha \times T$. Then, the operator family $A = (A_\Delta)_{\Delta\in\Lambda}$ is the quantum Poisson process on $\alpha \times T$ in Hudson and Parthasarathy's terms [9, 7, 20]. Let us remind that every operator $\mathcal{A}_h$, $h \in H$, can be represented as a quantum stochastic integral with respect to the quantum process A: $\mathcal{A}_h = \int_\alpha \int_T h(s,t)\, dA_{(s,t)}$.

Now, for any $g \in G \equiv L^2(T) \cap L^1(T) \cap L^\infty(T)$, we put $\mathcal{B}_g \equiv \mathcal{A}_{s\otimes g}$. Let Υ be the set of Borel bounded subsets of T. For any $\varepsilon \in \Upsilon$, we set also $B_\varepsilon \equiv \mathcal{B}_{\chi_\varepsilon}$, where χ_ε is the indicator of ε defined on T. As will be seen below, the operator family $B = (B_\varepsilon)_{\varepsilon\in\Upsilon}$ is a quantum compound Poisson process on T. In the special case when $T = \mathbb{R}_+$, we get the quantum process $(B_t)_{t\geq 0}$, where

$$B_t \equiv B_{[0,t]} = \int_\alpha \int_0^t s\, dA_{(s,t)}.$$

Evidently, for every $g \in G$,

$$B_g = \int_T g(t)\, dB_t = \int_\alpha \int_T sg(t)\, dA_{(s,t)}.$$

From [16], we have:

Lemma 3.1. *The operators $\mathcal{A}_h$, $h \in H$, are essentially self-adjoint and their closures $\mathcal{A}_h^\sim$ constitute a family of self-adjoint, commuting operators in $\Gamma(L^2(\alpha \times T))$.*

We will study the operator family $\mathcal{B} = (\mathcal{B}_\zeta)_{\zeta\in S(T)}$ $(S(T) \subset G)$. Define the set

$$\Xi = \text{l.s.}\big(\mathcal{B}_{\zeta_1}^{m_1} \cdots \mathcal{B}_{\zeta_n}^{m_n}\Omega\big), \qquad \zeta_1,\dots,\zeta_n \in S(T),\ \ m_1,\dots,m_n \in \mathbb{Z}_+,\ \ n \in \mathbb{N},$$

where l.s. stands for the linear span and Ω is the vacuum in $\Gamma(L^2(\alpha \times T))$.

As easily seen, $\Xi \subset \Gamma_{\mathrm{fin}}(S(\alpha\times T))$. Denote by $\Pi_{\mathrm{fin}}(S(\alpha\times T))$ and $\Pi_{q,\kappa}(S_p(\alpha\times T))$ the subspaces of $\Gamma_{\mathrm{fin}}(S(\alpha\times T))$ and $\Gamma_{q,\kappa}(S_p(\alpha\times T))$, respectively, that are defined as the corresponding closures of the set Ξ. The space $\Pi_{\mathrm{fin}}(S(\alpha\times T))$, as a subspace of a nuclear space, is itself a nuclear space. The dual $\Pi^*_{\mathrm{fin}}(S(\alpha\times T))$ of $\Pi_{\mathrm{fin}}(S(\alpha\times T))$ can be identified with the factor space

$$\Gamma^*_{\mathrm{fin}}(S(\alpha\times T))/\{\Phi\in\Gamma^*_{\mathrm{fin}}(S(\alpha\times T))\colon \langle\!\langle \Phi,\phi\rangle\!\rangle=0,\ \phi\in\Pi_{\mathrm{fin}}(S(\alpha\times T))\}.$$

Analogously, the dual $\Pi_{-q,-\kappa}(S_{-p}(\alpha\times T))$ of $\Pi_{q,\kappa}(S_p(\alpha\times T))$ is identified with the corresponding factorization of $\Gamma_{-q,-\kappa}(S_{-p}(\alpha\times T))$. Thus, in what follows, the writing $F\in\Pi^*_{\mathrm{fin}}(S(\alpha\times T))$, where F belongs really to $\Gamma^*_{\mathrm{fin}}(S(\alpha\times T))$, means the factor class the F belongs to. As will be seen later, $\Pi_{0,0}(L^2(\alpha\times T))$ *coincides* with $\Gamma(L^2(\alpha\times T))$. But, for each $p\geq 1$, $\Pi_{q,\kappa}(S_p(\alpha\times T))$ is a *principal* subspace of $\Gamma_{q,\kappa}(S_p(\alpha\times T))$.

The next lemma follows directly from [16].

Lemma 3.2.
a) *Every operator* $\mathcal{B}_\zeta$, $\zeta\in S(t)$, *acts continuously on* $\Pi_{\mathrm{fin}}(S(\alpha\times T))$.
b) *For any fixed* $f\in\Pi_{\mathrm{fin}}(S(\alpha\times T))$, *the following mapping is linear and continuous:*

$$S(T)\ni\zeta\to\mathcal{B}_\zeta f\in\Pi_{\mathrm{fin}}(S(\alpha\times T)).$$

By using Lemmas 3.1 and 3.2, analogously to the case of the usual PWN [16], one proves the following theorem.

Theorem 3.3.
1) *For each* $\omega\in S'(T)$, *there is a unique generalized joint eigenvector* $\mathcal{R}(\omega)=(\mathcal{R}^{(n)}(\omega))_{n=0}^\infty\in\Pi^*_{\mathrm{fin}}(S(\alpha\times T))$ *of the family* $\mathcal{B}$ *such that* $\mathcal{R}^{(0)}(\omega)=1$:

$$\langle\!\langle \mathcal{R}(\omega),\mathcal{B}_\zeta f\rangle\!\rangle=\langle\omega,\zeta\rangle\langle\!\langle\mathcal{R}(\omega),f\rangle\!\rangle,\qquad \zeta\in S(T),\ f\in\Pi_{\mathrm{fin}}(S(\alpha\times T)).$$

2) *For each* $x\in S'(\alpha\times T)$ *and* $n\in\mathbb{Z}_+$, *define the n-th (Poisson) Wick power of* x *as an element* $:x^{\otimes n}:$ *of* $S'^{\hat{\otimes}n}(\alpha\times T)$ *by the recursion relation*

$$\begin{aligned} :x^{\otimes 0}:=1,\quad :x^{\otimes 1}: &= x-1,\\ \langle :x^{\otimes(n+1)}:,\xi^{\otimes(n+1)}\rangle &= \langle :x^{\otimes 1}:,\xi\rangle\,\langle :x^{\otimes n}:,\xi^{\otimes n}\rangle\\ &\quad -n\,\langle :x^{\otimes n}:,\xi^{\otimes(n-1)}\hat{\otimes}(\xi^2)\rangle\\ &\quad -n\,\langle\xi,\xi\rangle\,\langle :x^{\otimes(n-1)}:,\xi^{\otimes(n-1)}\rangle, \end{aligned}$$

for $\xi\in S(\alpha\times T)$, $n\in\mathbb{N}$. *Now, for each* $x\in S'(\alpha\times T)$, *define* $R(x)=(R^{(n)}(x))_{n=0}^\infty\in\Gamma^*_{\mathrm{fin}}(S(\alpha\times T))$ *by* $R^{(n)}(x)=(n!)^{-1}:x^{\otimes n}:$. *Then,* $\mathcal{R}(\omega)=R(\imath(\omega))$, *where* $\imath(\omega)$ *denotes an arbitrary element of* $S'(\alpha\times T)$ *such that* $\langle\imath(\omega),s\rangle=\omega$.

3) *Put*

$$\begin{aligned}\Pi_{\rm fin}(S(\alpha\times T))\ni f &= \left(f^{(n)}\right)_{n=0}^{\infty}\to Uf=(Uf)(\omega)=\langle\!\langle\mathcal{R}(\omega),f\rangle\!\rangle\\ &= \sum_{n=0}^{\infty}\langle\mathcal{R}^{(n)}(\omega),f^{(n)}\rangle n!=\sum_{n=0}^{\infty}\langle R^{(n)}(\imath(\omega)),f^{(n)}\rangle n!\,.\end{aligned}$$

Then, there exists a unique measure $\mu_{\rm CP}$ – the spectral measure of the family $\mathcal{B}$ – such that U can be extended to a unitary mapping between $\Gamma(L^2(\alpha\times T))$ and the L^2-space $(L^2_{\rm CP})\equiv L^2(S'(T),d\mu_{\rm CP})$.
4) *The $\mu_{\rm CP}$ is the measure of CPWN on T with Lévy measure $\rho\nu$ given by its Fourier transform* (1.1).
5) *The image of the operator $\mathcal{B}^{\sim}_{\zeta}$ under the unitary operator U is the operator of multiplication by the linear functional $\langle\omega,\zeta\rangle$,i.e., $U\mathcal{B}^{\sim}_{\zeta}U^{-1}=\langle\omega,\zeta\rangle\cdot$. Moreover, any operator $B^{\sim}_{\varepsilon}$, $\varepsilon\in\Upsilon$, goes over into the operator of multiplication by a linear functional $X_\varepsilon=X_\varepsilon(\omega)$, i.e., $UB^{\sim}_{\varepsilon}U^{-1}=X_\varepsilon(\omega)\cdot$, and $X=(X_\varepsilon)_{\varepsilon\in\Upsilon}$ is the compound Poisson random measure on T with Lévy measure $\rho\nu$, i.e., X is a random measure on T such that*

$$\mathbf{E}(e^{icX_\varepsilon})=\int_{S'(T)}e^{icX_\varepsilon(\omega)}\,d\mu_{\rm CP}(\omega)=\exp\left[\nu(\varepsilon)\int_T(e^{ics}-1)\,d\rho(s)\right],\qquad s\in\mathbb{R},$$

for each $\varepsilon\in\Upsilon$.

For each $\omega\in S'(T)$ and $n\in\mathbb{Z}_+$, define the (compound Poisson) n-th Wick power of ω as $:\imath(\omega)^{\otimes n}:$, which is *not* defined uniquely now. Nevertheless, the Wick polynomial

$$\sum_{n=0}^{\infty}\langle:\imath(\omega)^{\otimes n}:,f^{(n)}\rangle=(Uf)(\omega),\qquad f=\left(f^{(n)}\right)_{n=0}^{\infty}\in\Pi_{\rm fin}(S(\alpha\times T)),$$

as a function of ω is evidently independent of the special choice of $\imath(\omega)$ (provided, of course, that $\imath(\omega)$ is chosen the same for all the summands). Thus, taking Theorem 3.3 into account, we will write

$$\phi(\omega)=\sum_{n=0}^{\infty}\langle:\imath(\omega)^{\otimes n}:,f^{(n)}\rangle\qquad\text{for }\ \phi=U\left(\left(f^{(n)}\right)_{n=0}^{\infty}\right)\in(L^2_{\rm CP}).$$

By analogy with the case of the usual Poisson [16], one proves the following fact.

Proposition 3.4. *We have*

$$U\big(\Pi_{\rm fin}(S(\alpha\times T))\big)=\mathcal{P}(S'(T)),$$

where $\mathcal{P}(S'(T))$ denotes the set of continuous polynomials on $S'(T)$, i.e., the functions of the form

$$\langle\omega^{\otimes n},g^{(n)}\rangle+\langle\omega^{\otimes(n-1)},g^{(n-1)}\rangle+\cdots+g^{(0)},$$

where $g^{(i)}\in S_c^{\hat{\otimes}i}(S(T))$.

Remark 3.5. Since the set of continuous polynomials on $S'(T)$ is dense in (L^2_{CP}), one could try to construct the chaos decomposition of (L^2_{CP}) in the following way. Let $\mathcal{P}^{\sim}_n(S'(T))$ denote the (L^2_{CP}) closure of the set $\mathcal{P}_n(S'(T))$ of continuous polynomials on $S'(T)$ of power $\le n$, and let $(L^2_{\mathrm{CP}})_n \equiv \mathcal{P}^{\sim}_n(S'(T)) \ominus \mathcal{P}^{\sim}_{n-1}(S'(T))$. Then, evidently, one obtains the following orthogonal decomposition: $(L^2_{\mathrm{CP}}) = \bigoplus_{n=0}^{\infty}(L^2_{\mathrm{CP}})_n$. Let us consider the projection in (L^2_{CP}) of a continuous monomial $\langle \omega^{\otimes n}, g^{(n)}\rangle$ on the n-th homogeneous chaos $(L^2_{\mathrm{CP}})_n$. This projection, say $:\langle \omega^{\otimes n}, g^{(n)}\rangle:$, is, of course, an element of (L^2_{CP}). But one can verify that, even for $n=2$, $:\langle \omega^{\otimes 2}, g^{(2)}\rangle:$ is *not* a continuous polynomial on $S'(T)$. Thus, the procedure of the orthogonalization of polynomials is not applicable in case of the compound Poisson.

We finish this section with considering a multiple Wiener integral over a function $g^{(n)} \in \hat{L}^2(T^n)$ by the compensated compound Poisson random measure

$$Y = (Y_\varepsilon)_{\varepsilon\in\Upsilon}, \qquad Y_\varepsilon \equiv X_\varepsilon - \langle s\rangle \nu(\varepsilon) = \langle :\imath(\omega)^{\otimes 1}:, s\otimes \chi_\varepsilon\rangle,$$

where, given a function $f = f(s)$, $\langle f\rangle$ denotes $\int f(s)\,d\rho(s)$. By definition, we have, for arbitrary disjoint sets $\varepsilon_1, \dots, \varepsilon_n \in \Upsilon$,

$$\int_{T^n} (\chi_{\varepsilon_1}\hat{\otimes}\cdots\hat{\otimes}\chi_{\varepsilon_n})(t_1,\dots,t_n)\,dY_{t_1}\dots dY_{t_n} = Y_{\varepsilon_1}\cdots Y_{\varepsilon_n}.$$

By Theorem 3.3, one infers also that

$$\begin{aligned} Y_{\varepsilon_1}\cdots Y_{\varepsilon_n} &= U\big((B_{\varepsilon_1} - \langle s\rangle\nu(\varepsilon_1))\cdots(B_{\varepsilon_n} - \langle s\rangle\nu(\varepsilon_n))\,\Omega\big)\\ &= U\big(s^{\otimes n}\otimes(\chi_{\varepsilon_1}\hat{\otimes}\cdots\hat{\otimes}\chi_{\varepsilon_n})\big) = \langle :\imath(\omega)^{\otimes n}:, s^{\otimes n}\otimes(\chi_{\varepsilon_1}\hat{\otimes}\cdots\hat{\otimes}\chi_{\varepsilon_n})\rangle. \end{aligned}$$

Since ν is non-atomic, the set of the vectors $\chi_{\varepsilon_1}\hat{\otimes}\cdots\hat{\otimes}\chi_{\varepsilon_n}$ with disjoint $\varepsilon_1,\dots,\varepsilon_n$ is total in $\hat{L}^2(T^n)$, so that we have (cf. [6]):

Proposition 3.6. *For each $g^{(n)} \in \hat{L}^2(T^n)$,*

$$\int_{T^n} g^{(n)}(t_1,\dots,t_n)\,dY_{t_1}\dots dY_{t_n} = \langle :\imath(\omega)^{\otimes n}:, s^{\otimes n}\otimes g^{(n)}\rangle.$$

So, if, for example, $\langle s^2\rangle = 1$, one constructs the following isometry:

$$\begin{aligned} \Gamma(L^2(T)) \ni g = (g^{(n)})_{n=0}^{\infty} \to Ig &= g^{(0)} + \sum_{n=1}^{\infty}\int_{T^n} g^{(n)}(t_1,\dots,t_n)\,dY_{t_1}\dots dY_{t_n}\\ &= g^{(0)} + \sum_{n=1}^{\infty}\langle :\imath(\omega)^{\otimes n}:, s^{\otimes n}\otimes g^{(n)}\rangle \in (L^2_{\mathrm{CP}}). \end{aligned}$$

Remark 3.7. Proposition 3.6 shows, in particular, that the condition that the support of ρ consists of more than one point implies that the random measure Y does not possess the chaotic representation property (cf. [22, 6]).

4. Spaces of test and generalized functions

In this section, we will only introduce spaces of test and generalized functions centered at (L^2_{CP}) and consider some examples of generalized functions. The further study of the analysis on these spaces, as well as applications to mathematical physics will be carried out in our forthcoming papers, see also the recent preprint [18].

Thus, in the way described in the previous section, by using the rigging (2.5), we construct the following rigging of $\Gamma(L^2(\alpha \times T))$:

$$\begin{aligned}\Pi_{-\kappa}(S'(\alpha \times T)) &= \operatorname*{ind\,lim}_{p,q\to\infty} \Pi_{-q,-\kappa}(S_{-p}(\alpha \times T)) \supset \Gamma(L^2(\alpha \times T)) \\ &\supset \operatorname*{proj\,lim}_{p,q\to\infty} \Pi_{q,\kappa}(S_p(\alpha \times T)) \supset \Pi_\kappa(S(\alpha \times T)).\end{aligned}$$

Next, by applying U (or its extension by continuity) to this rigging, we get

$$(S_{\mathrm{CP}})^{-\kappa} = \operatorname*{ind\,lim}_{p,q\to\infty} (S_{\mathrm{CP}})^{-\kappa}_{-p,-q} \supset (L^2_{\mathrm{CP}}) \supset \operatorname*{proj\,lim}_{p,q\to\infty} (S_{\mathrm{CP}})^{\kappa}_{p,q} = (S_{\mathrm{CP}})^{\kappa}, \quad \kappa \geq 0,$$

(we used obvious notations). We recall that the spaces $(S_{\mathrm{CP}})^{-\kappa}$ are understood as factor spaces.

For an arbitrary $F = \left(F^{(n)}\right)_{n=0}^{\infty} \in \Pi_{-\kappa}(S'(\alpha \times T))$, we will write

$$UF = \Phi = \Phi(\omega) = \sum_{n=0}^{\infty} \langle :\imath(\omega)^{\otimes n}:, F^{(n)}\rangle.$$

Let us consider some examples of generalized functions.

1) *CPWN on T (a system of coordinate functions)*:

$$\dot{X}_t = \dot{X}_t(\omega) \equiv \omega(t) \equiv\ :\omega(t):\ + \langle s\rangle \equiv \langle :\imath(\omega)^{\otimes 1}:, s \otimes \delta_t\rangle + \langle s\rangle.$$

Let $T = \mathbb{R}_+$, $d\nu(t) = dt$, then

$$X_t = X_t(\omega) = \langle \omega, \chi_{[0,t]}\rangle = \langle :\imath(\omega)^{\otimes 1}:, s \otimes \chi_{[0,t]}\rangle + \langle s\rangle t, \qquad t \geq 0.$$

Since $(\Delta t)^{-1}\chi_{[t,t+\Delta t]}(\cdot) \to \delta_t$ in $S_{-1}(T)$ as $\Delta t \to 0$, we get

$$(\Delta t)^{-1}(X_{t+\Delta t} - X_t) \to \dot{X}_t \qquad \text{in } (S_{\mathrm{CP}})^{-0}_{-1,0} \text{ as } \Delta t \to 0,$$

i.e., $\dot{X}_t$ is the time derivative of the compound Poisson process, which itself is a generalized stochastic process in $t \in T$. Notice that each $\omega(t) \in S'(T)$ is a CPWN sample path.

2) *Generalized multiple Wiener integrals*: Taking into account Proposition 3.6, we set, for each $G^{(n)} \in S_c'^{\hat{\otimes} n}(T)$,

$$\int_{T^n} G^{(n)}(t_1, \dots, t_n)\, dY_{t_1} \dots dY_{t_n} \equiv \langle :\imath(\omega)^{\otimes n}:, s^{\otimes n} \otimes G^{(n)}\rangle \in (S_{\mathrm{CP}})^{-0}.$$

Acknowledgements

The author is grateful to Professors Yu.M. Berezansky, A.L. Rebenko, G.F. Us and Doctor G.V. Shchepan'uk for useful discussions.

References

[1] ALBEVERIO, S., DALETSKY, YU.L., KONDRATIEV, YU.G., STREIT, L.: Non-Gaussian infinite dimensional analysis; Preprint, BiBoS University, Bielefeld 1994; J. Funct. Anal., to appear.

[2] BEREZANSKY, YU.M.: Spectral approach to white noise analysis; in Bielefeld Encounters in Mathematical Physics VIII, 131–140, World Scientific, Singapore, New Jersey, London, Hong-Kong 1993.

[3] BEREZANSKY, YU.M., KONDRATIEV, YU.G.: Spectral Methods in Infinite Dimensional Analysis; Kluwer Academic Publishers, Dordrecht, Boston, London 1994.

[4] BEREZANSKY, YU.M., LIVINSKY, V.O. LYTVYNOV, E.W.: Spectral approach to white noise analysis; Ukrainian Math. J. 46 (1993), 177–197.

[5] BEREZANSKY, YU.M., LIVINSKY, V.O. LYTVYNOV, E.W.: A generalization of Gaussian white noise analysis; Methods Funct. Anal. Topology. 1 (1995), 28–55.

[6] DERMOUNE, A.: Distributions sur l'espace de P. Lévy et calcul stochastic; Ann. Inst. H. Poincaré Probab. Statist. 26 (1990), 101–119.

[7] DERMOUNE, A.: Une remarque sur le process $\alpha a(f) + \bar{\alpha} a^+(f) + \lambda a^0(f)$; Ann. Sci. Univ. Clermont–Ferrand II Probab. Appl. 9 (1991), 55–58.

[8] GIELERAK, R., REBENKO, A.L.: On the Poisson integral representation in the classical statistical mechanics of continuous systems; Preprint, BiBoS University, Bielefeld 1994.

[9] HUDSON, R.L., PARTHASARATHY, K.R.: Quantum Itô's formula and stochastic evolutions; Comm. Math. Phys. 93 (1984), 301–323.

[10] ITO, Y.: Generalized Poisson functionals; Probab. Theory Related Fields 77 (1988), 1–28.

[11] ITO, Y., KUBO, I.: Calculus on Gaussian and Poisson white noises; Nagoya Math. J. 111 (1988), 41–84.

[12] KALLENBERG, O.: Random Measures; Akademie Verlag, Berlin 1975.

[13] KONDRATIEV, YU.G., LEUKERT, P., STREIT, L.: Wick calculus in Gaussian analysis; Preprint, BiBoS University, Bielefeld 1994.

[14] KONDRATIEV, YU.G., STREIT, L.: Spaces of white noise distributions: constructions, descriptions, applications I; Rep. Math. Phys. 33 (1993), 341–366.

[15] KONDRATIEV, YU.G., STREIT, L., WESTERKAMP, W., YAN, J.: Generalized functions in infinite dimensional analysis; Preprint, BiBoS University, Bielefeld 1995.

[16] LYTVYNOV, E.W.: Multiple Wiener integrals and non-Gaussian white noises: a Jacobi field approach; Methods Funct. Anal. Topology. 1 (1995), 61–85.

[17] LYTVYNOV E.W.: White noise calculus for a class of processes with independent increments; submitted.

[18] LYTVYNOV, E.W., REBENKO, A.L., SHCHEPAN'UK, G.V.: Quantum compound Poisson processes and white noise calculus; Preprint, BiBoS University, Bielefeld 1995.

[19] LYTVYNOV, E.W., REBENKO, A.L., SHCHEPAN'UK, G.V.: Wick calculus on spaces of gereralized functions of compound Poisson white noise; in preparation.

[20] MEYER, P.A.: Quantum Probability for Probabilists; Lecture Notes in Math. 1538 (1993).

[21] SURGAILIS, D.: On multiple Poisson stochastic integrals and associated Markov semigroups; Probab. Math. Stat. 3 (1984), 217–239.

[22] SURGAILIS, D.: On L^2 and non-L^2 multiple stochastic integration; Lecture Notes in Control and Inform. Sci. 36 (1981), 212–226.

Institute of Mathematics
National Academy of Sciences of Ukraine
3 Tereshchenkivska St.
252601, Kiev
Ukraine
mathkiev@imat.gluk.apc.org

1991 Mathematics Subject Classification: Primary 60G20; Secondary 46F25

Submitted: March 26, 1996

Operator Theory:
Advances and Applications, Vol. 102

Invariant and hyperinvariant subspaces of direct sums of simple Volterra operators

M.M. Malamud

Let J be the integration operator defined on $L_p[0,1]$, let J^α, $\alpha > 0$, be its positive powers, and let B be a nonsingular $n \times n$ diagonal matrix. The lattices of invariant and hyperinvariant subspaces of the Volterra operator $J^\alpha \otimes B$ defined on $L_p[0,1] \otimes \mathbb{C}^n$ are described in geometric terms.

1. Introduction

It is well-known ([GK], [N1], [FR]) that the Volterra integral operator J defined on $L_p[0,1]$ by $J : f \mapsto \int_0^x f(t)\,dt$ is unicellular for $p \in [1,\infty)$ and its lattice of invariant subspaces is anti-isomorphic to the segment $[0,1]$. The same is also true (see [GK], [N1]) for the simple Volterra operators

$$J^\alpha : f \mapsto \int_0^x \frac{(x-t)^{\alpha-1}}{\Gamma(\alpha)} f(t)\,dt, \qquad \Re(\alpha) > 0,$$

being the real powers of the integral operator J.

The simple Volterra operators defined on the space $L_p[0,1] \otimes \mathbb{C}^n$ of vector-functions are the tensor products $A = J^\alpha \otimes B$, where B is an $n \times n$ nonsingular diagonal matrix, $B = \operatorname{diag}(\lambda_1, \dots, \lambda_n)$.

These operators are interesting for many ressons. For example, they arise in the investigation of inverse problems for systems of ordinary differential equations

$$\left(B \otimes \frac{d}{dx}\right) y + Q(x)y = \lambda y$$

(see [M2], [M3]).

For a special choice of λ_j and $\alpha < 1$ these operators provide counterexamples (see [M4]) to one of Gohberg and Krein's conjectures about the equivalence of unicellularity and cyclicity for Volterra operators (see [G-K, p. 421]).

Note also that the operator $J \otimes B$ admits an abstract characterization up to unitary equivalence (for $B = B^*$ this was done in [K]).

To obtain a description of Hyplat A we need a description of the commutant $\{A\}'$ of an operator A. This is provided by Proposition 4.6 and Theorem 4.10. Theorem 4.11, whose proof is based on Theorem 4.10, presents a simple description of the lattice Hyplat A (in geometric terms) for an arbitrary nonsingular diagonal matrix B.

It should be pointed out that the operators $A = J^\alpha \otimes B$ with $B = \mathrm{diag}(\lambda_1, \dots, \lambda_n)$ satisfying the condition

$$\arg \lambda_i \neq \arg \lambda_j, \qquad 1 \leq i \neq j \leq n, \tag{1.1}$$

play a special role in the sequel. It turns out that each of the relations

$$\{J^\alpha \otimes B\}' = \oplus_{j=1}^n \{\lambda_j J^\alpha\}', \tag{1.2}$$

$$\mathrm{Hyplat}\,(J^\alpha \otimes B) = \oplus_{j=1}^n \mathrm{Hyplat}\,\lambda_j J^\alpha, \tag{1.3}$$

$$\mathrm{Lat}\,(J^\alpha \otimes B) = \oplus_{j=1}^n \mathrm{Lat}\,\lambda_j J^\alpha \tag{1.4}$$

is equivalent to the condition (1.1).

In particular, the relation (1.4) provides a description of Lat A under the condition (1.1) and means that the splitting of the lattice Lat A is equivalent to (1.1). When the condition (1.1) is violated we present a simple a criterion (in geometric terms) for a subspace $E \subset L_p[0,1] \otimes \mathbb{C}^n$ to be a cyclic subspace for the operator $A = J^\alpha \otimes B$. In this case some description of Lat$A$ is contained in Proposition 2.5. But as distinct from the case $n = 1$ this description is not completely satisfactory.

Note also that several criteria for an integral Volterra operator to be similar to the operator J^α, $\alpha > 0$, have been obtained in [M1]. Some generalizatons of these results from the scalar to the vector case, that is criteria for similarity to the operators $J^\alpha \otimes B$, are announced in [M4]. Their proofs will be published elsewhere.

The main results of this paper have been announced in [M3], [M4].

Notations. 1) X_1, X_2 are Banach spaces; 2) $[X_1, X_2]$ is the space of bounded linear operators from X_1 to X_2; 3) $[X] = [X, X]$ is the space of bounded operators on the Banach space X; 4) Lat T and Hyplat T are the lattices of invariant and hyperinvariant subspaces of an operator $T \in [X]$, respectively; 5) span E is the closed linear span of a set $E \subset X$; 6) $L_p^n[0,1] := L_p[0,1] \otimes \mathbb{C}^n$; 7) supp f is the support of a function f.

2. Cyclic subspaces

2.1. Let R be any commutative ring, in general without unity, let $A = (a_{ij})$ be an arbitrary $n \times n$ matrix with entries $a_{ij} \in R$, and let $\widetilde{A}$ be the adjoint matrix. Many well-known properties of matrices over a field remain valid for matrices with entries from R. In particular, the well-known Binet-Cauchy formula for the minors of the product AB of two such matrices A and B is valid. We note also that the formula

$$A \cdot \widetilde{A} = \widetilde{A} \cdot A = \det A \cdot I_n =: \Delta_A \cdot I_n \tag{2.1}$$

holds. Here Δ_A is the determinant of the matrix A and I_n is the $n \times n$ identity matrix. It follows that the rows and the columns of A are linearly dependent over R iff $\det A$ is a zero divisor in R.

2.2. It is well-known that the space $L_p[0,1]$, $1 \le p \le \infty$, with the convolution product

$$(2.2) \qquad (f*g)(x) = \int_0^x f(x-t)g(t)\,dt = \int_0^x g(x-t)f(t)\,dt = (g*f)(x)$$

is a commutative Banach algebra without unity. A description of zero divisors in $L_p[0,1]$ is given by the Titchmarsh Convolution Theorem [B2], [GK]:

$$(2.3)\ \ (f*g)(x) = 0,\ x \in [0,1] \Longrightarrow \operatorname{supp} f \subset [\alpha, 1],\ \operatorname{supp} g \subset [\beta, 1] \text{ and } \alpha + \beta \ge 1.$$

In particular, $f(x)$ is not a zero divisor in $L_p[0,1]$ if and only if

$$(2.4) \qquad \int_0^\varepsilon |f(x)|^p\,dx > 0, \qquad \varepsilon > 0.$$

This condition will be called the ε-condition.

The determinant of a functional matrix $F(x) = (f_{ij}(x))_{i,j=1}^n$ ($f_{ij} \in L_p[0,1]$) calculated with respect to the convolution product (4.2) will be called the $*$-determinant and will be denoted by $*$-det $F(x)$. Similarly, the $*$-minors of $F(x)$ are the minors with respect to the convolution product. Finally $*$-rank $F(x)$ will be the highest order of $*$-minors of $F(x)$ satisfying the ε-condition.

2.3. Let B be a $n \times n$ diagonal matrix with eigenvalues λ_j, $1 \le j \le r$, of equal arguments:

$$(2.5)\ \ B = \oplus_{j=1}^r \lambda_j I_{n_j} \in [\mathbb{C}^n],\ \ \arg\lambda_j = \arg\lambda_1,\ \ j \in \{1,\dots,r\},\ n = n_1 + \cdots + n_r.$$

In the following theorem we describe the cyclic subspaces of the operator $A = J^\alpha \otimes B$. We need some definitions to state the theorem.

Definition 2.1. Recall that a subspace E of some Banach space X is called a *cyclic subspace* for an operator $T \in [X]$ if span $\{T^n E : n \ge 0\} = X$. A vector f ($\in X$) is called *cyclic* if span $\{T^n f : n \ge 0\} = X$. The set of all cyclic vectors of an operator T is denoted by Cyc(T).

Definition 2.2. We set

$$\mu_T := \inf\{\dim E : E \text{ is a cyclic subspace of the operator } T \in [X]\}.$$

μ_T is called the *spectral multiplicity* of an operator $T \in [X]$. Note that μ_T can be infinite.

It is clear that an operator T is cyclic iff $\mu_T = 1$.

Theorem 2.3. *Let* $\Re(\alpha) > 0$, $p \in [1,\infty)$, *let* B *be a matrix of the form* (2.5), *where* $\lambda_j = \lambda_1/s_j^\alpha$, $1 = s_1 < s_2 < \cdots < s_r$, *and let*

$$f_{ij} = \{f_{ij1}, \dots, f_{ijn_j}\} \in L_p[0,1] \otimes \mathbb{C}^{n_j}, \qquad 1 \le i \le N,\ 1 \le j \le r.$$

Then $\mu_A = n$ *and the system* $\{f_i\}_{i=1}^N$ *of vectors* $f_i = \oplus_{j=1}^r f_{ij}$ *generates a cyclic subspace in the space* $L_p[0,1] \otimes \mathbb{C}^n$ *for the operator*

$$A = J^\alpha \otimes B = \oplus_{j=1}^r \lambda_j J^\alpha \otimes I_{n_j} \tag{2.6}$$

if and only if $N \geq n$ *and the matrices*

$$F_p(x) = \begin{pmatrix} f_{11}(s_1 x) & f_{12}(s_2 x) & \dots & f_{1m}(s_m x) \\ \vdots & \vdots & & \vdots \\ f_{N1}(s_1 x) & f_{N2}(s_2 x) & \dots & f_{Nm}(s_m x) \end{pmatrix} \tag{2.7}$$

are of maximal $*$*-ranks for all* $m \in \{1, \dots, r\}$, *namely*

$$*\text{-rank}\, F_p(x) = n_1 + n_2 + \dots + n_m, \qquad m \in \{1, \dots, r\}. \tag{2.8}$$

Proof. Sufficiency: Let $N \geq n = n_1 + n_2 + \dots + n_r$ and assume that the conditions (2.8) are satisfied.
The resolvent $(I - \lambda\lambda_k J^\alpha)^{-1}$ of the operator $\lambda_k J^\alpha$ is

$$(I - \lambda\lambda_k J^\alpha)^{-1} h = h(x) + \lambda\lambda_k \int_0^x (x-t)^{\alpha-1} E_{1/\alpha}(\lambda\lambda_k (x-t)^\alpha; \alpha) h(t)\, dt \tag{2.9}$$

where $h = h(x) \in L_p[0,1]$ and

$$E_{1/\alpha}(\lambda(x-t)^\alpha; \alpha) = \sum_{k=0}^\infty \frac{\lambda^k (x-t)^{\alpha k}}{\Gamma(\alpha + \alpha k)} \tag{2.10}$$

is the Mittag-Leffler function.
Let $g = \oplus_{j=1}^r g_j$ with $g_j = (g_{j1}, \dots, g_{jn_j}) \in L_q[0,1] \otimes \mathbb{C}^{n_j}$ be a vector from the annihilator $M^\perp \subset L_q[0,1] \otimes \mathbb{C}^n$ of the subspace $M = \text{span}\{A^k f_i : k \geq 0\,, 1 \leq i \leq N\}$. Let us show that $g = 0$.
Actually it follows from (2.9) and the obvious formula

$$(I - \lambda A)^{-1} = \oplus_{k=1}^r \left((I - \lambda_k \lambda J^\alpha)^{-1} \otimes I_{n_k}\right)$$

that for arbitrary $i \in \{1, \dots, N\}$

$$\begin{aligned} 0 &= \left((I - \lambda A)^{-1} f_i, g\right) \\ &= \sum_{j=1}^r \sum_{k=1}^{n_j} \int_0^1 f_{ijk}(x) \overline{g_{jk}(x)} dx \\ &\quad + \sum_{j=1}^r \sum_{k=1}^{n_j} \lambda\lambda_j \int_0^1 \overline{g_{jk}(x)} dx \int_0^x (x-t)^{\alpha-1} E_{1/\alpha}(\lambda\lambda_j (x-t)^\alpha; \alpha) f_{ijk}(t) dt \\ &= \sum_{j=1}^r \sum_{k=1}^{n_j} \Bigg[\int_0^1 f_{ijk}(x-t) \overline{g_{jk}(x)} dx \\ &\quad + \lambda\lambda_j \int_0^1 t^{\alpha-1} E_{1/\alpha}(\lambda\lambda_j t^\alpha; \alpha) dt \int_t^1 f_{ijk}(x-t) g_{jk}(x) dx \Bigg]. \end{aligned}$$

Making use of the change of variables $t_j = t/s_j$, $x_j = x/s_j$, $1 \le j \le r$, and denoting $\mu = \lambda\lambda_1$, we can rewrite the last identity in the form

$$\sum_{j=1}^{r}\sum_{k=1}^{n_j}\left[\int_0^1 f_{ijk}(x-t)\overline{g_{jk}(x)}\,dx + \mu\int_0^1 t^{\alpha-1}E_{1/\alpha}(\mu t^{\alpha};\alpha)\,\varphi_{ijk}(t)\,dt\right] = 0 \tag{2.11}$$

for $i \in \{1,\dots,N\}$ where

$$\varphi_{ijk}(t) = \begin{cases} s_j^{\alpha+1}\displaystyle\int_t^{s_j^{-1}} f_{ijk}((x-t)s_j)g_{jk}(xs_j)dx, & t\in[0,s_j^{-1}],\\ 0, & t\in[s_j^{-1},1]. \end{cases} \tag{2.12}$$

It is easy to see that (2.11) is equivalent to

$$\sum_{j=1}^{r}\sum_{k=1}^{n_j}\varphi_{ijk}(t) = 0, \qquad t\in[0,s_1], \tag{2.13}$$

for $i \in \{1,\dots,N\}$. We will show step by step that (2.13) implies the equalities

$$g_{jk}(x) = 0, \qquad x\in[0,1],\ j\in\{1,\dots,r\},\ k\in\{1,\dots,n_j\}. \tag{2.14}$$

In the first step we consider the system (2.13) for $t \in [s_2^{-1}, 1]$ and prove that

$$g_{1k} = 0, \qquad x\in[s_2^{-1},1],\ k\in\{1,\dots,n_1\}. \tag{2.15}$$

Indeed, for $t \in [s_2^{-1}, 1]$ the system (2.13) takes the form

$$\sum_{k=1}^{n_1}\int_t^1 f_{i1k}(x-t)\,g_{1k}(x)\,dx = 0, \qquad t\in[s_2^{-1},1],\ 1\le i\le n_1. \tag{2.16}$$

According to the first of the conditions (2.8) there exists an $n_1 \times n_1$-submatrix $\widetilde{F}_1(x)$ of the matrix $F_1(x)$ such that $*$-rank $\widetilde{F}_1(x) = *$-rank $F_1(x) = n_1$, i.e., $\Delta_1(x) = *$-det $\widetilde{F}_1(x)$ is not a zero divisor in $L_p[0,1]$. Let for definiteness

$$\widetilde{F}_1(x) = \begin{pmatrix} f_{11}(x)\\ \vdots \\ f_{n_1 1}(x)\end{pmatrix} = \begin{pmatrix} f_{111}(x) & f_{112}(x) & \dots & f_{11n_1}(x)\\ \vdots & \vdots & & \vdots\\ f_{n_1 11}(x) & f_{n_1 12}(x) & \dots & f_{n_1 1 n_1}(x)\end{pmatrix}$$

be such a matrix. Denoting by $A_{i1l}(x)$ the $*$-cofactor of an element $f_{i1l}(x)$, $1 \le$

$l \le n_1$, of the matrix $\widetilde{F}_1(x)$, we obtain from (2.16) and (2.1)

$$
\begin{aligned}
0 &= \sum_{i=1}^{n_1}\sum_{k=1}^{n_1}\int_y^1 A_{i1l}(t-y)\,dt\int_t^1 f_{i1k}(x-t)\,\overline{g_{1k}}(x)\,dx \\
&= \sum_{i=1}^{n_1}\sum_{k=1}^{n_1}\int_y^1 \overline{g_{1k}}(x)\,dx\int_y^x A_{i1l}(t-y)f_{i1k}(x-t)\,dt \\
&= \sum_{k=1}^{n_1}\int_y^1 \overline{g_{1k}}(x)\,dx\int_0^{x-y}\sum_{i=1}^{n_1} A_{i1l}(x-y-t)f_{i1k}(t)\,dt \\
&= \int_y^1 \overline{g_{1l}}(x)\Delta_1(x-y)\,dx, \qquad y\in[s_2^{-1},1],\ 1\le l\le n_1.
\end{aligned} \tag{2.17}
$$

Since $\Delta_1(x)$ satisfies the ε-condition (2.4), it follows from the Titchmarsh Convolution Theorem and from (2.17) that the equalities (2.15) are fulfilled. Further, suppose by induction that in the $(m-1)$-th step the relations

$$
g_{jk}(x)=0, \qquad x\in[s_m^{-1}s_j^{-1},1],\ j\in\{1,\dots,m-1\},\ k\in\{1,\dots,n_j\}, \tag{2.18}
$$

are established. Then for $t\in[s_{m+1},s_m]$ the system (2.13) can be rewritten in the form

$$
\sum_{j=1}^{m}\sum_{k=1}^{n_j}(s_j^{\alpha+1})\int_l^{s_m^{-1}} f_{ijk}\left((x-t)s_j\right)\overline{g_{jk}}\left(xs_j\right)\,dx=0. \tag{2.19}
$$

Let $N_m=n_1+\cdots+n_m$, and let $\widetilde{F}_m(x)$ be an $N_m\times N_m$-submatrix of the matrix $F_m(x)$ such that $*$-rank $\widetilde{F}_m(x)=*$-rank $F_m(x)=N_m$. Without loss of generality we can assume that $\widetilde{F}_m(x)$ consists of the first N_m rows of the matrix $F_m(x)$. Denoting by $A_{il\kappa}(x)$ the $*$-cofactor of an element $f_{il\kappa}(xs_l)$, $1\le i\le N_m$, $1\le l\le m$, $1\le\kappa\le n_l$, of $\widetilde{F}_m(x)$ and repeating the arguments used above, we obtain from (2.19) and (2.11)

$$
\begin{aligned}
0 &= \sum_{i=1}^{N_m}\sum_{j=1}^{m}\sum_{k=1}^{n_j}\int_0^{s_m^{-1}} A_{il\kappa}(t-y)\,dt\int_t^{s_m^{-1}} s_j^{\alpha+1}f_{ijk}((x-t)s_j)\overline{g_{jk}}(xs_j)\,dx \\
&= \sum_{j=1}^{m}\sum_{k=1}^{n_j}\int_y^{s_m^{-1}} \overline{g}_{jk}(xs_j)\,dx\int_0^{x-y}\sum_{i=1}^{N_m} s_j^{\alpha+1}A_{il\kappa}(x-y-t)f_{ijk}(ts_j)\,dt \\
&= \int_y^{s_m^{-1}} \overline{g_{l\kappa}(xs_l)}\Delta_m(x-y)\,dy, \qquad y\in[s_{m+1}^{-1},s_m^{-1}].
\end{aligned} \tag{2.20}
$$

Since $\Delta_m=*$-det $\widetilde{F}_m(x)$ is not a zero divisor in $L_p[0,1]$, it follows from (2.20) that

$$
g_{jk}(x)=0, \qquad x\in[s_{m+1}^{-1}s_j^{-1},s_m^{-1}s_j^{-1}],\ 1\le j\le m,\ 1\le k\le n_j. \tag{2.21}
$$

The equalities (2.21), (2.18) prove the induction hypothesis.

Necessity: Let the subspace $E = \text{span}\{f_i : 1 \le i \le N\}$ be cyclic for the operator A. At first we suppose that $N \ge n$ and show by contradiction that the conditions (2.8) hold. Let one of the conditions (2.8) be violated. Assume, for example, that $*$-rank $F_r(x)$ is less then n and denote by $F_r\begin{pmatrix} i_1 & i_2 & \dots & i_n \\ 1 & 2 & \dots & n \end{pmatrix}$ the $n \times n$ $*$-minor of the $N \times n$-matrix $F_r(x)$ (here $i_1 < i_2 < \dots < i_n$ are the numbers of rows). Since each such minor is a zero divisor in $L_p[0,1]$, there exists $\beta \in (0,1]$ such that

$$(2.22)\quad F_r\begin{pmatrix} i_1 & i_2 & \dots & i_n \\ 1 & 2 & \dots & n \end{pmatrix}(x) = 0, \quad x \in [0,\beta],\ 1 \le i_1 \le i_2 \le \dots \le i_n \le N.$$

Further, for any n vectors $g_i \in M = \text{span}\{A^m f_k : m \ge 0,\ 1 \le k \le N\}$ and for arbitrary $\varepsilon > 0$ there exist $n \cdot N$ polynomials $P_{ik}^{\varepsilon}(t)$ in one variable such that

$$\left\| \sum_{k=1}^{N} P_{ik}^{\varepsilon}(A)\, f_k - g_i \right\|_{L_p^n[0,1]} < \varepsilon, \qquad P_{ik}^{\varepsilon}(t) = \sum_{m=0}^{d_{ik}} a_{ikm}\, t^m, \qquad 1 \le i \le n.$$

Setting

$$(2.23)\quad g_{ij}^{\varepsilon} = \sum_{k=1}^{N} P_{ik}^{\varepsilon}\left(s_j^{-\alpha} J^{\alpha} \otimes I_{n_j}\right) f_{kj}, \qquad \widetilde{P}_{ik}^{\varepsilon}(t) = \sum_{m=0}^{d_{ik}} a_{ikm}(t^{m\alpha-1}/\Gamma(m\alpha)),$$

we obtain

$$(2.24)\quad \begin{aligned} g_{ij}^{\varepsilon}(xs_j) &= \sum_{k=1}^{N}\sum_{m=0}^{d_{ik}} a_{ikm} s_j^{-\alpha m} \int_0^{xs_j} \frac{(xs_j - t)^{m\alpha-1}}{\Gamma(\alpha m)} f_{kj}(t)\, dt \\ &= \sum_{k=1}^{N}\sum_{m=0}^{d_{ik}} a_{ikm} s_j^{-\alpha m} \int_0^{x} \frac{(x - t)^{m\alpha-1} s_j^{m\alpha-1}}{\Gamma(\alpha m)} f_{kj}(ts_j) s_j dt \\ &= \sum_{k=1}^{N} \widetilde{P}_{ik}^{\varepsilon}(t) * f_{kj}(ts_j). \end{aligned}$$

Let $g_i^0 = g_j = \oplus_{j=1}^{r} g_{ij}$ and

$$(2.25)\qquad G_r^{\varepsilon}(x) = G_r(g_1^{\varepsilon}, \dots, g_n^{\varepsilon}) = \begin{pmatrix} g_{11}^{\varepsilon}(xs_1) & g_{12}^{\varepsilon}(xs_2) & \dots & g_{1r}^{\varepsilon}(xs_r) \\ \vdots & \vdots & & \vdots \\ g_{n1}^{\varepsilon}(xs_1) & g_{n2}^{\varepsilon}(xs_2) & \dots & g_{nr}^{\varepsilon}(xs_r) \end{pmatrix}$$

be the $n \times n$ matrix-function, consisting of the vectors

$$g_i^{\varepsilon} = \oplus_{j=1}^{r} g_{ij}^{\varepsilon} = \sum_{1 \le k \le N} P_{ik}(A) f_k,$$

and let $P^\varepsilon(x) = \left(\widetilde{P}^\varepsilon_{ik}(x)\right)$ be the $n \times N$ matrix-function with entries $\widetilde{P}^\varepsilon_{ik}(x)$. It follows from (2.24), (2.25) and (2.7) that

$$G^\varepsilon_r(x) = P^\varepsilon(x) * F_r(x),$$

and therefore in view of the Binet-Cauchy formula

$$*\text{-}\det G^\varepsilon_r(x) = \sum_{i_1<\dots<i_n} P^\varepsilon\begin{pmatrix} 1 & 2 & \dots & n \\ i_1 & i_2 & \dots & i_n \end{pmatrix} * F_r\begin{pmatrix} i_1 & i_2 & \dots & i_n \\ 1 & 2 & \dots & n \end{pmatrix}$$

where $P^\varepsilon\begin{pmatrix} 1 & 2 & \dots & n \\ i_1 & i_2 & \dots & i_n \end{pmatrix}$ are the $n \times n$ $*$-minors of the matrix $P^\varepsilon(x)$. Consequently, by (2.22) we have

$$(2.26) \qquad *\text{-}\det G^\varepsilon_r(x) = 0, \qquad x \in [0,\beta],\ \varepsilon > 0.$$

This property is preserved when ε tends to zero:

$$(2.27) \qquad *\text{-}\det G_r(g_1,\dots,g_n)(x) = 0, \qquad x \in [0,\beta].$$

Thus the invariant subspace M is not all of $L_p[0,1] \otimes \mathbb{C}^n$ because for example for vectors $g_i = \{\delta_{ij}\}_{j=1}^n$ (δ_{ij} is the Kronecker delta), we have $G_r(g_1,\dots,g_n) = \underbrace{1 * 1 * \dots * 1}_{n} = \frac{x^{n-1}}{(n-1)!}$.

It remains to note that, in the case $N < n$, we can adjoin zero vectors $f_{N+1} = \dots = f_n = 0$ to the system $\{f_i\}_1^N$ and, repeating the previous arguments, obtain the equality (2.27) for an arbitrary system $\{g_i\}_1^n$ of vectors $g_i \in L_p[0,1] \otimes \mathbb{C}^n$. □

Corollary 2.4. *Let $X = L_p[0,1]$ with $p \in [1,\infty)$ and $\Re(\alpha) > 0$. Then the system $\{f_i\}_{i=1}^N$ of vectors*

$$(2.28) \qquad f_i(x) = \{f_{i1}(x),\dots,f_{in}(x)\} \in L_p[0,1] \otimes \mathbb{C}^n, \qquad 1 \le i \le N,$$

generates a cyclic subspace in $L_p[0,1] \otimes \mathbb{C}^n$ for the operator $A = J^\alpha \otimes I_n = \underbrace{J^\alpha \oplus \dots \oplus J^\alpha}_{n}$ ($\in [X^n]$) if and only if the matrix function

$$(2.29) \qquad F_n(x) = \begin{pmatrix} f_{11}(x) & f_{12}(x) & \dots & f_{1n}(x) \\ \vdots & \vdots & & \vdots \\ f_{N1}(x) & f_{N2}(x) & \dots & f_{Nn}(x) \end{pmatrix}$$

is of maximal $*$-rank, that is, $*$-rank $F_n(x) = n$.
In particular, if $N = n$ the subspace $E = \operatorname{span}\{f_i(x) : 0 \le i \le n\}$ will be a cyclic subspace for $A = J^\alpha \otimes I_n$ iff $*$-det $F_n(x)$ satisfies the ε-condition (2.4).

One immediately obtains the following description of the lattice Lat A from Theorem 2.3.

Proposition 2.5. *Under the conditions of Theorem* 2.3 *a subspace* $M \subset X^n$ *is an invariant subspace for* A, *that is,* $M \in LatA$, *iff there exists* $\beta > 0$ *such that for each system* $\{f_i = \oplus_{j=1}^{r} f_{ij}\}_{i=1}^{n}$ *of* n *vectors* ($f_i \in M$)

$$*\text{-}\det F_r(x) = 0, \qquad x \in [0, \beta],$$

where $F_r(x)$ *is the matrix of the form* (2.7).

Remark 2.6. In the scalar case Theorem 2.3 (as well as Corollary 2.4) gives a description of the set $\mathrm{Cyc}(J^\alpha)$ of the operator J^α acting in $L_p[0,1]$:

$$f \in \mathrm{Cyc}(J^\alpha) \quad \longleftrightarrow \quad f \text{ satisfies the } \varepsilon\text{-condition (2.4)}.$$

This fact is well-known and is equivalent to the unicellularity of J^α (see [B2], [GK], [N1], [FR]). There are two essentially different proofs of this fact concerning J in the literature: one is based on the Titchmarsh Convolution Theorem (see, for example [GK]), the other, suitable only for the space $L_2[0,1]$, uses either Livsic's theory of characteristic functions or the Nagy-Foias functional model [N1], [SF], [S1] (see also [FR] and papers cited therein).

In the present paper we develop the first approach though Corollary 2.4 for the space $L_2[0,1]$ can also be proved just like it was done by Sarason in the proof of unicellularity of J([S1]). But the proof we give is preferable for the following reasons: a) it is suitable for J^α with $\alpha \neq 1$, for the operator (4.6) with distinct λ_j and even for more general convolution operators; b) it is suitable for Banach function spaces like $L_p[0,1]$, $1 \le p < \infty$, $C[0,1], W_p^k[0,1]$.

3. The lattices of invariant subspaces

3.1. We start with the following definition.

Definition 3.1. An operator $T \in [X]$ is called *unicellular* if its lattice of invariant subspaces Lat T is linearly ordered.

It is well-known that the operator J^α with $\Re(\alpha) > 0$ is unicellular in $L_p[0,1]$, $1 \le p < \infty$. More precisely,

$$\text{(3.1)} \qquad \mathrm{Lat}\, J^\alpha = \{E_a^p := \chi_{[a,1]} L_p[0,1] : \ 0 \le a \le 1\},$$

and in particular Lat J^α is anti-isomorphic to the segment $[0,1]$. In (3.1) $\chi_{[a,1]}$ indicates the characteristic function of the segment $[a,1]$.

In this section some cases of splitting of the lattice of invariant subspaces of an operator $A = J^\alpha \otimes B$ are investigated.

This result reduces the description of Lat A to the cases described in Theorem 2.3 and the relation (3.1).

Proposition 3.2. *Let $\alpha > 0$, and let B be a diagonal matrix,*

$$B = \oplus_{j=1}^{r} B_j, \qquad B_j = \operatorname{diag}(\lambda_{j1}, \dots, \lambda_{jn_j}), \qquad 1 \le j \le r, \tag{3.2}$$

where $\lambda_{jk} = \lambda_{j1}/s_{jk}^{\alpha}$, $1 = s_{j1} \le s_{j2} \le \cdots \le s_{jn_j}$ and $\arg\lambda_{j1} \ne \arg\lambda_{k1} \pmod{2\pi}$ for $j \ne k$. Let also $A = J^{\alpha} \otimes B \in [L_p[0,1] \otimes \mathbb{C}^n]$ with $n = n_1 + \cdots + n_r$. Then

$$\mu_A = \max_{1 \le j \le r} n_j \qquad \text{and} \qquad \operatorname{Lat} A = \oplus_{j=1}^{r} \operatorname{Lat}(J^{\alpha} \otimes B_j). \tag{3.3}$$

Proof. Let π_j be a projection from $L_p[0,1] \otimes \mathbb{C}^n$ onto $L_p[0,1] \otimes \mathbb{C}^{n_j}$ in the direct sum decomposition $L_p^n[0,1] = \oplus_{j=1}^{r}(L_p[0,1] \otimes \mathbb{C}_{n_j})$. For an arbitrary nontrivial invariant subspace $M(\subset L_p^n[0,1])$ of the operator A we introduce the subspaces

$$M_j = \operatorname{span}\{(J^{\alpha} \otimes B_j)^k f_j : k \ge 0,\ f_j \in \pi_j M\}, \qquad 1 \le j \le r. \tag{3.4}$$

It is clear that $M \subseteq \oplus_{j=1}^{r} M_j$. We shall establish the opposite inclusion.
For arbitrary vectors $f = \oplus_{j=1}^{r} f_j \in M$ and $g = \oplus_{j=1}^{r} g_j \in M^{\perp} \subset L_q^n[0,1]$, where $p^{-1} + q^{-1} = 1$, and

$$f_j = \{f_{j1}, \dots, f_{jn_j}\} \in L_p[0,1] \otimes \mathbb{C}^{n_j}, \quad g_j = \{g_{j1}, \dots, g_{jn_j}\} \in L_q[0,1] \otimes \mathbb{C}^{n_j},$$

let us define the functions

$$\varphi_{jk}(t) = \begin{cases} s_{jk}^{\alpha+1} \int_t^{s_{jk}^{-1}} f_{jk}((x-t)s_{jk}) g_{jk}(x s_{jk}) dx, & t \in [0, s_{jk}^{-1}], \\ 0, & t \in [s_{jk}^{-1}, 1], \end{cases} \tag{3.5}$$

and

$$\varphi_j(t) := \sum_{k=1}^{n_j} \varphi_{jk}(t), \qquad 1 \le j \le r.$$

Since $R_{\lambda}(A) f \in M$ for all $f \in M$ and

$$R_{\lambda}(A) = (I - \lambda A)^{-1} = \oplus_{j=1}^{n} (I - \lambda J^{\alpha} \otimes B_j)^{-1}, \tag{3.6}$$

it follows from (2.9), (3.5) and (3.6) that

$$\begin{aligned} 0 &= \sum_{j=1}^{r} \sum_{k=1}^{n_j} \int_0^1 g_{jk}(x)\, dx \int_0^x (x-t)^{\alpha-1} E_{1/\alpha}(\lambda\lambda_{jk}((x-t)^{\alpha}; \alpha) f_{jk}(t)\, dt \\ &= \sum_{j=1}^{r} \sum_{k=1}^{n_j} \int_0^1 t^{\alpha-1} E_{1/\alpha}(\lambda\lambda_j s_{jk}^{-\alpha} t^{\alpha}; \alpha)\, dt \int_t^1 f_{jk}(x-t) g_{jk}(x)\, dx \qquad (3.7) \\ &= \sum_{j=1}^{r} \int_0^1 t^{\alpha-1} E_{1/\alpha}(\lambda\lambda_j t^{\alpha}; \alpha) \varphi_j(t)\, dt =: \sum_{j=1}^{r} \Phi_j(\lambda). \end{aligned}$$

Here $\lambda_j := \lambda_{j1}$ for $1 \leq j \leq r$. We shall obtain the inclusion $M \supset \oplus_{j=1}^r M_j$ by proving the implication

$$\sum_{j=1}^{r} \Phi_j(\lambda) = 0 \implies \Phi_j(\lambda) = 0, \quad j \in \{1, \dots, r\}. \tag{3.8}$$

Assuming the contrary we establish that there exists a ray such that on this ray one of the summands in (3.7) is of larger order of growth than the others. To find this ray we use the asymptotic behavior of the function $E_{1/\alpha}(z;\alpha)$ considering the cases $\alpha \geq 2$ and $0 < \alpha < 2$ separately.
i) In the case $2 \leq \alpha < \infty$ the asymptotic formula

$$E_{1/\alpha}(z;\alpha) = \frac{1}{\alpha} \sum_{k} \left(z^{1/\alpha} \exp(i2\pi k \alpha^{-1}) \right)^{1-\alpha} \cdot \exp\left(z^{1/\alpha} e^{i\frac{2\pi k}{\alpha}} \right) + O(|z|^{-2}) \tag{3.9}$$

holds as $|z| \to \infty$ (see [Dj]). Here the summation is taken over those $k \in \mathbb{Z}$ for which $|\arg z + 2\pi k| \leq \pi\alpha/2$.
It is easily seen that the right-hand side of (3.9) does not depend on the choice of a branch of the multi-valued function $z^{1/\alpha}$.
Denoting by $[a_j, b_j]$ the minimal segment containing the support of the function $\varphi_j(t)$ we suppose for definiteness that

$$b_1|\lambda_1|^{1/\alpha} = \max\{b_j|\lambda_j|^{1/\alpha} : 1 \leq j \leq r\}. \tag{3.10}$$

We will show that the ray $\Lambda = \{\lambda \in \mathbb{C} : \arg\lambda = -\arg\lambda_1\}$ is the desired one. Letting

$$\beta_1 = 0, \qquad \beta_j = \arg \lambda_j \lambda_1^{-1} \ (\neq 0), \quad 2 \leq j \leq r, \tag{3.11}$$

we note that

$$\varepsilon := 1 - \max_{j,k} \left\{ \cos \frac{\beta_j + 2\pi k}{\alpha} \ : \ |\beta_j + 2\pi k| \leq \pi\alpha/2 \, , \, 2 \leq j \leq r \right\} > 0. \tag{3.12}$$

Now it follows from (3.11) and (3.12) that

$$\begin{aligned} &\int_0^1 t^{\alpha-1} E_{1/\alpha}((xt)^\alpha \lambda_j \lambda_1^{-1}; \alpha) \varphi_j(t)\, dt \\ &= \int_{a_j}^{b_j} t^{\alpha-1} E_{1/\alpha}((xt)^\alpha \lambda_j \lambda_1^{-1}; \alpha) \varphi_j(t)\, dt \\ &= O\left(x^{1-\alpha} \sum_k \exp\left(x b_j \left| \frac{\lambda_j}{\lambda_1} \right|^{1/\alpha} \cos\left(\frac{\beta_j + 2\pi k}{\alpha} \right) \right) \right) \\ &= O(\exp(b_1(1-\varepsilon)x), \qquad 2 \leq j \leq r, \end{aligned} \tag{3.13}$$

where the summation is taken over those $k \in \mathbb{Z}$ for which $|\beta_j + 2\pi k| \leq \pi\alpha/2$. At the same time from (3.11) we see that asymptotically as $x \to \infty$

$$\begin{aligned} &\int_{a_1}^{b_1} t^{\alpha-1} E_{1/\alpha}((xt)^\alpha;\alpha)\varphi_1(t)dt \\ &= \sum_{|k|\leq\alpha/4} x^{1-\alpha} \int_{a_1}^{b_1} \exp\left(xt\cos\frac{2\pi k}{\alpha}\right)\varphi(t)dt + o(1) \\ &= x^{1-\alpha}\int_{a_1}^{b_1} e^{xt}\varphi_1(t)dt + O\left(e^{b_1 cx}\right) \end{aligned} \tag{3.14}$$

where

$$c := \max\{\cos 2\pi k/\alpha : k \in \mathbb{Z}\setminus\{0\},\ |k| \leq \alpha/4\} < 1. \tag{3.15}$$

Since both functions $\varphi_1(t-a_1)$ and $\varphi_2(b_1-t)$ satisfy the ε-condition (2.4), the indicator diagram of the function

$$F(z) = \int_{a_1}^{b_1} e^{zt}\varphi_1(t)\,dt \tag{3.16}$$

is the segment $[a_1, b_1] \subseteq [0,1]$ (see [L]). Therefore the indicator function $h_F(\theta) = \limsup_{r\to\infty} r^{-1}\ln|F(re^{i\theta})|$ is of the form

$$h_F(\theta) = 2^{-1}(b_1-a_1)|\cos\theta| + 2^{-1}(a_1+b_1)\cos\theta,$$

and consequently

$$\limsup_{r\to+\infty}\frac{\ln|F(r)|}{r} = h_F(0) = b_1. \tag{3.17}$$

On the other hand it follows from (3.7) and (3.13)-(3.16) that

$$\limsup_{r\to+\infty}\frac{\ln|F(r)|}{r} \leq b_1(1-\varepsilon). \tag{3.18}$$

Comparing the relations (3.17) and (3.18) we obtain $b_1 = 0$, and bearing in mind (3.10), we see that $b_j = 0$ for all $j \in \{1,\dots,r\}$. This means that $\varphi_j(t) = 0$, $j \in \{1,\dots,n\}$, and hence

$$\left((I-\lambda J^\alpha \otimes B_j)^{-1} f_j, g_j\right) = 0, \qquad f \in \pi_j M,$$

that is, $g_j \in M_j^\perp$. Thus $M = \oplus_{j=1}^r M_j$.

ii) Now let $0 < \alpha < 2$. Then for any γ, $\pi\alpha/2 < \gamma < \min\{\pi, \pi\alpha\}$, we have the following asymptotic formulas as $|z| \to \infty$ [Dj]:

$$\begin{aligned} E_{1/\alpha}(z;\alpha) &= \frac{1}{\alpha} z^{(1-\alpha)\alpha^{-1}} \exp\left(z^{1/\alpha}\right) + O\left(\frac{1}{|z|^2}\right), \quad |\arg z| \leq \gamma, \\ E_{1/\alpha}(z;\alpha) &= O\left(\tfrac{1}{|z|^2}\right), \quad \gamma \leq |\arg z| \leq \pi. \end{aligned} \tag{3.19}$$

Taking (3.19) into account one can repeat the previous arguments using the estimates

$$t^{\alpha-1}E_{1/\alpha}(\lambda_1^{-1}\lambda_j(xt)^\alpha;\alpha) = O\left(\frac{1}{x^2}\right)$$

instead of (3.11) when $|\beta_j| \in [\gamma, \pi]$. One can complete the proof as in case i). □

Remark 3.3. a) If $\alpha = p$ ($\in \mathbb{Z}_+$) is an integer, the proof immediately follows from a simple fact about the sum of entire functions. Indeed, in this case setting $\lambda = \mu^p$ one can easily obtain from (2.10) the explicit formula

$$t^{p-1}E_{1/p}((\mu t)^p;p) = \frac{1}{\mu^{p-1}}\sum_{k=0}^{\infty}\frac{(\mu t)^{kp+p-1}}{(kp+p-1)!} = \frac{1}{p\mu^{p-1}}\sum_{k=1}^{p}\varepsilon_k e^{\varepsilon_k \mu t} \tag{3.20}$$

instead of the asymptotic formulas (3.8), (3.19). Here $\varepsilon_k = \exp(i2\pi k/p)$ are the distinct k-th roots of unity. It follows from (3.20) and the well-known Polya Theorem [L] that each entire function

$$\Phi_j(\mu) = \int_0^1 t^{p-1}E_{1/p}(\lambda_j(\mu t)^p;p)\varphi(t)dt = \frac{1}{p\mu^{p-1}}\sum_{k=1}^{p}\varepsilon_k\int_{a_j}^{b_j} e^{\varepsilon_k\mu t}\varphi_j(t)dt$$

is of exponential type b_j and its indicator diagram is the polygon with vertices

$$b_j|\lambda_j|\exp(-i2\pi k/p)\ , \quad 1 \le k \le p.$$

To obtain the implication

$$\sum_{1\le j\le n}\Phi_j(\mu) \equiv 0 \implies \Phi_j(\mu) = 0, \quad j \in \{1,\dots,r\},$$

it remains to note that, as it follows from Polya's Theorem, the indicator diagram of a sum of entire functions of exponential type contains the convex hull of those extreme points of each of the indicator diagrams that do not lie in any of the other diagrams [L].

b) We also mention the papers [CW] and [N2] devoted to the splitting of $\mathrm{Lat}\,(T_1 \oplus T_2)$. In [CW] the splitting of $\mathrm{Lat}\,(T_1 \oplus T_2)$ is investigated for Hilbert space contractions of class C_0, or, what is the same, for dissipative operators T_1 and T_2. It should be pointed out that the operators $\lambda_1 J$ and $\lambda_2 J$ are simultaneously dissipative if and only if $\arg\lambda_1 = \arg\lambda_2$.

In [N2] the splitting of the lattice $\mathrm{Lat}\,(T_1 \oplus T_2)$ is studied under the spectral independence condition: $\sigma(T_1) \cap \sigma(T_2) = \emptyset$ (see, for example, Theorem 1.75 and Corollaries 1.69, 1.71, 1.77 in [N2]).

3.2. Now we are ready to formulate the main result of the section.

Theorem 3.4. *Let $\alpha_1 > \alpha_2 > \cdots > \alpha_n > 0$, and let B_{kj} be the $n_{kj} \times n_{kj}$ diagonal matrices with eigenvalues λ_{kji} of equal arguments:*

$$B_{kj} = \mathrm{diag}\,(\lambda_{kj1},\dots,\lambda_{kjn_j}), \quad \arg\lambda_{kj1} = \arg\lambda_{kji}, \quad 1 \le i \le n_{kj},$$

and $\arg \lambda_{kj_1 1} \neq \arg \lambda_{kj_2 1}$ *if* $j_1 \neq j_2$. *Let also*

$$B_k = \oplus_{j=1}^{n_k} B_{kj}, \quad A_k = J^{\alpha_k} \otimes B_k, \quad A = \oplus_{k=1}^{n} A_k. \tag{3.21}$$

Then

$$\operatorname{Lat} A = \oplus_{k=1}^{n} \oplus_{j=1}^{n_k} \operatorname{Lat}\,(J^{\alpha_k} \otimes B_{kj}). \tag{3.22}$$

Proof. Let

$$N_k = \sum_{1 \le j \le n_k} n_{kj}, \qquad N = \sum_{1 \le k \le n} N_k,$$

and let π_k be the projection from $L_p[0,1] \otimes \mathbb{C}^N$ onto $L_p[0,1] \otimes \mathbb{C}^{N_k}$ in the direct sum decomposition $L_p^N[0,1] = \oplus_{k=1}^n L_p^{N_k}[0,1]$. In just the same way as in the proof of Proposition 3.2 we introduce the subspaces

$$M_k = \operatorname{span}\{(J^{\alpha_k} \otimes B_k)^m f_k : m \ge 0,\ f_k \in \pi_k M\}, \qquad 1 \le k \le n,$$

for an arbitrary invariant subspace M ($\subset L_p^N[0,1]$) of the operator A.
The equality $M = \oplus_{k=1}^n M_k$ amounts to the inclusion $M \supset \oplus_{k=1}^n M_k$ because the opposite inclusion is obvious.
Let $f = \oplus_{k=1}^n f_k \in M$, $g = \oplus_{k=1}^n g_k \in M^{\perp} \subset L_q^n[0,1]$, where $f_k \in M_k$, $g_k \in L_q^{N_k}[0,1]$. Because of

$$(I - \lambda A)^{-1} = \oplus_{k=1}^n (I - \lambda(J^{\alpha_k} \otimes B_k))^{-1}, \tag{3.23}$$

we have

$$\sum_{k=1}^{n} ((I - \lambda(J^{\alpha_k} \otimes B_k)) f_k, g_k) = 0. \tag{3.24}$$

Since each non-zero summand in the left-hand side of the equality (3.24) is an entire function $\Phi_k(\lambda)$ of order α_k, the sum $\sum_k \Phi_k(\lambda)$ has the same order and type as the summand of highest order. Therefore the equality (3.24) is equivalent to the system

$$\Phi_k(\lambda) := \big((I - \lambda(J^{\alpha_k} \otimes B_k))^{-1} f_k, g_k\big) = 0, \qquad 1 \le k \le n. \tag{3.25}$$

Thus $g = \oplus_{k=1}^n g_k \in M^{\perp}$ iff $g_k \in M_k^{\perp}$ for every $k \in \{1, \dots, n\}$, and hence $M = \oplus_{k=1}^n M_k$. To complete the proof it remains to apply Proposition 3.2 to each of the operators $J^{\alpha_k} \otimes B_k$ acting in the space $L_p[0,1] \otimes \mathbb{C}^{N_k}$, $1 \le k \le n$. □

Let π_{kj} be the projection from $L_p^N = L_p[0,1] \otimes \mathbb{C}^N$ onto $L_p^{n_{kj}} = L_p[0,1] \otimes \mathbb{C}^{n_{kj}}$ in the decomposition

$$L_p^N = \oplus_{(k,j) \in E} L_p^{n_{kj}}, \quad E := \{(k,j) : 1 \le k \le n,\ 1 \le j \le n_k\}, \quad N = \operatorname{card} E. \tag{3.26}$$

The following corollaries follow immediately from Theorem 2.3 and Theorem 3.4.

Corollary 3.5. *Under the conditions of Theorem 3.4 the following statements hold:*
1) *The spectral multiplicity μ_A satisfies*

$$\mu_A = \max\{n_{kj} : 1 \leq k \leq n,\ 1 \leq j \leq n_k\}. \tag{3.27}$$

2) *A subspace M in $L_p[0,1] \otimes \mathbb{C}^N$ is a cyclic subspace for the operator A of the form (4.21) if and only if for each pair $(k,j) \in E$ the subspace $M_{kj} = \pi_{kj} M$ is a cyclic subspace for the operator $A_{kj} = J^{\alpha_k} \otimes B_{kj}$.*
3) *A system $\{f_i\}_{i=1}^r$ of vectors $f_i = \oplus_{k=1}^n \oplus_{j=1}^{n_k} f_{ikj}$, where $f_{ikj} \in L_p[0,1] \otimes \mathbb{C}^{n_{kj}}$, generates a cyclic subspace for the operator A iff $r \geq \mu_A$ and for each pair $(k,j) \in E$ the system $\{f_{ikj}\}_{i=1}^r$ satisfies the conditions of Theorem 2.3.*

Corollary 3.6. *Let $A = \oplus_{k=1}^r J^{\alpha_k} \otimes B_k$, $B_k = \operatorname{diag}(\lambda_{k1}, \dots, \lambda_{kn_k})$, $n = n_1 + \cdots + n_r$, and $\arg \lambda_{ki} \neq \arg \lambda_{kj}$ for $i \neq j$. Then the following statements are true:*
1) *The operator A is cyclic in $L_p[0,1] \otimes \mathbb{C}^n$, $1 \leq p < \infty$.*
2) *The lattice $\operatorname{Lat} A$ is anti-isomorphic to the cube $\square_n = \{x \in R^n :\ 0 \leq x_i \leq 1\}$:*

$$\operatorname{Lat} A = \{\oplus_{i=1}^n \chi_{[a_i,1]} L_p[0,1] : 0 \leq a_i \leq 1\} = \oplus_{i=1}^n \operatorname{Lat} J^{\alpha_i}. \tag{3.28}$$

3) *A vector $f = \oplus_{k=1}^n \oplus_{j=1}^{n_k} f_{kj}$, where $f_{kj} \in L_p[0,1]$, is cyclic ($f \in Cyc(A)$) iff for each $(k,j) \in E$ the vector $f_{kj}(x)$ satisfies the ε-condition (2.4).*

Remark 3.7. Setting $r = 1$ and $\alpha = \alpha_1$, from Corollary 3.6 we obtain a description of the lattice $\operatorname{Lat}\,(J^\alpha \otimes B_1)$. Namely, $\operatorname{Lat}\,(J^\alpha \otimes B_1)$ is of the form (3.28) with $n = n_1$. For the operator $J \otimes B_1$ (i.e., for $\alpha = 1$) a quite different proof of this fact based on Lavrentyev's Approximation Theorem was obtained in [OS1], [OS2].

3.3. To state the next corollary we recall the following definition.

Definition 3.8. Let S be an involution [GK] in a Hilbert space H.
a) An operator T ($\in [H]$) is called *S-real* if $AS = SA$.
b) An S-real operator T is said to be *S-unicellular* if $\operatorname{Lat}_S T := \operatorname{Lat} T \cap \operatorname{Lat} S$ is totally ordered by set inclusion.

We will consider only the involution S of complex conjugation in $L_p[0,1]$ and the involution $S_n := S \otimes I_n$ in $L_p[0,1] \otimes \mathbb{C}^n$ given by

$$S_n = S \otimes I_n : (f_1, \dots, f_n) \mapsto (\overline{f}_1, \dots, \overline{f}_n). \tag{3.29}$$

Corollary 3.9. *Let S_2 be the involution in $L_p[0,1] \otimes \mathbb{C}^2$ defined by (3.29), and let $B = \begin{pmatrix} b & d \\ -d & b \end{pmatrix}$. Then the operator $A = J^\alpha \otimes B$ is S_2-unicellular in $L_p \otimes \mathbb{C}^2$ and*

$$\operatorname{Lat}_S A = \{\chi_{[a,1]} L_p[0,1] \oplus \chi_{[a,1]} L_p[a,1] : a \in [0,1]\}.$$

Proof. Since

$$U^{-1}BU = D := \begin{pmatrix} b+id & 0 \\ 0 & b-id \end{pmatrix}, \qquad U = \begin{pmatrix} 1 & 1 \\ i & -i \end{pmatrix},$$

it is clear that

$$M \in \mathrm{Lat}\,(J^\alpha \otimes B) \iff (I \otimes U^{-1})M \in \mathrm{Lat}\,(J^\alpha \otimes D).$$

In accordance with Proposition 3.2 there exist $a_1, a_2 \in [0,1]$ such that

$$M = \{\{f_1, f_2\} : f_1(x) - if_2(x) = 0,\ x \in [0, a_1],\ f_1(x) + if_2(x) = 0,\ x \in [0, a_2]\}.$$

It is easily seen that $a_1 = a_2$ if in addition $M \in \mathrm{Lat}\,S$. Let for example $a_1 < a_2$. Then one can find a pair of functions $\{f_1^0, f_2^0\} \in M$ such that

$$\text{(3.30)} \qquad \mathrm{mes}\{x \in [a_1, a_2] : f_1^0(x) - if_2^0(x) \neq 0\} > 0.$$

But since $M \in \mathrm{Lat}\,S$, $f_1^0(x) = -if_2^0(x)$, $\overline{f_1^0(x)} = -i\overline{f_2^0(x)}$, $x \in [0, a_2]$, that is, $f_1^0(x) = f_2^0(x) = 0$, $x \in [0, a_2]$, which contradicts (3.30). □

Remark 3.10. Quite different proofs of the S_2-unicellularity of the operator $J \otimes B$ have been obtained earlier in [B1] for $B = \begin{pmatrix} 0 & 1 \\ -1 & 0 \end{pmatrix}$ and in [O-S2] for $B = \begin{pmatrix} \cos\alpha & \sin\alpha \\ -\sin\alpha & \cos\alpha \end{pmatrix}$.

4. The lattices of hyperinvariant subspaces

4.1. We recall that an invariant subspace E of an operator $A \in [H]$ is called hyperinvariant for A if E is invariant for any bounded operator that commutes with A, that is, it is invariant for all the operators from the commutant $\{A\}'$. The lattice of all the hyperinvariant subspaces of an operator A will be denoted by Hyplat A. To describe them we need the descriptions of the commutants $\{A\}'$.

One can obtain the following proposition in the same way as Proposition 3.2.

Proposition 4.1. *Let $X = L_p[0,1]$, $p \in [1, \infty)$, $X^n = \overbrace{X \oplus \ldots \oplus X}^{n}$, and let B be a diagonal matrix with simple nonzero eigenvalues of different arguments:*

$$\text{(4.1)} \qquad B = \mathrm{diag}\,(\lambda_1, \ldots, \lambda_n), \qquad \arg\lambda_i \neq \arg\lambda_j,\ i \neq j.$$

Then the commutant $\{A\}'$ of the operator $A = J^\alpha \otimes B$ $(\in [X^n])$ with $\alpha > 0$ is

$$\text{(4.2)} \qquad \{A\}' = \{K : K = \mathrm{diag}\,(K_1, \ldots, K_n),\ K_j \in [X]\},$$

all the K_j being of the form

$$K_j f = \frac{d}{dx}\int_0^x k_j(x-t)f(t)\,dt, \qquad k_j(x) \in L_q[0,1], \tag{4.3}$$

where $q^{-1}+p^{-1}=1$.

Proof. Let $K=(K_{ij})_{i,j=1}^n$ be the block-matrix partition of the operator K with respect to the direct sum decomposition $X^n = \underbrace{X\oplus\ldots\oplus X}_{n}$. Then the equality $KJ^\alpha = J^\alpha K$ is equivalent to the system

$$\lambda_j K_{ij} J^\alpha = \lambda_i J^\alpha K_{ij}, \qquad 1\le i,j\le n. \tag{4.4}$$

It was shown in [M1] that there are no bounded operators intertwining the operators J^α and cJ^α for $c\in\mathbb{C}$ and $\arg c\neq 0$, that is, $K_{ij}=0$ for $i\neq j$.
A brief direct proof can be easily derived from (2.9), (3.8) and (3.19) together with the implication (3.7). In fact, it follows from (4.4) that

$$K_{ij}(I-\lambda_j J^\alpha)^{-1} = (I-\lambda_i J^\alpha)^{-1}K_{ij}, \qquad i\neq j. \tag{4.5}$$

Therefore for arbitrary vectors $f\in L_p[0,1]$, $g\in L_q[0,1]$ we have

$$\int_0^1 t^{\alpha-1}E_{1/\alpha}(\lambda\lambda_1 t^\alpha;\alpha)\varphi_1(t)\,dt + \int_0^1 t^{\alpha-1}E_{1/\alpha}(\lambda\lambda_2 t^\alpha;\alpha)\varphi_2(t)\,dt = 0, \tag{4.6}$$

where

$$\varphi_1(t) = \int_t^1 f(x-t)g_1(x)\,dx, \quad \varphi_2(t) = -\int_t^1 f_1(x-t)g(x)\,dx$$

and $g_1 = K_{ij}^* g$, $f_1 = K_{ij}f$. Denote by $[a_j,b_j]$, $j=1,2$, the minimal segment containing the support of the function $\varphi_j(t)$ and suppose that $b_1|\lambda_1|^{1/\alpha}\ge b_2|\lambda_2|^{1/\alpha}>0$. The asymptotic formulas (3.8) and (3.9) yield that on the ray $\lambda_1^{-1}R_+ = \{\lambda_1^{-1}x : x\ge 0\}$ the first summand in (4.6) is of larger order of growth than the second one. It means that $b_1=b_2=0$ that is $\varphi_1(t)=\varphi_2(t)=0$, $t\in[0,1]$, and due to Titchmarsh's Theorem $K_{ij}=0$ for $i\neq j$.
It suffices now to prove that each bounded operator $K\in[H]$ commuting with J^α has the form (4.3). This fact is well-known but we provide here a brief and simple proof.
Since $P := JK$ maps $L_p[0,1]$ into $C[0,1]$, it is an integral operator (see [KA]) with a kernel $P(x,t)\in L_\infty \hat{\otimes} L_q$:

$$Pf = \int_0^1 P(x,t)f(t)\,dt, \qquad C(P) = \operatorname*{ess\,sup}_{0\le x\le 1}\int_0^1 |P(x,t)|^q\,dt < \infty.$$

Let $R(x,t)$ be the kernel of the operator $R := JPJ$. Then

$$R(x,t) = \int_0^x d\xi \int_t^1 P(\xi,\eta)\,d\eta \tag{4.7}$$

and $R(x,t)$ is an absolutely continuous function with respect to x and t for fixed $t \in [0,1]$ and $x \in [0,1]$, respectively. Since $RJ^\alpha = J^\alpha R$, we have $RJ^{\alpha m} = J^{\alpha m} R$ for all $m \in \mathbb{Z}_+$ and the Muntz-Szasz's Theorem ([PW], part 2) yields $RJ = JR$, that is,

$$\int_t^1 R(x,s)\,ds = \int_0^x R(s,t)\,ds. \tag{4.8}$$

Applying the operator $D_x D_t$ to the equation (4.8) we obtain $(D_x + D_t)R(x,t) = 0$. This means that $R(x,t) = R(x-t)$, i.e., $R(x,t)$ depends only on the difference of the variables. We can now rewrite (4.8) in the form

$$0 = \int_{x-1}^{x-t} R(s)\,ds - \int_{-t}^{x-t} R(s)\,ds = \int_{x-1}^{-t} R(s)\,ds.$$

It follows that $R(s) = 0,\ s \in [-1,0]$. Taking (4.7) into account we have

$$P(x,t) = D_x D_t R(x,t) = -R''(x-t) = P(x-t), \qquad P(x) = 0,\ x \in [-1,0].$$

Thus $JKf = P(x) * f(x)$, and therefore K is of the form (4.3) and $C(P) = \|P(x)\|_{L_q[0,1]}$. □

Remark 4.2. i) In the case $p = 2$ one concludes that $P = JK$ admits an integral representation (4.6) because P is a Hilbert-Schmidt operator. Another description of the commutant $\{J\}'$ in $L_2[0,1]$ was obtained by Sarason [S2].
ii) Dixmier [D] has found that in the space $L_1[0,1]$ the commutant $\{J\}'$ consists precisely of the convolution operators induced by measures on $[0,1)$. This fact can be established also from the representation (4.3).
One can consider an operator K_j of the form (4.3) as a convolution operator: $f \mapsto f * k_j'$ with k_j' being a distribution. However, not all the operators in the commutant $\{J\}'$ of J on $L_p[0,1]$ for $p > 1$ are generated by measures on $[0,1)$. For example $k(x) = x^i$ generates a bounded operator of the form (4.3), although x^i is not a function of bounded variation on $[0,1]$.

Remark 4.3. It is obvious that Lat $(A_1 \oplus A_2) \supset$ Lat $A_1 \oplus$ Lat A_2. The opposite inclusion, which is equivalent to the splitting of Lat$(A_1 \oplus A_2)$, is closely connected with the existence of an operator $T \in [X_1, X_2]$ intertwining the operators $A_i \in [X_i]$, $i = 1,2$. In fact, if there exists a bounded operator $(O \neq)\ T \in [X_1, X_2]$ such that $TA_1 = A_2 T$ then

$$\text{Lat}\,(A_1 \oplus A_2) \neq (\text{Lat}\,A_1 \oplus \text{Lat}\,A_2).$$

Indeed, the subspace $M = \operatorname{gr} T = \{(h, Th) : h \in X_1\} \subset X_1 \oplus X_2$ is invariant for $A_1 \oplus A_2$ because of

$$\begin{pmatrix} h \\ Th \end{pmatrix} \in M \Rightarrow (A_1 \oplus A_2)\begin{pmatrix} h \\ Th \end{pmatrix} = \begin{pmatrix} A_1 h \\ A_2 Th \end{pmatrix} = \begin{pmatrix} A_1 h \\ TA_1 h \end{pmatrix} \in M,$$

but $M \notin \operatorname{Lat} A_1 \oplus \operatorname{Lat} A_2$.
Therefore the first part of Proposition 4.1, i.e., the diagonal form of an operator $K \in \{A\}'$, immediately follows from Proposition 3.2.

The following two results are simple corollaries of the previous one.

Proposition 4.4. *Under the conditions of Proposition* 4.1

(4.9) $$\operatorname{Hyplat} A = \operatorname{Lat} A = \{\oplus_{i=1}^{n} \chi_{[a_i,1]} L_p[0,1] : 0 \le a_i \le 1\} = \oplus_{j=1}^{n} \operatorname{Lat}(\lambda_j J^{\alpha}),$$

and consequently the lattice Hyplat A *is anti-isomorphic to the unit cube* $\square_n = \{x \in R^n : 0 \le x_i \le 1\}$.

The following Proposition may be proved by analogy with the proof of Proposition 3.2.

Proposition 4.5. *Let* $A = \oplus_{k=1}^{r} J^{\alpha_k} \otimes B_k$, $\alpha_1 > \cdots > \alpha_r$ *and* $B_k = \operatorname{diag}(\lambda_{k1}, \ldots, \lambda_{kn_k})$ *a nonsingular diagonal matrix,* $n = n_1 + \cdots + n_r$ *and* $\arg \lambda_{ki} \ne \arg \lambda_{kj}$ *for* $i \ne j$. *Then:*

1) *The commutant* $\{A\}'$ *of the operator* A *is of the form* (4.2), (4.3).

2) Hyplat A *is of the form* (4.9) *and therefore it is anti-isomorphic to the unit cube* $\square_n = \{x \in R^n : 0 \le x_i \le 1\}$.

4.2. To state the next results we need some notations. Corresponding to any $a \in (0,1)$ we define an isometric operator $U_a : L_p[0,1] \to L_p[0,a^{-1}]$ given by

(4.10) $$U_a : L_p[0,1] \ni f(x) \mapsto \widetilde{f}(x) = a^{1/p} f(ax) \in L_p[0,a^{-1}].$$

As usual for $a < l$ we will identify $L_p[a,l]$ and $L_p[0,a]$ with the subspaces E_a^p and $\widetilde{E}_a^p$ of $L_p[0,l]$:

(4.11) $$\begin{aligned} E_a^p &= \{f(x) \in L_p[0,l] : f(x) = 0,\ x \in [0,a]\} = \chi_{[a,1]} L_p[0,l], \\ \widetilde{E}_a^p &= \{f(x) \in L_p[0,l] : f(x) = 0,\ x \in [a,l]\} = \chi_{[0,a]} L_p[0,l], \end{aligned}$$

respectively.

Let P_a be the projection

(4.12) $$P_a : L_p[0,a^{-1}] \to \widetilde{E}_1^p \cong L_p[0,1]$$

of $L_p[0,a^{-1}]$ onto $\widetilde{E}_1^p$ in the decomposition $L_p[0,a^{-1}] = E_1^p \oplus \widetilde{E}_1^p$.

According to the decomposition $L_p[0,1] = E_a^p \oplus \widetilde{E}_a^p$ we introduce the embedding operator $i_a : L_p[0,a] \to L_p[0,1]$ given by

$$i_a : f(x) \mapsto g(x) = \begin{cases} 0, & x \in [0, 1-a], \\ f(x-1+a), & x \in [1-a, 1]. \end{cases} \tag{4.13}$$

Proposition 4.6. *Let $X = L_p[0,1]$, $1 \le p < \infty$, and let $B = \operatorname{diag}(\lambda_1, \dots, \lambda_n)$ be a nonsingular diagonal matrix with eigenvalues of equal arguments:*

$$\lambda_j = \lambda_1 s_j^\alpha, \qquad 1 = s_1 \le s_2 \le \cdots \le s_n,$$

and assume in addition

a) $P_{ij} = P_{a_{ij}}$, $U_{ij} = U_{a_{ij}}$ *and* $a_{ij} = s_i^{-1} s_j < 1$ *if* $i > j$,

b) $i_{ji} = i_{a_{ji}}$ *if* $i < j$,

c) $U_{jj} = i_{jj} = I$ *(I is the identity operator on $L_p[0,1]$).*

Then the commutant $\{A\}'$ of the operator $A = J^\alpha \otimes B \in [X^n]$ with $\Re(\alpha) > 0$ is

$$\{A\}' = \{K : K = (K_{ij})_{i,j=1}^n,\ K_{ij} \in [X]\},$$

where $K_{jj} \in \{J^\alpha\}'$, that is, K_{jj} is of the form (4.3) for each $j \in \{1, \dots, n\}$,

$$K_{ij} = C_{ij} P_{ij} U_{ij}, \qquad C_{ij} \in \{J^\alpha\}', \qquad i > j, \tag{4.14}$$

$$K_{ij} = i_{ji} U_{ij}^{-1} C_{ij}, \qquad C_{ij} \in \{J^\alpha\}', \qquad i < j. \tag{4.15}$$

Proof. In view of (4.4) the equality $KA = AK$ is equivalent to the system

$$a_{ij}^\alpha J^\alpha K_{ij} = K_{ij} J^\alpha, \qquad a_{ij} = s_i^{-1} s_j. \tag{4.16}$$

We consider the two cases $i > j$ and $i < j$ separately.

i) Let $i > j$. Then $a_{ij} = s_i^{-1} s_j < 1$. To begin with, we shall verify that K_{ij} of the form (4.14) satisfies (4.16). Setting for an arbitrary $f(x) \in L_p[0,1]$

$$\varphi(x) = \begin{cases} a_{ij}^{1/p}((a_{ij}x)^{\alpha-1} * f)(a_{ij}x), & x \in [0,1], \\ 0, & x \in [1, a_{ij}^{-1}], \end{cases}$$

we have

$$\begin{aligned} \Gamma(\alpha) a_{ij}^\alpha J^\alpha P_{ij} U_{ij} f &= a_{ij}^\alpha \int_0^x (x-t)^{\alpha-1} a_{ij}^{1/p} f(a_{ij}t)\,dt \\ &= a_{ij}^{1/p} \int_0^{a_{ij}x} (a_{ij}x - t)^{\alpha-1} f(t)\,dt \\ &= (P_{ij}\varphi)x = \Gamma(\alpha) P_{ij} U_{ij} J^\alpha f. \end{aligned}$$

Thus for $i > j$ the operator

$$N_{ij} = P_{ij} U_{ij} \tag{4.17}$$

satisfies the commutation relation (4.16). Since $C_{ij} \in \{J^\alpha\}'$, the same is true for $K_{ij} = C_{ij}N_{ij}$.
Conversely, let K_{ij} satisfy (4.16). To prove (4.14) we set $f_1 = K_{ij}f$, $g_1 = K_{ij}^* g$, $\psi_1(t) = \int_t^1 f(x-t)g_1(x)\,dx$ and

$$\psi_2(t) = \begin{cases} \dfrac{1}{a_{ij}^{\alpha+1}} \displaystyle\int_t^{a_{ij}} f_1\left(\frac{x-t}{a_{ij}}\right) g\left(\frac{x}{a_{ij}}\right) dx, & t \in [0, a_{ij}], \\ 0, & t \in [a_{ij}, 1]. \end{cases} \tag{4.18}$$

Bearing in mind that $\lambda_j = a_{ij}^\alpha \lambda_j$ we obtain from (4.5) and (4.6)

$$\int_0^1 t^\alpha E_{1/\alpha}(\lambda t^\alpha; \alpha)[\psi_1(t) - \psi_2(t)]\,dt = 0$$

This means that $\psi_1(t) = \psi_2(t)$, $t \in [0,1]$, and in view of (3.18)

$$\psi_1(t) = \int_t^1 f(x-t)g_1(x)\,dx = 0, \qquad t \in [a_{ij}, 1].$$

It follows that

$$g_1(x) = (K_{ij}^* g)(x) = 0, \qquad x \in [a_{ij}, 1], \tag{4.19}$$

and consequently $E_{a_{ij}} = \chi_{[a_{ij},1]} L_p[0,1] \subseteq \ker K_{ij}$. According to (4.17) $\Re(N_{ij}^*) = E_{a_{ij}}$ and in view of Douglas' Lemma [N1], the inclusion $\Re(K_{ij}^*) \subseteq E_{a_{ij}} = \Re(N_{ij}^*)$ is equivalent to the representation

$$K_{ij} = C_{ij}N_{ij}. \tag{4.20}$$

Here C_{ij} is a bounded operator on $L_p[0,1]$ which is uniquely determined by the additional condition $\ker C_{ij} \supset \ker N_{ij}$.
Substituting the operator K_{ij} from (4.20) into (4.16) and using (4.17) we obtain

$$(J^\alpha C_{ij} - C_{ij}J^\alpha)N_{ij} = 0. \tag{4.21}$$

Since $\Re(N_{ij}) = L_p[0,1]$, one concludes from (4.21) that $C_{ij} \in \{J^\alpha\}'$ for $i > j$, and according to Proposition 4.1 there exist $C_{ij} \in [L_p[0,1]]$ such that

$$C_{ij}f = \frac{d}{dx}\int_0^x C_{ij}(x-t)f(t)\,dt. \tag{4.22}$$

ii) Now let $i < j$. Then $a_{ji} = a_{ij}^{-1} = s_j^{-1}s_i < 1$ and

$$a_{ji}^\alpha K_{ij}J^\alpha = J^\alpha K_{ij}. \tag{4.23}$$

We prove that for each pair (i,j) with $i < j$ the operator

$$M_{ij} = i_{a_{ji}} U_{ij}^{-1} \tag{4.24}$$

satisfies the commutation relation (4.22). Indeed, setting ($a < 1$)

$$\varphi(x) = \begin{cases} 0, & x \in [0, 1-a], \\ a \displaystyle\int_0^{(a+x-1)/a} (x - 1 + a - at)^{\alpha-1} f(t)\, dt, & x \in [1-a, 1], \end{cases}$$

we have

$$\Gamma(\alpha) a^\alpha i_a U_a^{-1} J^\alpha f = a i_a \left(\int_0^{xa^{-1}} (xa^{-1} - t)^{\alpha-1} f(t)\, dt \right) = \varphi(x). \tag{4.25}$$

On the other hand,

$$\Gamma(\alpha) J^\alpha i_a U_a^{-1} f = \Gamma(\alpha) J^\alpha i_a f\left(\frac{x}{a}\right) = \varphi(x). \tag{4.26}$$

Comparing (4.25) and (4.26) one obtains $a_{ji}^\alpha M_{ij} J^\alpha = J^\alpha M_{ij}$, that is, M_{ij} of the form (4.24) is a solution of the equation (4.23). Because of $C_{ij} \in \{J^\alpha\}'$, the same is true for $K_{ij} = M_{ij} C_{ij}$.

Next, let K_{ij} be an arbitrary bounded solution of the relation (4.23). We observe that

$$(K_{ij} f)(x) = 0, \qquad x \in [0, 1-a],$$

and consequently $\Re(K_{ij}) \supseteq \chi_{[1-a,1]} L_p[0,1] = E_{1-a}^p$. This fact can be deduced from (4.5) and (4.6) in the same way as the relation (4.19), but it immediately follows from (4.19) because of

$$a_{ij} (J^\alpha)^* K_{ij}^* = K_{ij}^* (J^\alpha)^*.$$

Using Douglas' Lemma we can deduce from the inclusion $\Re(K_{ij}) \supseteq E_{1-a}^p = \Re(M_{ij})$ that $K_{ij} = M_{ij} C_{ij}$ with some $C_{ij} \in [X]$. One can conclude now using (4.23) that $M_{ij}(C_{ij} J^\alpha - J^\alpha C_{ij}) = 0$. Since $\ker M_{ij} = \{0\}$, the last identity is equivalent to the relation $C_{ij} J^\alpha - J^\alpha C_{ij} = 0$, that is, $C_{ij} \in \{J^\alpha\}'$ and consequently C_{ij} is of the form (4.22). □

Remark 4.7. It follows from (4.14) and (4.15) that there is no quasiaffinity T (i.e., an operator T such that $\ker T = \{0\}$ and $\overline{\Re(T)} = H$) intertwining the operators J^α and $sJ^\alpha$with $s \in R \setminus \{1\}$. Indeed, if $sJ^\alpha T = TJ^\alpha$ and $s < 1$ then $\ker T \supseteq E_s^p = \chi_{[s,1]} L_p[0,1]$), otherwise if $sJ^\alpha T = TJ^\alpha$ and $s > 1$ then $\Re(T) \subset E_{1-s}^p = \chi_{[1-s,1]} L_p[0,1]$). In particular, the operators J^α and sJ^α are not quasisimilar in the sence of B. Nagy and C. Foias [SF]. This assertion was obtained earlier in [M1].

Proposition 4.8. *Let* $X = L_p[0,1]$, $1 \le p < \infty$, $\Re(\alpha) > 0$, *and let* B *satisfy the conditions of Proposition* 4.6. *Then the lattice* Hyplat A *of the operator* $A = J^\alpha \otimes B \in [X^n]$ *has the form*

$$\operatorname{Hyplat} A = \{\oplus_{j=1}^n E_{a_j}^p : (a_1, \ldots, a_n) \in P(s_1, \ldots, s_n)\}, \tag{4.27}$$

where

$$P(s_1,\dots,s_n)=\{\,(a_1,\dots,a_n)\in\square_n:\\ s_ja_{j+1}\le s_{j+1}a_j\le s_{j+1}-s_j+s_ja_{j+1},\ 1\le j\le n-1\}. \tag{4.28}$$

Therefore the lattice Hyplat A *is anti-isomorphic to the oblique parallelepiped* $P(s_1,\dots,s_n)$. *The dimension* $\dim P(s_1,\dots,s_n)$ *of this parallelepiped is equal to the number of different* s_j, $1\le j\le n$.

Proof. Let $\lambda_j\in\mathbb{C}$ and $\arg\lambda_j\ne\arg\lambda_i$ for $i\ne j$. Since $\oplus_{j=1}^n\lambda_jJ^\alpha\in\{A\}'$, it follows from (4.9) that

$$\text{Hyplat}\,A\subset\text{Lat}\,(\oplus_{j=1}^n\lambda_jJ^\alpha)=\{\oplus_{j=1}^nE_{a_j}^p:0\le a_j\le 1\}.$$

The following implications can be easily derived from (4.14) and (4.15):

$$f_j(x)\in E_{a_j}^p\quad\Rightarrow\quad K_{ij}f_j\in E^p_{a_js_is_j^{-1}},\qquad i>j\ge 1,$$
$$f_j(x)\in E_{a_j}^p\quad\Rightarrow\quad K_{ij}f_j\in E^p_{1-s_j^{-1}s_i(1-a_j)},\qquad i+1\le j\le n.$$

To complete the proof it suffices to observe that the system of inequalities

$$s_ja_k\le s_ka_j\le s_k-s_j+s_ja_k,\qquad 1\le j\le n-1,\ j+1\le k\le n,$$

is equivalent to the reduced one described in (4.28). □

Corollary 4.9. *Let* $X=L_p[0,1]$, $\Re(\alpha)>0$, *and let* $A=J^\alpha\otimes I\in[X^n]$. *Then* Hyplat A *is of the form*

$$\text{Hyplat}\,(J^\alpha\otimes I)=\{\oplus_{j=1}^nE_{a_j}^p:a_1=a_2=\dots=a_n\in[0,1]\}, \tag{4.29}$$

and therefore Hyplat A *is anti-isomorphic to the segment* $[0,1]$.

4.3. Now we are ready to state the main results of this section. Combining Proposition 4.1 and Proposition 4.6 we obtain the first of them.

Theorem 4.10. *Let* $X=L_p[0,1]$, $\alpha>0$, *and let* $B=\oplus_{p=1}^rB_m$ *with* $B_m=\text{diag}\,(\lambda_{m1},\lambda_{m2},\dots,\lambda_{mn_m})$ *being a nonsingular diagonal matrix with eigenvalues* λ_{mk}, $1\le k\le n_m$, *of equal arguments:*

$$\lambda_{mk}=\lambda_{m1}s_{mk}^\alpha,\qquad 1=s_{m1}\le s_{m2}\le\dots\le s_{mn_m},$$

and such that $\arg\lambda_{m1}\ne\arg\lambda_{k1}$ *for* $m\ne k$. *Then*

$$\{J^\alpha\otimes B\}'=\oplus_{m=1}^r\{J^\alpha\otimes B_m\}', \tag{4.30}$$

that is, $T\in\{J^\alpha\otimes B\}'$ *if and only if*

$$T=\text{diag}\,(T_{11},T_{22},\dots,T_{rr}), \tag{4.31}$$

where $T_{mm} = (K_{mij})_{i,j=1}^{n_m} \in \{J^\alpha \otimes B_m\}'$, $1 \leq m \leq r$, *is an* $n_m \times n_m$ *operator-matrix with entries* K_{mij} *as described in Proposition* 4.6.

Combining Proposition 4.4, Proposition 4.8 and Theorem 4.10 we obtain the final result of this section, namely the description of the hyperinvariant subspaces of an operator $J^\alpha \otimes B$ for an arbitrary nonsingular diagonizable matrix B.

Theorem 4.11. *Let* $X = L_p[0,1]$, $1 \leq p < \infty$, $\alpha > 0$, *and let* B *satisfy the conditions of Theorem* 4.10 *and* $n = n_1 + \cdots + n_r$. *Then the lattice* Hyplat A *of the operator* $A = J^\alpha \otimes B \in [X^n]$ *is of the form*

$$\begin{aligned} \text{Hyplat}\,(J^\alpha \otimes B) &= \oplus_{m=1}^r \text{Hyplat}\,(J^\alpha \otimes B_m) \\ &= \oplus_{m=1}^r \{\oplus_{j=1}^{n_m} E_{a_{mj}}^p : (a_{m1}, .., a_{mn_m}) \in P(s_{m1}, .., s_{mn_m})\} \end{aligned} \tag{4.32}$$

where $P(s_{m1}, \ldots, s_{mn_m})$ *is the parallelepiped defined by the relation* (4.28).

Acknowledgements

This paper was partially supported by the ISSEP under grant No. APU 051060 and by the INTAS Project 93-0249.

References

[B1] Brodskii, M.S.: On unicellularity of real Volterra operators; Dokl. Acad. Nauk. 147:5 (1962), 1010–1012 (in Russian).

[B2] ______: Triangular and Jordan Representations on Linear operators; Transl. Math. Monographs 32, Amer. Math. Soc., Providence RI 1971.

[CW] Conway, G.B., Wu, P.Y.: The splitting of $A(T_1 \oplus T_2)$ and related questions; Indiana Univ. Math. J. 26:1 (1977), 41–56.

[D] Dixmier, J.: Les opérateurs permutables à l'opérateur intégral; Portugal Math. 8 (1949), 73–84.

[Dj] Djarbashan, M.M.: Integral Transformations and Functions Representations in Complex Domain; Nauka, Moscow 1966 (in Russian).

[FR] Frankfurt, R., Rovnyak, J.: Recent results and unsolved problems of finite convolution operators; in: Linear Spaces and Approximation, Butzer, P.L., Sz.-Nagy, B. (eds.), Internat. Ser. Numer. Math. 40 (1978), 133–150.

[GK] Gohberg, I.C., Krein, M.G.: Theory and Applications of Volterra operators in Hilbert space; Transl. Math. Monographs 24, Amer. Math. Soc., Providence RI 1970.

[K] Kalish, G.K.: Characterization of direct sums and commuting sets of Volterra operators; Pacific J. Math. 18:3 (1966), 545–552.

[KA] Kantorovich, L.V., Akilov, G.P.: Functional Analysis; Nauka, Moscow 1977 (in Russian).

[L] LEVIN, B.JA.: Distribution of Zeros of Entire Functions; Transl. Math. Monographs 5, Amer. Math. Soc., Providence RI 1964.

[M1] MALAMUD, M.M.: Similarity of Volterra operators and related questions of the theory of differential equations of fractional order; Trans. Moscow Math. Soc. 55 (1994), 57–122.

[M2] ———: The connection between a potential matrix of a Dirac system and its Wronskian; Dokl. Acad. Nauk. 344:5 (1995), 601–604 (in Russian).

[M3] ———: Inverse problems for some systems of ordinary differential equations; Uspekhi Mat. Nauk. 50:4 (1995), 145–146 (in Russian).

[M4] ———: On cyclic subspaces of Volterra operators; Dokl. Acad. Nauk. 349:3 (1996), 454–458 (in Russian).

[N1] NIKOLSKII, N.K.: Treatise on the Shift Operator; Springer Verlag, Berlin 1986.

[N2] ———: Multicyclicity phenomenon. I. An introduction and maxi-formulas; Operator Theory: Adv. Appl. 42 (1989), 9–57.

[OS1] OSILENKER, B.P., SHUL'MAN, V.S.: On lattices of invariant subspaces of some operators; Funktsional Anal. i Prilozhen. 17:1 (1983), 81–82 (in Russian).

[OS2] ———: On lattices of invariant subspaces of some operators; in: Investigations on functions theory of several real variables, Yaroslavl' 1984, 105–113 (in Russian).

[PW] PALEY, R., WIENER, N.: Fourier transforms in the complex domain; Amer. Math. Soc., New York 1934.

[S1] SARASON, D.: A remark on the Volterra operator; J. Math. Anal. Appl. 12 (1965), 244–246.

[S2] ———: Generalized interpolation in H^∞; Trans. Amer. Math. Soc. 127 (1967), 179–203.

[SF] SZ.-NAGY, B., FOIAS, C.: Harmonic Analysis of Operators on Hilbert Space; Akademiai Kiado, Budapest 1970.

Department of Mathematics
Donetsk State University
Universitetskaya str. 24
Donetsk 340055
Ukraine
mmm@univ.donetsk.ua

1991 Mathematics Subject Classification: Primary 2T47

Submitted: May 21, 1996

Operator Theory:
Advances and Applications, Vol. 102

Some interior and exterior boundary-value problems for the Helmholtz equation in a quadrant

E. MEISTER, F. PENZEL, F.-O. SPECK and F.S. TEIXEIRA

Boundary-transmission problems for two-dimensional Helmholtz equations in a quadrant Q_1 and its complement Q_1^c, respectively, are considered in a Sobolev space setting. The first problem of a quadrant with Dirichlet condition on one face and transmission condition on the other is solved in closed form for the case where all quadrants are occupied by the same medium. Unique sovlability can also be shown in the case of two different media up to exceptional cases of wave numbers, while Fredholm property holds in general. In the second problem transmission conditions are prescribed on both faces. These problems are reduced to equivalent integral equations in $L_2^+(\mathbb{R})$ of Wiener-Hopf-Hankel type which can be solved explicitly by obtaining canonical generalized factorizations of certain non-rational 2×2-matrix-valued symbols.

1. The Dirichlet problem D_{Q_1}

Find the solution $u \in H_1(Q_1)$ of

$$(1.1) \qquad (\Delta + k^2)u = 0 \quad \text{in} \quad Q_1$$

with the boundary conditions

$$(1.2) \qquad T_{0,\Gamma_1}u := u|_{\Gamma_1} = f_1, \qquad T_{0,\Gamma_2}u := u|_{\Gamma_2} = f_2$$

for some given $(f_1, f_2) \in H_{1/2}(\partial Q_1)$, i.e.,

$$(1.3) \qquad f_1, f_2 \in H_{1/2}(\mathbb{R}^+) \quad \text{such that} \quad f_1 - f_2 \in \widetilde{H}_{1/2}(\mathbb{R}_+).$$

Here always $k \in \mathbb{C}_{++}$ is assumed.

Theorem 1.1. *The problem D_{Q_1} is uniquely solvable by*

$$(1.4) \qquad u(x_1, x_2) = \mathcal{F}^{-1}_{\xi \to x_1}\left\{\widehat{\ell^e f_1}(\xi)e^{-t(\xi)x_2}\right\} + \mathcal{F}^{-1}_{\eta \to x_2}\left\{\widehat{\ell^o \widetilde{f}_2}(\eta)e^{-t(\eta)x_1}\right\}$$

for $(x_1, x_2) \in Q_1$ with

$$(1.5) \qquad \widetilde{f}_2(x_2) := f_2(x_2) - T_{0,\Gamma_2}\mathcal{F}^{-1}_{\xi \to x_1}\left\{\widehat{\ell^e f_1}(\xi)e^{-t(\xi)x_2}\right\}$$

and

$$(1.6) \qquad t(\xi) := (\xi^2 - k^2)^{1/2} \quad \textit{such that} \quad \Re(t(\xi)) \geq 0$$

for $\xi \in \mathbb{R}$.

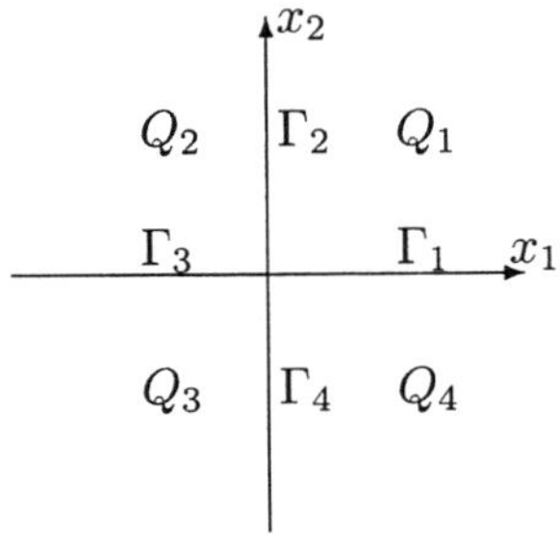

Figure 1: Quadrants

Sketch of proof. Find a solution $v \in H_1(\mathbb{R}^2_+)$ of (1.1) with trace

(1.7) $$T_0 v := v|_{x_2=0} = \ell^e f_1 \in H_{1/2}(\mathbb{R}).$$

This problem is uniquely solvable (cf. [4, 8]) by

(1.8) $$v(x_1,x_2) = \mathcal{F}^{-1}_{\xi \to x_1} \left\{ \widehat{\ell^e f_1}(\xi) e^{-t(\xi)x_2} \right\}$$

for $x_2 > 0$ having as trace $T_{0,\Gamma_2} v \in H_{1/2}(\mathbb{R}_+)$ such that

(1.9) $$f_2 - T_{0,\Gamma_2} v \in \widetilde{H}_{1/2}(\mathbb{R}_+).$$

By use of the decomposition

(1.10) $$u = w + v|_{Q_1}$$

the Dirichlet problem D_{Q_1} is reduced to

(1.11) $$(\Delta + k^2) w = 0 \quad \text{in} \quad Q_1$$

with the boundary conditions

(1.12) $$T_{0,\Gamma_1} w = 0 \quad \text{and} \quad T_{0,\Gamma_2} w = \widetilde{f}_2 \in \widetilde{H}_{1/2}(\mathbb{R}_+),$$

i.e., $\ell^o \widetilde{f}_2 \in H_{1/2}(\mathbb{R})$ such that

(1.13) $$w \equiv w_{\text{odd}}(x_1,x_2) := \mathcal{F}^{-1}_{\eta \to x_2} \left\{ \widehat{\ell^o \widetilde{f}_2}(\eta)\, e^{-t(\eta)x_1} \right\}$$

is the unique solution to the Helmholtz equation in $\mathbb{R}^2_r$, i.e., $x_1 > 0$, with 0-trace on Γ_1 since w is odd with respect to x_2. □

Theorem 1.2. *The solution $u \in H_1(Q_1)$ to the problem D_{Q_1} is also represented by the symmetric formula*

(1.14) $$u(x_1,x_2) = \mathcal{F}^{-1}_{\xi \to x_1} \left\{ \widehat{\ell^o f_1}(\xi) e^{-t(\xi)x_2} \right\} + \mathcal{F}^{-1}_{\eta \to x_2} \left\{ \widehat{\ell^o f_2}(\eta) e^{-t(\eta)x_1} \right\}$$

for $(x_1,x_2) \in Q_1$.

Remark 1.3. Each separate term does not represent bounded operators from $H_{1/2}(\mathbb{R}_+)$ into $H_1(Q_1)$ but the sum does!

$$w_1(x_1,x_2) := \mathcal{F}^{-1}_{\xi\to x_1}\left\{\widehat{\ell^o f_1}(\xi)e^{-t(\xi)x_2}\right\} \in H_{1-\epsilon}(\mathbb{R}^2_+) \tag{1.15}$$

for $x_2 > 0$ and

$$w_2(x_1,x_2) := \mathcal{F}^{-1}_{\eta\to x_2}\left\{\widehat{\ell^o f_2}(\eta)e^{-t(\eta)x_1}\right\} \in H_{1-\epsilon}(\mathbb{R}^2_r) \tag{1.16}$$

for $x_1 > 0$ and any $\epsilon \in (0,1)$ by taking

$$f_1, f_2 \in H_{1/2-\epsilon}(\mathbb{R}_+) \quad \text{such that} \quad \ell^o f_1, \ell^o f_2 \in H_{1/2-\epsilon}(\mathbb{R})$$

by estimating C^∞-functions $s \in \mathcal{S}$ with the corresponding Sobolev norm. Alternatively it can be shown with help of Bessel potential operators on Lipschitz domains (cf. [6, 1, 2]) directly that $w_1 + w_2|_{Q_1} \in H_{1-\epsilon}(Q_1)$.

The Neumann problem N_{Q_1} with the boundary conditions

$$T_{1,\Gamma_1}u := -\frac{\partial}{\partial x_2}u|_{\Gamma_1} = g_1, \qquad T_{1,\Gamma_2}u := -\frac{\partial}{\partial x_1}u|_{\Gamma_2} = g_2 \tag{1.17}$$

where $(g_1, g_2) \in H_{-1/2}(\partial Q_1)$, i.e.,

$$g_1, g_2 \in H_{-1/2}(\mathbb{R}_+) \quad \text{such that} \quad g_1 - g_2 \in \widetilde{H}_{-1/2}(\mathbb{R}_+)$$

is uniquely solvable by the symmetric formula

$$u(x_1,x_2) = \mathcal{F}^{-1}_{\xi\to x_1}\left\{\widehat{\ell^e g_1}(\xi)t^{-1}(\xi)e^{-t(\xi)x_1}\right\} + \mathcal{F}^{-1}_{\eta\to x_2}\left\{\widehat{\ell^e g_2}(\eta)t^{-1}(\eta)e^{-t(\eta)x_2}\right\} \tag{1.18}$$

for $(x_1, x_2) \in Q_1$.

2. The mixed problem M_{Q_1}

Find the solution $u \in H_1(Q_1)$ of

$$(\Delta + k^2)u = 0 \quad \text{in} \quad Q_1 \tag{2.1}$$

with the boundary conditions

$$T_{0,\Gamma_1}u = f_1, \qquad T_{1,\Gamma_2}u = g_2 \tag{2.2}$$

for given $f_1 \in H_{1/2}(\mathbb{R}_+)$, $g_2 \in H_{-1/2}(\mathbb{R}_+)$.

Theorem 2.1. *The problem M_{Q_1} is uniquely solvable by*

$$u(x_1,x_2) = \mathcal{F}^{-1}_{\xi\to x_1}\left\{\widehat{\ell^e f_1}(\xi)\, e^{-t(\xi)x_2}\right\} + \mathcal{F}^{-1}_{\eta\to x_2}\left\{\widehat{\ell^o g_2}(\eta)\, t^{-1}(\eta)\, e^{-t(\eta)x_1}\right\} \tag{2.3}$$

for $(x_1, x_2) \in Q_1$.

Remark 2.2. The first term solves the Dirichlet problem (D) in $\mathbb{R}^2_+$. The second term solves the Neumann problem (N) in $\mathbb{R}^2_r$ with corresponding even and odd data, respectively!

Now with the potential operators

$$\begin{aligned} K_{D,Q_j} &: H_{1/2}(\partial Q_j) \to H_1(Q_j), \\ K_{N,Q_j} &: H_{-1/2}(\partial Q_j) \to H_1(Q_j), \\ K_{M,Q_j} &: H_{1/2}(\Gamma_j) \times H_{-1/2}(\Gamma_{j+1}) \to H_1(Q_j; \Delta; k) \end{aligned}$$

for $j = 1, \ldots, 4$ and j mod 4 given by the afore mentioned interior boundary value problem, we can solve the exterior quadrant problems for $H_1(\mathbb{R}^2 \setminus \overline{Q}_3)$ e.g. by reduction to an equivalent system of boundary pseudodifferential equations first and then to a scalar Wiener-Hopf-Hankel integral equation.

3. The complement problem $D_{Q_3^c}$

Let $\Omega := \mathbb{R}^2 \setminus \overline{Q}_3 = Q_3^c$. Find the solution $u \in H_1(\Omega)$ of

$$(\Delta + k^2)u = 0 \quad \text{in} \quad \Omega \tag{3.1}$$

with the boundary conditions

$$T_{0,\Gamma_3}u = f_3, \qquad T_{0,\Gamma_4}u = f_4 \tag{3.2}$$

such that the compatibility condition

$$f_3 - f_4 \in \widetilde{H}_{1/2}(\mathbb{R}_-) \tag{3.3}$$

holds. Then $u_j := u|_{Q_j}$, $j = 1, 2, 4$, are the solutions in $H_1(Q_j)$ whose traces on Γ_1, Γ_2 and those of their normal derivatives fulfill homogeneous transmission conditions:

$$\begin{aligned} T_{0,\Gamma_1}(u_1 - u_4) &= 0, \\ T_{1,\Gamma_1}(u_1 + u_4) &= 0, \\ T_{0,\Gamma_2}(u_1 - u_2) &= 0, \\ T_{1,\Gamma_2}(u_1 + u_2) &= 0. \end{aligned} \tag{3.4}$$

If conversely u_j, $j = 1, 2, 4$, are the solutions of Helmholtz equations with transmission conditions (3.4) and boundary conditions

$$T_{0,\Gamma_3}u_2 = f_3, \qquad T_{0,\Gamma_4}u_4 = f_4, \tag{3.5}$$

then

$$u := \begin{cases} u_1 & \text{in} \quad Q_1, \\ u_2 & \text{in} \quad Q_2, \\ u_4 & \text{in} \quad Q_4, \end{cases} \tag{3.6}$$

is a solution of $D_{Q_3^c}$.

The three functions u_j have representation formulae like (1.14) in Theorem 1.2. The remaining transmission conditions (3.4) lead to two boundary integral equations for f_1, f_2:

$$\begin{aligned}
&\int_{\mathbb{R}} e^{-i\xi x_1}\, t(\xi)\, \widehat{\ell^o f_1}(\xi)\, d\xi + i \int_{\mathbb{R}} \xi\, \widehat{\ell^o f_2}(\xi)\, e^{-t(\xi)x_1}\, d\xi \\
&= \int_{\mathbb{R}} e^{-i\xi x_1}\, t(\xi)\, \widehat{\ell^o f_3}(\xi)\, d\xi + i \int_{\mathbb{R}} \xi\, \widehat{\ell^o f_4}(\xi)\, e^{-t(\xi)x_1}\, d\xi
\end{aligned} \tag{3.7}$$

for $x_1 > 0$ and

$$\begin{aligned}
&i \int_{\mathbb{R}} \xi \widehat{\ell^o f_1}(\xi)\, e^{-t(\xi)x_2}\, d\xi + \int_{\mathbb{R}} e^{-i\xi x_2}\, t(\xi)\, \widehat{\ell^o f_2}\, d\xi \\
&= i \int_{\mathbb{R}} \xi \widehat{\ell^o f_3}(\xi)\, e^{-t(\xi)x_2}\, d\xi + \int_{\mathbb{R}} e^{-i\xi x_2}\, t(\xi)\, \widehat{\ell^o f_4}\, d\xi
\end{aligned} \tag{3.8}$$

for $x_2 > 0$ with the invertible convolution operator

$$A := \mathcal{F}^{-1} t^{-1} \mathcal{F} : H_{-1/2}(\mathbb{R}) \to H_{1/2}(\mathbb{R}) \tag{3.9}$$

and the "operator around the corner"

$$C : H_{1/2}(\mathbb{R}) \to H_{-1/2}(\mathbb{R}_+) \tag{3.10}$$

given by

$$Cf(x) := \frac{i}{2\pi} \int_{\mathbb{R}} \xi \hat{f}(\xi)\, e^{-t(\xi)x}\, d\xi \tag{3.11}$$

for $x > 0$.

Due to the compatibility conditions

$$f_1 - f_{4(-)} \in \widetilde{H}_{1/2}(\mathbb{R}_+), \tag{3.12}$$

$$f_2 - f_{3(-)} \in \widetilde{H}_{1/2}(\mathbb{R}_+) \tag{3.13}$$

with $f_{j(-)}(x) := f_j(-x)$, the equations (3.7) and (3.8) then give

$$r_+ A^{-1} \ell^o f_1 + C \ell^o f_2 = -r_+ A^{-1} \ell^o f_3 + C \ell^o f_4, \tag{3.14}$$

$$C \ell^o f_1 + r_+ A^{-1} \ell^o f_2 = C \ell^o f_3 - r_+ A^{-1} \ell^o f_4 \tag{3.15}$$

with continuous operators from $H_{1/2}(\partial Q_1)$ into $H_{1/2}(\mathbb{R}_+)$ as composition of T_{1,Γ_2} with potential operator K_{D,Q_1}. The right-hand side operator in (3.14) is continuous from $H_{1/2}(\partial Q_4)$ into $H_{-1/2}(\mathbb{R}_+)$ and analogous in equation (3.15).

Without any restrictive assumptions

$$\varphi_1 := f_1 - f_{4(-)}, \tag{3.16}$$

$$\varphi_2 := f_2 - f_{3(-)} \tag{3.17}$$

lead to a (2×2)-system of pseudodifferential equations:

$$\begin{pmatrix} 2r_+A^{-1}\ell^o & C\ell^o \\ C\ell^o & 2r_+A^{-1}\ell^o \end{pmatrix} \begin{pmatrix} \varphi_1 \\ \varphi_2 \end{pmatrix} = \begin{pmatrix} h_1 \\ h_2 \end{pmatrix}. \tag{3.18}$$

Applying the odd extension ℓ^o and denoting

$$\ell^o r_+ A^{-1}\ell^o = A^{-1}\ell^o, \tag{3.19}$$

from the second equation it is obtained

$$\ell^o\varphi_2 = \frac{1}{2}A\ell^o(h_2 - C\ell^o\varphi_1). \tag{3.20}$$

Introducing this in the first equation and by restriction it follows

$$2r_+A^{-1}\ell^o\varphi_1 - \frac{1}{2}CA\ell^oC\ell^o\varphi_1 = h_1 - \frac{1}{2}CA\ell^o h_2 \tag{3.21}$$

where in $H_{-1/2}(\mathbb{R}_+)$

$$h_1 := (r_+A^{-1}\ell^o f_4 + C\ell^o f_4) + (r_+A^{-1}\ell^o f_3 + C\ell^o f_3), \tag{3.22}$$

$$h_2 := (r_+A^{-1}\ell^o f_3 + C\ell^o f_3) + (r_+A^{-1}\ell^o f_4 + C\ell^o f_4) \tag{3.23}$$

with $J\varphi(x) := \varphi(-x)$ for $x \in \mathbb{R}$.

Theorem 3.1. *The operator* $CA\ell^oC\ell^o : \widetilde{H}_{1/2}(\mathbb{R}_+) \to H_{-1/2}(\mathbb{R}_+)$ *is a product of the Hankel operator*

$$\mathcal{H} := -2r_+A^{-1}J|_{\widetilde{H}^+_{1/2}(\mathbb{R})} \to H_{-1/2}(\mathbb{R}^+) \tag{3.24}$$

with the 0*-extension operator* ℓ_o.

Sketch of proof. Let $\varphi \in C_o^\infty(\overline{\mathbb{R}_+})$ and use $\mathcal{F}\ell^o = (I - J)\mathcal{F}\ell_o$. Then

$$\begin{aligned} (\mathcal{F}\ell^oC\ell^o\varphi)(\xi) &= -\frac{1}{\pi}\int_{\mathbb{R}} \eta\,\xi\,(\xi^2 + t^1(\eta))^{-1}\widehat{\ell^o\varphi}\,d\eta \\ &= -2i\xi \cdot \widehat{\ell_o\varphi}(it(\xi)) \end{aligned} \tag{3.25}$$

due to the oddness of last kernel function and residue calculus with respect to η.

$$\begin{aligned} (CA\ell^oC\ell^o\varphi)(x) &= -\frac{1}{\pi}\int_{\mathbb{R}} e^{iyx}\,t(y)\,\widehat{\ell_o\varphi}(y)\,dy \\ &= -2(r_+A^{-1}J\ell_o\varphi)(x) \\ &= (\mathcal{H}\ell_o\varphi)(x) \end{aligned} \tag{3.26}$$

for $x > 0$. Thus equation (3.21) is reduced to a scalar equation of Wiener-Hopf-Hankel type

$$2r_+ A^{-1}\ell_o\varphi_1 - r_+ A^{-1}J\ell_o\varphi_1 = h_1 - \frac{1}{2}CA\ell^o h_2 \tag{3.27}$$

(cf. [9]). The last equation is equivalent to a Hankel integral equation in $L_2^+(\mathbb{R})$. Using the invertible Bessel potential operators

$$\begin{aligned} A_+ &:= \mathcal{F}^{-1}\, t_+^{-1}\, \mathcal{F} : L_2(\mathbb{R}) \to H_{1/2}(\mathbb{R}), \\ A_- &:= \mathcal{F}^{-1}\, t_-^{-1}\, \mathcal{F} : H_{-1/2}(\mathbb{R}) \to L_2(\mathbb{R}), \end{aligned} \tag{3.28}$$

where

$$t_\pm(\xi) := (\xi \pm k)^{1/2}, \tag{3.29}$$

equation (3.27) is equivalent to

$$2A^{-1}\ell_o\varphi_1 - A^{-1}J\ell_o\varphi_1 = \ell^o(h_1 - \frac{1}{2}CA\ell^o h_2) + \varphi^- \in H^-_{-1/2}(\mathbb{R}) \tag{3.30}$$

with

$$\ell_o\varphi_1 = A_+\Psi^+, \quad \Psi^+ \in L_2^+(\mathbb{R}). \tag{3.31}$$

Then the Hankel equation

$$\mathcal{K}_2\Psi^+ := 2\Psi^+ - \mathcal{P}^+\mathcal{K}J\Psi^+ = f^+ \tag{3.32}$$

is obtained where

$$\mathcal{K} := \mathcal{F}^{-1}\,(-it_+t_-^{-1})\,\mathcal{F} \tag{3.33}$$

and

$$f^+ := \mathcal{P}^+A_-\ell^o(h_1 - \frac{1}{2}CA\ell^o h_2). \tag{3.34}$$

The operator $\mathcal{K}_2 := 2I - P^+\mathcal{K}J$ can be inverted explicitly (cf. [9, 5]). Thus there is a unique solution

$$\varphi_1 = r_+A_+\mathcal{K}_2^{-1}f^+, \tag{3.35}$$

$$\varphi_2 = \frac{1}{2}r_+A\ell^o(h_2 - C\ell^o\varphi_1). \tag{3.36}$$

□

Theorem 3.2. *The problem* $D_{Q_3^c}$ *is uniquely solvable for any* $(f_3, f_4) \in H_{1/2}(\partial\Omega)$ *given by* (3.6) *with* (1.14) *in Theorem* 1.2. *with*

$$f_1 := r_+A_+\mathcal{K}_2^{-1}f^+ + f_{4(-)}, \tag{3.37}$$

$$f_2 := \frac{1}{2}A\ell^o(h_2 - C\ell^o r_+A_+\mathcal{K}_2^{-1}f^+). \tag{3.38}$$

Remark 3.3. The exterior Neumann boundary value problem $N_{Q_3^c}$ can be solved in a similar way using K_{N,Q_j}!

The exterior mixed boundary value problem $M_{Q_3^c}$ is as follows:

Find the solution $u \in H_1(\Omega)$, $\Omega := \mathbb{R}^2 \setminus \overline{Q}_3 = Q_3^c$, of the Helmholtz equation with the (mixed) boundary conditions

$$T_{0,\Gamma_3} u = f_3, \qquad T_{1,\Gamma_4} u = g_4 \tag{3.39}$$

where $(f_3, g_4) \in H_{1/2}(\mathbb{R}_-) \times H_{-1/2}(\mathbb{R}_-)$.

Theorem 3.4. ("Representation formula") *The function*

$$u := \begin{cases} u_1 & in \quad Q_1, \\ u_2 & in \quad Q_2, \\ u_4 & in \quad Q_4, \end{cases} \tag{3.40}$$

is a solution to $M_{Q_3^c} = DN_{Q_3^c}$ *iff*

$$u_1(x_1, x_2) = \mathcal{F}^{-1}_{\xi \to x_1} \left\{ \widehat{\ell^o f_1}(\xi)\, e^{-t(\xi)x_2} \right\} + \mathcal{F}^{-1}_{\eta \to x_2} \left\{ \widehat{\ell^o g_2}(\eta)\, t^{-1}(\eta)\, e^{-t(\eta)x_1} \right\} \tag{3.41}$$

for $(x_1, x_2) \in Q_1$,

$$u_2(x_1, x_2) = \mathcal{F}^{-1}_{\xi \to x_1} \left\{ \widehat{\ell^e f_3}(\xi)\, e^{-t(\xi)x_2} \right\} - \mathcal{F}^{-1}_{\eta \to x_2} \left\{ \widehat{\ell^o g_2}(\eta)\, t^{-1}(\eta)\, e^{+t(\eta)x_1} \right\} \tag{3.42}$$

for $(x_1, x_2) \in Q_2$ *and*

$$u_4(x_1, x_2) = \mathcal{F}^{-1}_{\xi \to x_1} \left\{ \widehat{\ell^e f_1}(\xi)\, e^{t(\xi)x_2} \right\} + \mathcal{F}^{-1}_{\eta \to x_2} \left\{ \widehat{\ell^o g_4}(\eta)\, t^{-1}(\eta)\, e^{-t(\eta)x_1} \right\} \tag{3.43}$$

for $(x_1, x_2) \in Q_4$.

The homogeneous transmission conditions

$$T_{0,\Gamma_2}(u_1 - u_2) = 0, \tag{3.44}$$

$$T_{1,\Gamma_1}(u_1 + u_4) = 0 \tag{3.45}$$

lead to a (2×2)-system of boundary pseudodifferential equations for the unknowns f_1 and g_2:

$$\begin{pmatrix} C_0 \ell^o & 2r_+ A \ell^e \\ 2r_+ A^{-1} \ell^e & C_1 \ell^o \end{pmatrix} \begin{pmatrix} f_1 \\ g_2 \end{pmatrix} = \begin{pmatrix} C_0 \ell^e f_3 \\ C - 1 \ell^o g_4 \end{pmatrix} \tag{3.46}$$

with the "operators around the corner"

$$C_0 f(x) = \frac{1}{2\pi} \int_{\mathbb{R}} \hat{f}(\xi)\, e^{-t(\xi)x}\, d\xi, \tag{3.47}$$

$$C_1 g(x) = \frac{i}{2\pi} \int_{\mathbb{R}} \xi\, t^{-1}(\xi)\, \hat{g}(\xi)\, e^{-t(\xi)x}\, d\xi \tag{3.48}$$

for $x > 0$ where

$$\begin{array}{rcl} C_0 & : & H_{1/2}(\mathbb{R}) \quad \rightarrow \quad H_{1/2}(\mathbb{R}_+), \\ C_1 & : & H_{-1/2}(\mathbb{R}) \quad \rightarrow \quad H_{-1/2}(\mathbb{R}_+). \end{array}$$

Theorem 3.5. *The operator*

$$C_1 A^{-1} \ell^o C_0 \ell^e = -\mathcal{H} \ell_o \tag{3.49}$$

maps $\widetilde{H}_{1/2}(\mathbb{R}_+)$ *into* $H_{-1/2}(\mathbb{R}_+)$.

Sketch of proof. The relation $\ell^o + \ell^e = 2\ell_o$ and residue calculus lead to

$$(\mathcal{F}\ell^o C_0 \ell^e \varphi)(\xi) = 2i\xi\, t^{-1}(\xi)\, \widehat{\ell_o \varphi}(it(\xi)) \tag{3.50}$$

and to the scalar Wiener-Hopf-Hankel equation in $\widetilde{H}^+_{1/2}(\mathbb{R})$,

$$2r_+ A^{-1} \ell_o \varphi + r_+ A^{-1} J \ell_o \varphi = C_1 \ell^o g_4 - 2r_+ A^{-1} \ell^e f_3 \tag{3.51}$$

with

$$\mathcal{K}_{-2} \Psi^+ := -2\Psi^+ - \mathcal{P}^+ \mathcal{K} J \Psi^+ = f^+, \tag{3.52}$$

where

$$\Psi^+ := A_+^{-1} \ell_o \varphi_1 \in L_2^+(\mathbb{R}) \tag{3.53}$$

and

$$f^+ := \mathcal{P}^+ A_- \ell^o (2r_+ A^{-1} \ell^e f_3 - C_1 \ell^o g_4). \tag{3.54}$$

Herein the operator $\mathcal{K}_{-2} := -2I - \mathcal{P}^+ \mathcal{K} J$ is invertible (cf. [9]). □

Theorem 3.6. *The problem* $M_{Q_3^c}$ *is uniquely solvable for any* $(f_3, g_4) \in H_{1/2}(\mathbb{R}_-) \times H_{-1/2}(\mathbb{R}_-)$ *given by* (3.41)–(3.43) *with*

$$f_1 := r_+ A_+ \mathcal{K}_{-2}^{-1} f^+ + f_{3(-)}, \tag{3.55}$$

$$g_2 := -\frac{1}{2} r_+ A^{-1} \ell^o C_0 \ell^e \varphi_1 \tag{3.56}$$

with

$$\varphi_1 := r_+ A_+ \mathcal{K}_{-2}^{-1} f^+ \tag{3.57}$$

and f^+ *given by* (3.54).

Acknowledgements

The authors want to thank the DFG for a research grant under grant number KO 634/32-1 and also the JNICT (Portugal), grant number 87422/MATM.

References

[1] DUDUČAVA, R. SPECK, F.-O.: Bessel Potential Operators for the Quarter-Plane; Sem. Anal. und Anw., Bericht Nr. 25, Mathematisches Institut A, Universität Stuttgart 1990.

[2] DUDUČAVA, R.: Wiener-Hopf Equations with the Transmission Property; Integral Equations Operator Theory 15 (1992), 412–426.

[3] DUDUČAVA, R., SPECK, F.-O.: Pseudodifferential Operators on Compact Manifolds with Lipschitz Boundary; Math. Nachr. 160 (1993), 149–191.

[4] MEISTER, E., SPECK, F.-O.: Modern Wiener-Hopf Methods in Diffraction Theory; in: Ordinary and Partial Differential Equations, Vol. 2, Proceedings of a Conference in Dundee, Eds. B.D. Sleeman and R.J. Jarvis, Research Notes in Mathematics (1989), 130–171.

[5] MEISTER, E., SPECK, F.-O., TEIXEIRA, F. S.: Wiener-Hopf-Hankel Operators for Some Wedge Diffraction Problems with Mixed Boundary Conditions; J. Integral Equations Appl. 4:2 (1992), 229–255.

[6] SCHNEIDER, R.: Reduction of Order for Pseudodifferential Operators on Lipschitz Domains; Preprint No. 1253, Technische Hochschule Darmstadt 1989.

[7] SCHNEIDER, R.: Bessel Potential Operators for Canonical Lipschitz Domains; Math. Nachr. 150 (1991), 277–299.

[8] SPECK, F.-O.: Mixed Boundary Value Problems of the Type of Sommerfeld's Half-plane Problem; Proc. Roy. Soc. Edinburgh Sect. A 104 (1986), 261–277.

[9] TEIXEIRA, F. S.: Diffraction by a Rectangular Wedge: Wiener-Hopf-Hankel formulation; Integral Equations Operator Theory 14 (1991), 436–455.

Fachbereich Mathematik, AG 12
TH Darmstadt
Schloßgartenstr. 7
64289 Darmstadt
Germany
meister@mathematik.th-darmstadt.de

Departamento de Matematica
Instituto Superior Tecnico
Av. Rovisco Pais
1096 Lisbon
Portugal

1991 Mathematics Subject Classification: Primary 78A45; Secondary 45E10, 45F15

Submitted: April 29, 1996

Operator Theory:
Advances and Applications, Vol. 102

Interpolation of some function spaces and indefinite Sturm-Liouville problems

S.G. PYATKOV

We consider self-adjoint Sturm-Liouville problems of the form $Lu = \lambda g(x)u$, where L is an ordinary differential operator of order $2m$, defined on the interval $(0,1)$, and g is a real-valued function assuming both positive and negative values. For our problem, we prove under some assumptions that the eigenvectors and associated vectors constitute a Riesz basis in the space L_2 with the weight $|g|$. To study the problem, we consider the question of interpolation of some Sobolev spaces with weight.

1. Introduction

We consider the problem

$$Lu = \lambda g(x)u, \quad x \in (0,1), \tag{1.1}$$

where L is an ordinary differential operator of order $2m$ which is defined by the differential expression

$$Lu = \sum_{i,j=1}^{m} \frac{d^i}{dx^i} a_{ij} \frac{d^j u}{dx^j}, \qquad x \in (0,1), \tag{1.2}$$

and the boundary conditions

$$B_k u = \sum_{i=0}^{2m-1} \left(\alpha_{ik} u^{(i)}(0) + \beta_{ik} u^{(i)}(1) \right) = 0, \qquad k = \overline{1,m}. \tag{1.3}$$

The operator L is assumed to be self-adjont in $L_2(0,1)$.

Sturm-Liouville problems with an indefinite weight function (and elliptic eigenvalue problems of this kind) have been the subject of many investigations. These problems arise in many areas of engineering, physics, and applied mathematics. To begin with, we should mention the early works of Hilbert [13], who proved the existence of infinitely many positive and negative eigenvalues for the case in which $m = 1$ and L is a positive ordinary differential operator (i.e., $n = 1$) and considered the corresponding eigenfunctions expansion. Similar questions were considered also in [10, 11, 12, 20]. The first results in the multidimensional case are due to Holmgren [14]. He considered the Dirichlet problem $\Delta u + \lambda g(x)u = 0$, $x \in G \subset \mathbb{R}^n$, when g is a continuous function assuming both positive and negative values. In this case he proved the existence of an infinite number of positive and negative eigenvalues which can be characterized by a "min-max principle". The asymptotic distribution of these eigenvalues has been established by Pleijel [15]. Then his results have been generalized by many mathematicians (see, for example, [2, 3, 7, 8]). The main problem under consideration here is whether the root functions constitute an unconditional basis in the space $L_{2,g}((0,1)\setminus G^0)$, where $G^0=\{x\in(0,1):g(x)=0\}$, with the norm defined by the equality

$$\|u\|^2_{L_{2,g}((0,1)\setminus G^0)}=\int_{(0,1)\setminus G^0}|g||u|^2\,dx.$$

The first results devoted to this problem appeared only recently (see [1, 6, 4, 5, 16, 17, 18, 19, 22, 23]). The main assumption used in these articles is the condition that the function g behaves like a power of $|x-x_0|$ in some neighborhood of the "turning point" x_0 (at this point g changes its sign). The exceptions are the article [22] and the book [23]. In this article, we do not use this condition on the function g. To study the problem, we exploit some properties of the Sobolev spaces with weight. We consider the question of interpolation of some weighted Sobolev spaces and apply the results presented in [19]. Almost all notations are conventional (see [21]).

2. Interpolation of some Sobolev spaces

By $W_2^m(a,b)$, $a,b\in\mathbb{R}$, we mean the Sobolev space with the norm

$$\|u\|^2_{W_2^m(a,b)}=\int_a^b|u^{(m)}|^2+|u|^2\,dx,$$

where $u^{(m)}=\frac{d^mu}{dx^m}$ is the generalized derivative of the function $u(x)$. Let $\overset{\circ}{W}{}_2^m(a,b)$ denote the subspace of $W_2^m(a,b)$ consisting of the functions whose derivatives of order less than m and the functions themselves vanish at $x=a$ and $x=b$. The space $\overset{\circ}{W}{}_2^m(a,b)$ is equipped with the same norm as the space $W_2^m(0,1)$. Assume that ω is a continuous and positive function on (a,b) and $\omega\in L_1(a,b)$. We denote by $L_{2,\omega}(a,b)$ the space of measurable functions $u(x)$ such that

$$\|u\|^2_{L_{2,\omega}(a,b)}=\int_a^b\omega|u|^2\,dx\ <\infty.$$

If A and B are Banach spaces then by $(A,B)_{\theta,2}$ we mean the space obtained by the real interpolation method (see [21]). The symbol $L(A,B)$ denotes the space of linear bounded operators from A into B. If $A=B$ then we write $L(A)$ rather than $L(A,A)$. Assign

$$H^s(a,b)=(W_2^m(a,b),L_{2,\omega}(a,b))_{1-\frac{s}{m},2},\quad \overset{\circ}{H}{}^s(a,b)=(\overset{\circ}{W}{}_2^m(a,b),L_{2,\omega}(a,b))_{1-\frac{s}{m},2}.$$

The main question that we consider in this section is the question of coincidence of the spaces $H^s(a,b)$ and $\overset{\circ}{H}{}^s(a,b)$.

Two positive functions g and f will be referred to as *equivalent* ($f \sim g$) if for some constant $M > 0$ we have

$$\frac{1}{M}g(x) \le f(x) \le Mg(x), \qquad x \in (a,b).$$

If $g \sim \omega$ then the spaces $L_{2,g}(a,b)$ and $L_{2,\omega}(a,b)$ coincide. If g is a positive continuous function on $[c,d] \subset (a,b)$, then we put

$$\operatorname*{osc}_{[c,d]} g = \max_{x\in[c,d]} g(x) / \min_{x\in[c,d]} g(x).$$

Given ω, there exists a positive continuous function ρ with the following properties:

$$\begin{aligned} &\rho(a) = \rho(b) = 0, \\ &|\rho(x) - \rho(y)| \le |x-y|, \quad \operatorname*{osc}_{[x-\rho(x),x+\rho(x)]} \omega \le M, \quad x,y \in [a,b], \end{aligned} \tag{2.1}$$

where $M > 1$ is a fixed constant. In particular, for a given constant $M > 1$, we can put

$$\rho(x) = \rho_0(x) = \max\{\rho : \operatorname*{osc}_{[x-\rho,x+\rho]} \omega \le M,\ (x-\rho, x+\rho) \subset (a,b)\}, \quad x \in (a,b).$$

It is easy to find that this function ρ_0 meets (2.1). We assume the following conditions to be fulfilled:

(A) There exists a constant $\delta > 0$ ($\delta < b - a$) such that

$$\sup_{x\in(a,a+\delta)} \frac{\int_a^x \omega(\xi)\rho^{2k}(\xi)\,d\xi}{\omega(x)\rho^{2k+1}(x)} < \infty, \qquad k = \overline{0, m-1}. \tag{2.2}$$

(B) There exists a constant $\delta > 0$ ($\delta < b - a$) such that

$$\sup_{x\in(b-\delta,b)} \frac{\int_x^b \omega(\xi)\rho^{2k}(\xi)\,d\xi}{\omega(x)\rho^{2k+1}(x)} < \infty, \qquad k = \overline{0, m-1}. \tag{2.3}$$

Below, in Lemmas 2.5 and 2.6 we shall present sufficient conditions which ensure that the conditions (A), (B) are fulfilled.

We now obtain some auxiliary statements. Without loss of generality, we can assume that $|\rho'| \le q < 1$ almost everywhere on (a,b); otherwise, we consider the function $q\rho(x)$ rather than $\rho(x)$. Let $k \in \mathbb{Z}$ (the set of all integers). Take $x_0 \in (a,b)$ and put $I_0 = (x_0 - \rho(x_0), x_0 + \rho(x_0))$. We can determine the points x_k from the equations $x_k = x_{k+1} - \rho(x_{k+1})$, $k = -1, -2, \dots$, $x_{k+1} = x_k + \rho(x_k)$,

$k = 0, 1, \ldots$. Then we put $\rho_k = \rho(x_k)$, $a_k = x_k - \rho_k$, $b_k = x_k + \rho_k$, $I_k = (a_k, b_k)$. We have $\cup_{k=-\infty}^{\infty} I_k = (a, b)$. Moreover, there exists a constant $c_0 = c_0(q)$ such that $\rho_{k-1}/c_0 \leq \rho_k \leq c_0\rho_{k-1}$. As usually, we can construct a partition of unity $\{\psi_k\}$ on (a, b) subject to the covering $\{I_k\}$ and such that $\operatorname{supp}\psi_k \cap \operatorname{supp}\psi_{k+2} = \emptyset$ and $\psi_k^{(l)} \leq c/(\rho_k)^l$, $k \in \mathbb{Z}$, $x \in (a, b)$. There exists a system of polynomials $\{p_k\}_{k=0}^{m-1}$ with the following properties:

$$\int_0^1 p_k(x)\overline{p_s(x)}\, dx = \delta_{ks}, \qquad \deg p_k = k, \tag{2.4}$$

where δ_{ks} is the Kronecker symbol. Assign

$$S_k(u)(x) = \sum_{k=0}^{m-1} \frac{1}{2\rho_k} \int_{a_k}^{b_k} u(\tau) p_k\left(\frac{\tau - a_k}{2\rho_k}\right) d\tau \; p_k\left(\frac{x - a_k}{2\rho_k}\right). \tag{2.5}$$

We can introduce the operator

$$S(u) = \sum_{k=-\infty}^{\infty} S_k(u)\psi_k.$$

The definitions of the operators S_k and S imply that $S_k(p) = p$ and $S(p) = p$ for each polynomial p whose degree is less than m. Using the system $\{\psi_k\}$, we can construct functions $\widetilde{\omega}, \widetilde{\rho} \in C^\infty(a, b)$ such that $\widetilde{\omega} \sim \omega$, $\widetilde{\rho} \sim \rho$, and

$$|\widetilde{\omega}^{(l)}(x)| \leq c\omega(x)/\rho^l(x), \qquad |\widetilde{\rho}^{(l)}(x)| \leq c\rho^{1-l}, \quad x \in (a, b), \quad l = 1, 2, \ldots,$$

where c is some positive constant.

Lemma 2.1. *If conditions* (A), (B) *hold, then there exist constants* θ_0 *and* $M_0 > 0$ *such that, for all* $\theta \in (\theta_0, 1]$ *and some constant* $\delta < b - a$, *we have*

$$\int_a^b \rho^{2m(\theta-1)}\omega^\theta(\xi)\, d\xi < \infty,$$

and

$$\frac{\int_a^x \rho^{2m(\theta-1)+2k}\omega^\theta(\xi)\, d\xi}{\rho^{2m(\theta-1)+2k+1}(x)\omega^\theta(x)} < M_0, \qquad x \in (a, a+\delta], \tag{2.6}$$

$$\frac{\int_x^b \rho^{2m(\theta-1)+2k}\omega^\theta(\xi)\, d\xi}{\rho^{2m(\theta-1)+2k+1}(x)\omega^\theta(x)} < M_0, \qquad x \in [b-\delta, b), \tag{2.7}$$

where $k = 0, 1, \ldots, m-1$.

Proof. The proofs of (2.6) and (2.7) are similar. Show, for example, inequality (2.6). We take a sequence of functions $\{\varphi_n\}$ with the following properties:

$\varphi_n \in C^\infty([a,b])$, $\varphi_n(x) = 1$ for $x \in [a+1/n, b]$, $\varphi_n(x) = 0$ for $x \in [a, a+1/2n]$, $\varphi_n(x), \varphi_n'(x) \geq 0$ for all $x \in [a,b]$, and $\varphi_n(x) \leq \varphi_{n+1}(x)$ for all $x \in [a,b]$ and $n = 1, 2, \ldots$.

Fix some constant $\delta < b - a$. For $x \in (a, a+\delta]$, we have

$$\begin{aligned}
&\int_a^x \widetilde{\rho}^{2m(\theta-1)}(\xi)\widetilde{\omega}^\theta(\xi)\varphi_n(\xi)\,d\xi \\
&= \int_a^x \widetilde{\rho}^{2m(\theta-1)}\widetilde{\omega}^{\theta-1}(\xi)\varphi_n(\xi)\frac{d}{d\xi}\int_a^\xi \widetilde{\omega}(\eta)\,d\eta\,d\xi \\
&\leq \varphi_n(x)\int_a^x \widetilde{\omega}(\eta)\,d\eta\,\widetilde{\rho}^{2m(\theta-1)}\widetilde{\omega}^{\theta-1}(x) \\
&\quad - \int_a^x (\theta-1)[\widetilde{\rho}^{2m(\theta-1)}\widetilde{\omega}^{\theta-2}\widetilde{\omega}' + \widetilde{\omega}^{\theta-1}\widetilde{\rho}^{2m(\theta-1)-1}\widetilde{\rho}' 2m]\int_a^\xi \widetilde{\omega}(\eta)\,d\eta\,\varphi_n(\xi)\,d\xi.
\end{aligned}$$

Using condition (A), we obtain

$$\begin{aligned}
\int_a^x \widetilde{\rho}^{2m(\theta-1)}(\xi)\widetilde{\omega}^\theta(\xi)\varphi_n(\xi)\,d\xi \;\leq\;& \varphi_n(x)M\widetilde{\rho}^{2m(\theta-1)+1}(x)\widetilde{\omega}^\theta(x) \\
&+ M(\theta-1)\int_a^x \widetilde{\rho}^{2m(\theta-1)}(\xi)\widetilde{\omega}^\theta(\xi)\varphi_n(\xi)\,d\xi,
\end{aligned}$$

where M is some constant. If $M(\theta-1) \leq 1/2$, then

$$\int_a^x \widetilde{\rho}^{2m(\theta-1)}(\xi)\widetilde{\omega}^\theta(\xi)\varphi_n(\xi)\,d\xi \leq \varphi_n(x)2M\widetilde{\rho}^{2m(\theta-1)+1}(x)\widetilde{\omega}^\theta(x).$$

Levi's Theorem implies that

$$\int_a^\delta \widetilde{\rho}^{2m(\theta-1)}(\xi)\widetilde{\omega}^\theta(\xi)\,d\xi < \infty$$

and

$$\int_a^x \widetilde{\rho}^{2m(\theta-1)}(\xi)\widetilde{\omega}^\theta(\xi)\,d\xi \leq 2M\widetilde{\rho}^{2m(\theta-1)+1}(x)\widetilde{\omega}^\theta(x), \qquad x \in (a, a+\delta].$$

Thus, we have proved the first assertion of the lemma and (2.6) for $k = 0$. To prove (2.6) for $k > 0$ and (2.7), we use similar arguments. □

The next lemma is an analog of the Hardy inequality. We assume here that the conditions (A), (B) hold and $\theta \in (\theta_0, 1]$.

Lemma 2.2. *For all $v \in C_0^\infty(a,b)$ and some $\varepsilon > 0$, the following estimate holds:*

$$\begin{aligned}
\int_a^b \rho^{2m(\theta-1)}(x)\omega^\theta(x)|v|^2(x)\,dx \;\leq\;& \int_a^b \rho^{2m(\theta-1)+2m}(x)\omega^\theta(x)|v^{(m)}|^2(x)\,dx \\
&+ \int_{a+\varepsilon}^{b-\varepsilon} \omega^\theta(x)\sum_{k=0}^{m-1}|v^{(k)}|^2(x)\,dx.
\end{aligned} \tag{2.8}$$

Proof. First, we put $\varepsilon = \delta/2$ ($\delta < b-a$) and find a function ψ with the properties $\psi \in C^\infty([a,b])$, $\psi(x) = 1$ for $x \in [a, a+\varepsilon]$, and $\psi(x) = 0$ for $x \in [a+2\varepsilon, b]$. Next, we consider the integral

$$\int_a^{a+2\varepsilon} |v^{(k)}|^2(x)\rho^{2m(\theta-1)+2k}(x)\omega^\theta(x)\psi(x)\,dx =: I.$$

Integrating by parts, we obtain

$$\begin{aligned} I &= -\int_a^{a+2\varepsilon}\int_a^x \rho^{2m(\theta-1)+2k}(\xi)\omega^\theta(\xi)\,d\xi\ \psi(x)[v^{(k)}\overline{v^{(k+1)}} + v^{(k+1)}\overline{v^{(k)}}](x)\,dx \\ &\quad -\int_a^{a+2\varepsilon}\int_a^x \rho^{2m(\theta-1)+2k}(\xi)\omega^\theta(\xi)\,d\xi\ \psi'(x)|v^{(k)}|^2(x)\,dx. \end{aligned}$$

Conditions (A), (B) and this inequality yield the estimate

$$\begin{aligned} I &\le c_1\int_a^{a+2\varepsilon} \rho^{2m(\theta-1)+2k+2}(x)\omega^\theta(x)\psi(x)|v^{(k+1)}|^2(x)\,dx \\ &\quad + c_2\int_{a+\varepsilon}^{a+2\varepsilon} \omega^\theta(x)|v^{(k)}|^2(x)\,dx, \end{aligned} \tag{2.9}$$

where the constants c_1 and c_2 are independent of v. Repeating the arguments in the neighborhood $[b-\delta, b)$ of the point b and using the inequalities obtained for $k = 0, 1, 2, \ldots$, we obtain the assertion of the lemma. □

Lemma 2.3. *For all $u \in W_2^m(a,b)$ and $\theta \in (\theta_0, 1]$, the following inequality holds:*

$$\begin{aligned} \int_a^b \rho^{2m(\theta-1)}(x)\omega^\theta(x)|u|^2\,dx &\le c\int_a^b \omega^\theta(x)|u|^2(x)\,dx \\ &\quad + c\sum_{k=-\infty}^{\infty} \rho_k^{2m(\theta-1)}\omega_k^\theta\|u - S_k(u)\|^2_{L_2(a_k,b_k)}, \end{aligned}$$

where $\omega_k = \omega(x_k)$ and c is some positive constant.

Proof. We use Lemma 2.2, where we take $v = S(u)$ ($u \in C_0^\infty(a,b)$). The definition of the operator S implies the inequality

$$\int_{a+\varepsilon}^{b-\varepsilon} \omega^\theta(x)\sum_{k=0}^{m-1}|S(u)^{(k)}|^2(x)\,dx \le c\int_a^b |u|^2(x)\omega^\theta(x)\,dx, \tag{2.10}$$

where the constant c is independent of u. We have that

$$S^{(m)}(u) = \sum_{k=-\infty}^{\infty}\sum_{l=0}^{m-1} c_l^m S_k^{(l)}(u)\psi_k^{(m-l)},$$

where c_l^m are binomial coefficients. Consider the expression

$$S_k^{(l)}(u)(x)\psi_k^{(m-l)}(x) + S_{k+1}^{(l)}(u)(x)\psi_{k+1}^{(m-l)}(x) = J_k$$

for $x \in \operatorname{supp}\psi_k \cap \operatorname{supp}\psi_{k+1}$. Since $\psi_k + \psi_{k+1} = 1$ for these points x, we have

$$J_k = \left[S_{k+1}^{(l)}(u) - S_k^{(l)}(u)\right]\psi_{k+1}^{(m-l)}(x) = S_{k+1}^{(l)}(u - S_k(u))(x)\psi_{k+1}^{(m-l)}(x).$$

Therefore, taking account of the definition of the operator S_k, we obtain

$$\|J_k\|_{L_2(a_k,b_k)} \leq \frac{c}{\rho_k^m}\|u - S_k(u)\|_{L_2(a_k,b_k)}.$$

This inequality implies

$$\begin{aligned}
&\int_a^b \rho^{2m(\theta-1)+2m}(x)\omega^\theta(x)|S^{(m)}(u)|^2(x)\,dx \\
&\leq c_0 \sum_{k=-\infty}^{\infty} \rho_k^{2m(\theta-1)+2m}\omega_k^\theta\|S^{(m)}(u)\|^2_{L_2(a_k,b_k)} \\
&\leq c_1 \sum_{k=-\infty}^{\infty} \rho_k^{2m(\theta-1)+2m}\omega_k^\theta\left[\|J_k\|^2_{L_2(a_k,b_k)} + \|J_{k-1}\|^2_{L_2(a_k,b_k)}\right] \\
&\leq c_2 \sum_{k=-\infty}^{\infty} \rho_k^{2m(\theta-1)}\omega_k^\theta\|u - S_k(u)\|^2_{L_2(a_k,b_k)},
\end{aligned}$$

where c_i, $i = 0, 1, 2$ are some constants. The last inequality and (2.10) yield the estimate

$$\begin{aligned}
\int_a^b \rho^{2m(\theta-1)}(x)\omega^\theta(x)|S(u)|^2(x)\,dx \;\leq\;& c_3\int_a^b \omega^\theta(x)|u|^2(x)\,dx \\
&+ c_3 \sum_{k=-\infty}^{\infty} \rho_k^{2m(\theta-1)}\omega_k^\theta\|u - S_k(u)\|^2_{L_2(a_k,b_k)}
\end{aligned}$$

for all $u \in C_0^\infty(a,b)$. This inequality and the representation

$$u = u - S(u) + S(u) = \sum_{k=-\infty}^{\infty} [u - S_k(u)]\psi_k + S(u)$$

imply the inequality

$$\begin{aligned}
\int_a^b \rho^{2m(\theta-1)}(x)\omega^\theta(x)|u|^2(x)\,dx \;\leq\;& c\int_a^b \omega^\theta(x)|u|^2(x)\,dx \\
&+ c \sum_{k=-\infty}^{\infty} \rho_k^{2m(\theta-1)}\omega_k^\theta\|u - S_k(u)\|^2_{L_2(a_k,b_k)}
\end{aligned}$$

for all $u \in C_0^\infty(a,b)$. Next, we take $u \in W_2^m(a,b)$. There exists a sequence of functions $u_n \in C^\infty([a,b])$ such that $\|u-u_n\|_{W_2^m(a,b)} \to 0$ as $n \to \infty$. In particular, $\|u-u_n\|_{C([a,b])} \to 0$ as $n \to \infty$. Pick up a function $\psi_\gamma \in C_0^\infty(a,b)$ $(\gamma < (b-a)/2)$ with the properties: $\psi_\gamma(x) = 1$ for $x \in [a+\gamma, b-\gamma]$ and $0 \le \psi_\gamma \le 1$. We consider the function $u_\gamma^n = \psi_\gamma u_n \in C_0^\infty(a.b)$. Taking some sequence $\varepsilon_m \to 0$ as $m \to \infty$ and choosing suitable $\gamma = \gamma_m$ and $n = n(m)$, we can construct a sequence of functions u_m such that

$$\int_a^b \rho^{2m(\theta-1)}(x)\omega^\theta(x)|u_m|^2(x)\,dx \to \int_a^b \rho^{2m(\theta-1)}(x)\omega^\theta(x)|u|^2(x)\,dx$$

as $m \to \infty$ and

$$\sum_{k=-\infty}^{\infty} \rho_k^{2m(\theta-1)}\omega_k^\theta \|u_m - S_k(u_m)\|^2_{L_2(a_k,b_k)} \to \sum_{k=-\infty}^{\infty} \rho_k^{2m(\theta-1)}\omega_k^\theta \|u - S_k(u)\|^2_{L_2(a_k,b_k)}$$

as $m \to \infty$. This fact proves the lemma. □

Theorem 2.4. *Let the conditions* (A), (B) *be fulfilled. Then there exists* $s_0 > 0$ *such that* $H^s(a,b) = \overset{\circ}{H}{}^s(a,b)$ *for all* $s \in [0, s_0)$.

Proof. To simplify the proof, we assume in the proof of the theorem that $s \in [0,1/2)$ and $\theta = 1-s/m \in (\theta_0, 1]$. We use the K-method. Introduce the functionals

$$K_i^2(t,u) = \inf_{v \in H_i} \left(\|v\|^2_{H_i} + t^2 \|\sqrt{\omega}(v-u)\|^2_{L_2(a,b)} \right), \qquad i = 1,2,$$

where $H_1 = W_2^m(a,b)$ and $H_2 = \overset{\circ}{W}{}_2^m(a,b)$. Then $H^s(a,b)$ consists of the functions $u \in L_{2,\omega}(a,b)$ such that

$$\|u\|^2_{H^s(a,b)} = \int_0^\infty K_1^2(t,u) t^{-2\theta} \frac{dt}{t} < \infty$$

Similarly, we can define the space $\overset{\circ}{H}{}^s(a,b)$. For $u \in C_0^\infty(a,b)$, we estimate the functional $K_2(t,u)$ from above. We consider the problem

$$(2.11)\quad (-1)^m w^{(2m)} + (2\rho_k)^{2m}(1+t^2\omega_k)w = (2\rho_k)^{2m} t^2 \omega_k u_k(\tau)\psi_k(2\rho_k\tau + a_k),$$

$$(2.12)\quad w^{(l)}(0) = w^{(l)}(1) = 0, \quad l = 0,1,\dots,m-1, \qquad \tau \in (0,1),$$

where $u_k(\tau) = u(2\rho_k\tau + a_k)$. Let $\{\varphi_k\}_{k=0}^\infty$ be the eigenfunctions of the operator $Lu = (-1)^m u^{(2m)}$ whose domain is the set of functions $u \in W_2^{2m}(0,1)$ satisfying (2.12). By $(\cdot,\cdot)$ we mean the inner product in $L_2(0,1)$, i.e.,

$$(u,v) = \int_0^1 u(x)\overline{v(x)}\,dx.$$

Assume that $(\varphi_i, \varphi_j) = \delta_{ij}$ and let $\{\lambda_k\}$ be the nondecreasing sequence of eigenvalues. A solution v_k to the problem (2.11), (2.12) can be written in the form

$$v_k(\tau) = \sum_{i=0}^{\infty} \frac{(u_k\psi_k(2\rho_k\tau + a_k), \varphi_i)\varphi_i(\tau)(2\rho_k)^{2m}t^2\omega_k}{\lambda_i + (2\rho_k)^{2m}(1 + t^2\omega_k)}.$$

Extend the functions $v_k(\frac{x-a_k}{2\rho_k})$ by zero on the whole interval (a, b) and assign

$$w(x) = \sum_{k=-\infty}^{\infty} v_k\left(\frac{x - a_k}{2\rho_k}\right).$$

Obviously, $w \in \overset{\circ}{W}{}_2^m(a, b)$. We have that

$$\begin{aligned} K_2^2(t,u) &\leq \|w\|^2_{W_2^m(a,b)} + t^2\|\sqrt{\omega}(w-u)\|^2_{L_2(a,b)} \\ &\leq c_0 \sum_{k=-\infty}^{\infty} \|v_k\|^2_{W_2^m(a_k,b_k)} + t^2\omega_k\|(v_k - \psi_k u)\|^2_{L_2(a_k,b_k)} \\ &= \sum_{k=-\infty}^{\infty} J_k(v_k), \end{aligned}$$

where c_0 is a positive constant. The last expression can be written in the form

$$J_k(v_k) = (2\rho_k)^{1-2m} \int_0^1 \left(|v_k^{(m)}|^2 + (2\rho_k)^{2m}|v_k|^2 + t^2(2\rho_k)^{2m}\omega_k|v_k - u_k\psi_k|^2\right) d\tau.$$

The definition of the function v_k yields that

$$J_k(v_k) = (2\rho_k)^{1-2m} \sum_{i=0}^{\infty} \frac{|(u_k\psi_k, \varphi_i)|^2(2\rho_k)^{2m}t^2\omega_k(\lambda_i + (2\rho_k)^{2m})}{(\lambda_i + (2\rho_k)^{2m}(1 + t^2\omega_k))}.$$

Using the fact that the domain of the operator $L^{(1-\theta)/2}$ is the space $W_2^{m(1-\theta)}$ (see [21]), we obtain

$$\begin{aligned} &\int_0^{\infty} t^{-2\theta} K_2^2(t,u)\,\frac{dt}{t} \\ &\leq c(\theta) \sum_{k=-\infty}^{\infty} \sum_{i=0}^{\infty} \frac{|(u_k\psi_k, \varphi_i)|^2(2\rho_k)^{2m(\theta-1)+1}\omega_k^{\theta}}{(\lambda_i + (2\rho_k)^{2m})^{\theta-1}} \\ &\leq c_1 \sum_{k=-\infty}^{\infty} \omega_k^{\theta}(2\rho_k)^{2m(\theta-1)+1} \sum_{i=0}^{\infty} |(u_k\psi_k(2\rho_k\tau + a_k), \varphi_i)|^2(\lambda_i)^{1-\theta} \\ &\leq c_2 \sum_{k=-\infty}^{\infty} \omega_k^{\theta}(2\rho_k)^{2m(\theta-1)+1}\|u_k\psi_k(2\rho_k\tau + a_k)\|^2_{W_2^{m(1-\theta)}(0,1)}, \end{aligned} \tag{2.13}$$

where c, c_1, and c_2 are positive constants. The norm in the space $W_2^{m(1-\theta)}(0,1)$ can be defined by the equality

$$\|v\|^2_{W_2^{m(1-\theta)}(0,1)} = \int_0^1 \int_0^1 \frac{|v(x)-v(y)|^2}{|x-y|^{1+2m(1-\theta)}}\,dx\,dy + \int_0^1 |v(x)|^2\,dx.$$

This equality, inequality (2.13), and the properties of the functions ψ_k imply that

$$\begin{aligned}
&\int_0^\infty t^{-2\theta} K_2^2(t,u)\,\frac{dt}{t}\\
&\le c_3 \sum_{k=-\infty}^{\infty} \left(\omega_k^\theta \int_{a_k}^{b_k}\int_{a_k}^{b_k} \frac{|u(x)-u(y)|^2}{|x-y|^{1+2m(1-\theta)}}\,dx\,dy + \omega_k^\theta (2\rho_k)^{2m(\theta-1)} \int_{a_k}^{b_k} |u(x)|^2\,dx \right)\\
&\le c_4 \sum_{k=-\infty}^{\infty} \omega_k^\theta \int_{a_k}^{b_k}\int_{a_k}^{b_k} \frac{|u(x)-u(y)|^2}{|x-y|^{1+2m(1-\theta)}}\,dx\,dy + \int_a^b \omega^\theta(x)\rho^{2m(\theta-1)}|u(x)|^2\,dx\\
&< \infty.
\end{aligned} \tag{2.14}$$

As in the proof of Lemma 3, we can establish that inequality (2.14) holds for all $u \in W_2^m(a,b)$. Therefore, $W_2^m(a,b) \subset \overset{\circ}{H}{}^s(a,b)$.
Next, we estimate the functional $K_1^2(t,u)$ from below. For $u \in W_2^m(a,b)$, we have the inequality

$$\begin{aligned}
K_1^2(t,u) &\ge c_0 \inf_{v\in H_1} \sum_{k=-\infty}^{\infty} \|v\|^2_{W_2^m(a_k,b_k)} + t^2 \|\sqrt{\omega}(v-u)\|^2_{L_2(a_k,b_k)}\\
&\ge c_1 \inf_{v_k \in W_2^m(a_k,b_k)} \sum_{k=-\infty}^{\infty} \|v_k\|^2_{W_2^m(a_k,b_k)} + t^2\omega_k \|(v_k-u)\|^2_{L_2(a_k,b_k)}\\
&= c_1 \sum_{k=-\infty}^{\infty} \inf_{v_k\in W_2^m(a_k,b_k)} J_k(v_k),
\end{aligned}$$

where c_0 and c_1 are some positive constants. Put $u_k(\tau) = u(2\rho_k\tau + a_k)$, $\tau = \frac{x-a_k}{2\rho_k}$. Then the expression $J_k(v_k)$ takes the form

$$J_k(v_k) = (2\rho_k)^{1-2m} \int_0^1 \left(|v_k^{(m)}|^2 + (2\rho_k)^{2m}|v_k|^2 + t^2(2\rho_k)^{2m}\omega_k |v_k - u_k|^2 \right) d\tau.$$

The definition of the functional J_k yields that

$$\inf_{v_k\in W_2^m(0,1)} J_k(v_k) = J_k(w),$$

where w is a solution to the problem

$$(-1)^m w^{(2m)} + (2\rho_k)^{2m}(1+t^2\omega_k)w = (2\rho_k)^{2m} t^2 \omega_k u_k, \tag{2.15}$$

$$w^{(l)}(0) = w^{(l)}(1) = 0, \qquad l = m, m+1, \dots, (2m-1). \tag{2.16}$$

Let $\{\varphi_k\}_{k=0}^{\infty}$ be the eigenfunctions of the operator $Lu = (-1)^m u^{(2m)}$ whose domain is the set of functions $u \in W_2^{2m}(0,1)$ satisfying (2.16). Assume that $(\varphi_i, \varphi_j) = \delta_{ij}$ and let $\{\lambda_k\}$ be the nondecreasing sequence of eigenvalues. The first m eigenfunctions $\varphi_0, \varphi_1, \dots, \varphi_{m-1}$ are the polynomials $p_0, p_1, \dots, p_{m-1}$. The solution to the problem (2.15), (2.16) can be written in the form

$$w(x) = \sum_{i=0}^{\infty} \frac{(u_k, \varphi_i)(2\rho_k)^{2m} t^2 \omega_k}{\lambda_i + (2\rho_k)^{2m}(1 + t^2\omega_k)}.$$

Hence, we obtain the equality

$$J_k(w) = (2\rho_k)^{1-2m} \sum_{i=0}^{\infty} \frac{|(u_k, \varphi_i)|^2 (2\rho_k)^{2m} t^2 \omega_k (\lambda_i + (2\rho_k)^{2m})}{(\lambda_i + (2\rho_k)^{2m}(1 + t^2\omega_k))}.$$

Note that this series converges whenever $u \in W_2^m(a,b)$. With the use of Levi's theorem and the fact that the domain of the operator $L^{(1-\theta)/2}$ is the space $W_2^{m(1-\theta)}$, we obtain

$$\begin{aligned}
\infty &> \int_0^{\infty} t^{-2\theta} K_1^2(t,u)\,\frac{dt}{t} \\
&\geq c(\theta) \sum_{k=-\infty}^{\infty} \sum_{i=0}^{\infty} |(u_k, \varphi_i)|^2 (2\rho_k)^{2m(\theta-1)+1} \omega_k^{\theta} / (\lambda_i + (2\rho_k)^{2m})^{\theta-1} \\
&\geq \delta_1 \sum_{k=-\infty}^{\infty} \omega_k^{\theta} (2\rho_k)^{2m(\theta-1)+1} \sum_{i=0}^{\infty} |(u_k - S_k(u)(2\rho_k\tau + a_k), \varphi_i)|^2 (\lambda_i)^{1-\theta} \\
&\quad + \delta_2 \sum_{k=-\infty}^{\infty} \rho_k \omega_k^{\theta} \sum_{i=0}^{\infty} |(u_k, \varphi_i)|^2 \\
&\geq \delta_3 \sum_{k=-\infty}^{\infty} \omega_k^{\theta} (2\rho_k)^{2m(\theta-1)+1} \|u_k - S_k(u)(2\rho_k\tau + a_k)\|^2_{W_2^{m(1-\theta)}(0,1)} \\
&\quad + \delta_4 \sum_{k=-\infty}^{\infty} \rho_k \omega_k^{\theta} \|u_k\|^2_{L_2(0,1)},
\end{aligned}$$

where δ_i are some positive constants. From this inequality we infer

$$\begin{aligned}
&\int_0^{\infty} t^{-2\theta} K_1^2(t,u)\,\frac{dt}{t} \\
&\geq \delta_5 \sum_{k=-\infty}^{\infty} \Big(\omega_k^{\theta} \rho_k^{2m(\theta-1)} \|u_k - S_k(u)\|^2_{L_2(a_k,b_k)} \\
&\quad + \omega_k^{\theta} \int_{a_k}^{b_k} \int_{a_k}^{b_k} \frac{|u(x) - S_k(u)(x) - u(y) + S_k(u)(y)|^2}{|x-y|^{1+2m(1-\theta)}}\,dx\,dy \Big) \\
&\quad + \delta_6 \int_a^b \omega^{\theta}(x) |u(x)|^2\,dx,
\end{aligned} \tag{2.17}$$

where δ_i are some positive constants. It is easy to find that

$$\int_{a_k}^{b_k}\int_{a_k}^{b_k}\frac{|S_k(u)(x)-S_k(u)(y)|^2}{|x-y|^{1+2m(1-\theta)}}\,dx\,dy \le c\rho_k^{2m(\theta-1)}\|S_k(u)\|^2_{L_2(a_k,b_k)}. \tag{2.18}$$

Lemma 2.3 together with inequalities (2.17) and (2.18) yield the estimate

$$\begin{aligned}&\int_0^\infty t^{-2\theta}K_1^2(t,u)\,\frac{dt}{t}\\ &\ge \delta\sum_{k=-\infty}^{\infty}\omega_k^\theta\int_{a_k}^{b_k}\int_{a_k}^{b_k}\frac{|u(x)-u(y)|^2}{|x-y|^{1+2m(1-\theta)}}\,dx\,dy+\int_a^b\omega^\theta(x)\rho^{2m(\theta-1)}|u(x)|^2\,dx,\end{aligned} \tag{2.19}$$

where the constant $\delta>0$ is independent of $u\in W_2^m(a,b)$. Inequalities (2.14) and (2.19) complete the proof of the theorem. □

Next, we present some examples and some sufficient conditions which ensure that the conditions (A), (B) are fulfilled. Put

$$\rho_1(x)=\frac{\int_a^x\omega(\xi)\,d\xi}{\omega(x)},\qquad \rho_2(x)=\frac{\int_a^x\omega^{1/(2m-1)}(\xi)\,d\xi}{\omega^{1/(2m-1)}(x)},$$

$$\rho_3(x)=\frac{\int_x^b\omega(\xi)\,d\xi}{\omega(x)},\qquad \rho_4(x)=\frac{\int_x^b\omega^{1/(2m-1)}(\xi)\,d\xi}{\omega^{1/(2m-1)}(x)}.$$

The simplest sufficient conditions can be described as follows:

(a_1) The functions $\rho_i, i=1,2$, satisfy the Lipschitz condition on $[a,a+\delta]$ for some $0<\delta<b-a$.

(b_1) The functions ρ_i, $i=3,4$, satisfy the Lipschitz condition on $[b-\delta,b]$ for some $0<\delta<b-a$.

(a_2) The function ρ_1 satisfies the Lipschitz condition on $[a,a+\delta]$ for some $0<\delta<b-a$ and $\rho_1'\ge 0$ almost everywhere on $[a,a+\delta]$.

(b_2) The function ρ_3 satisfies the Lipschitz condition on $[b-\delta,b]$ for some $0<\delta<b-a$ and $\rho_3'\le 0$ almost everywhere on $[b-\delta,b]$.

(a_3) The function ρ_1 satisfies the Lipschitz condition on $[a,a+\delta]$ for some $0<\delta<b-a$ and $\|\rho_1'\|_{L_\infty(a,x)}\to 0$ as $x\to a$.

(b_3) The function ρ_3 satisfies the Lipschitz condition on $[b-\delta,b]$ for some $0<\delta<b-a$ and $\|\rho_3'\|_{L_\infty(x,b)}\to 0$ as $x\to b$.

(a_4) $\omega\in C^1(a,a+\delta]$ for some $0<\delta<b-a$, $\omega'>0$ on $(a,a+\delta]$, $\omega/\omega'\in C^1(a,a+\delta)$, $(\omega/\omega')'\ge -q$ $(0<q<1/(2m-1))$ on $(a,a+\delta]$, and

$$\overline{\lim}_{x\to a}(\omega^{2m/(2m-1)}(x)/\omega'(x))=0.$$

(b_4) $\omega \in C^1(b-\delta, b)$ for some $0 < \delta < b-a$, $\omega' < 0$ on $[b-\delta, b)$, $\omega/\omega' \in C^1[b-\delta, b)$, $(\omega/\omega')' \geq -q$ $(0 < q < 1/(2m-1))$ on $[b-\delta, b)$, and

$$\overline{\lim}_{x \to b}(-\omega^{2m/(2m-1)}(x)/\omega'(x)) = 0.$$

Lemma 2.5. *Condition* (A) *holds under one of the conditions* (a_i), $i = 1, 2, 3, 4$.

Proof. Let condition (a_1) hold. We take $\rho = \rho_0$; the function ρ_0 was defined at the beginning of the section. If the function ρ_1 satisfies the Lipschitz condition, then there exists $\varepsilon > 0$ such that $(\varepsilon\rho_1)' \leq q < 1$ almost everywhere on $(a, a+\delta)$, where q is some fixed constant. Hence, $\operatorname*{osc}_{[x-\varepsilon\rho_1(x), x+\varepsilon\rho_1(x)]} \rho_1 \leq c$ for all $x \in (a, a+\delta_1)$ $(\delta_1 + \varepsilon\rho_1(\delta_1) = \delta)$ and some constant c depending on q only. From the definition of ρ_1, we obtain

$$\omega(x) = c_1 \frac{e^{-\int_x^{a+\delta_1} 1/\rho_1(\xi)\, d\xi}}{\rho_1(x)},$$

where c_1 is a positive constant. This equality implies that

$$\operatorname*{osc}_{[x-\varepsilon\rho_1(x), x+\varepsilon\rho_1(x)]} \omega \leq M_1 < \infty$$

for some constant M_1 and all $x \in (a, a+\delta_1)$. The definition of the function ρ_0, where $M \geq M_1$, yields the inequality

$$\varepsilon\rho_1(x) \leq \rho_0(x), \qquad x \in (a, a+\delta_1). \tag{2.20}$$

Similar arguments and the definition of the function ρ_2 imply that

$$\varepsilon_1\rho_2(x) \leq \rho_0(x) \tag{2.21}$$

for some $\varepsilon_1 > 0$ and all $x \in (a, a+\delta_2)$ $(\delta_2 > 0)$. Thus, on some interval $(a, a+\delta_3)$ we have the inequality

$$\rho_i(x) \leq c\rho_0(x), \qquad i = 1, 2, \tag{2.22}$$

where c is a positive constant. Consider the functions $\widetilde{\rho}$ $(\widetilde{\rho} \sim \rho_0)$ and $\widetilde{\omega}$ $(\widetilde{\omega} \sim \omega)$ which were defined at the beginning of the section. We have

$$|(\widetilde{\rho}\widetilde{\omega})'| \leq |\widetilde{\rho}'\widetilde{\omega} + \widetilde{\rho}\widetilde{\omega}'| \leq c\widetilde{\omega}(x)$$

for some constant $c > 0$. The inequality $\widetilde{\rho}(x) \leq c(x-a)$ for some c and the inclusion $\widetilde{\omega} \in L_1(a, b)$ imply that $\lim_{x \to a} \widetilde{\rho}\widetilde{\omega} = 0$. Integrating the above inequality over the interval (a, x) $(x \leq a+\delta_3)$, we obtain the inequality

$$\widetilde{\rho}\widetilde{\omega} \leq c \int_a^x \widetilde{\omega}(\xi)\, d\xi.$$

Therefore, there exists a constant c such that

$$\rho(x) = \rho_0(x) \le c\frac{\int_a^x \omega(\xi)\,d\xi}{\omega(x)} = c\rho_1(x).$$

In view of (2.22)((2.20)), the functions ρ_0 and ρ_1 are equivalent on the interval $(a, a+\delta_3)$. Similarly, using (2.22) we can show that the functions ρ_0 and ρ_2 are equivalent on the same interval. From the definitions of the functions ρ_1 and ρ_2, we obtain the inequalities

$$\frac{\displaystyle\int_a^x \omega(\xi)\,d\xi}{\omega(x)\rho(x)} \le R < \infty, \qquad \frac{\displaystyle\int_a^x \rho^{2m-2}\omega(\xi)\,d\xi}{\omega(x)\rho^{2m-1}(x)} \le R < \infty,$$

where R is some constant and $x \in (a, a+\delta_3)$. Using the Hölder inequality, we arrive at the bounds

$$\frac{\displaystyle\int_a^x \rho^{2k}\omega(\xi)\,d\xi}{\omega(x)\rho^{2k+1}(x)} \le R < \infty, \qquad x \in (a, a+\delta_3), \quad k = 0, 1, \dots, m-1,$$

i.e., condition (A) holds.

Let condition (a_2) hold. We take $\rho = \rho_0$. By similar arguments we can show that $\rho_0 \sim \rho_1$ on some interval $(a, a+\delta_1)$ ($\delta_1 \le \delta$). Monotonicity of the function ρ_1 implies the inequality

$$\frac{\displaystyle\int_a^x \rho_1^{2k}\omega(\xi)\,d\xi}{\omega(x)\rho_1^{2k}(x)} \le \frac{\displaystyle\int_a^x \omega(\xi)\,d\xi\,\rho_1^{2k}(x)}{\omega(x)\rho_1^{2k}(x)} \le \rho_1(x), \qquad x \in (a, a+\delta_1).$$

If the condition (a_3) holds, then we use the arguments of Lemma 2.1 in order to prove the lemma.

Let condition (a_4) hold. In this case condition (a_1) is fulfilled. Indeed, there exists a constant $c > 0$ such that $(\omega/\omega')' \ge -(c-1)/c$. From this inequality it follows that

$$c\omega(x) + c\omega(x)(\omega(x)/\omega'(x))' \ge \omega(x), \qquad x \in (a, a+\delta).$$

Thus,

$$c(\omega^2(x)/\omega'(x))' \ge \omega(x), \qquad x \in (a, a+\delta).$$

Integrating this inequality, we obtain

$$\frac{\displaystyle\int_a^x \omega(\xi)\,d\xi\,\omega'(x)}{\omega^2(x)} \le c, \qquad x \in (a, a+\delta).$$

Therefore, the derivative of the function $\int_a^x \omega(\xi)\, d\xi/\omega(x)$ is bounded on the interval $(a, a+\delta]$. In a similar way, it can be proved that the derivative of the function $\int_a^x \omega^{1/(2m-1)}(\xi)\, d\xi/\omega^{1/(2m-1)}(x)$ is bounded. Therefore, condition (a_1) holds. □

By similar arguments we can show the following lemma:

Lemma 2.6. *Condition* (B) *holds under one of the conditions* (b_i), $i = 1, 2, 3, 4$.

Remark 2.7. As it can be seen from the proofs of Lemmas 2.5 and 2.6, for validity of these lemmas it suffices to demand that one of the conditions (a_i) (or (b_i)) is valid for some function g equivalent to the function ω.

Now we present some examples. We consider the interval $(0,1)$. The functions

$$\omega(x) = x^s \ (s > -1), \qquad \omega(x) = e^{-1/x^s} \ (s > 0), \qquad \omega(x) = e^{-e^{1/x^s}} \ (s > 0)$$

satisfy condition (a_4). We can say that all functions, regular in some sense, satisfy one of the conditions (a_i). The exceptions are oscillating functions. But it is not difficult to construct an oscillating function which also satisfies the conditions of Lemma 2.5 or Lemma 2.6. For example, we put $\rho(x) = 2(x^\alpha(\sin(1/x^\delta)+1) + x^\beta(-\sin(1/x^\delta)+1))/\delta$, where $\min(\alpha,\beta) = \delta + 1$ and $\delta > 0$. Then the function

$$\omega(x) = \frac{e^{-\int_x^1 1/\rho(\xi)\, d\xi}}{\rho(x)}$$

satisfies condition (a_1) with $m = 1$.

3. Indefinite Sturm-Liouville problems

For simplicity, we assume that the coefficients of the operator L are smooth functions, i.e., $a_{ij} = (-1)^{i+j}\overline{a_{ji}} \in C^{\max(i,j)}([0,1])$. The case of measurable coefficients satisfying some natural conditions can be considered by analogy with [4].

We assume that the operator L with the domain

$$D(L) = \{u \in W_2^{2m}(0,1) : B_k u = 0, \ \ k = \overline{1,m}\}$$

is self-adjoint in $L_2(0,1)$ and, for some constants $c_1, c_3 > 0$, and c_2, we have the inequality

$$\|u\|^2_{W_2^m(0,1)} \geq (Lu,u) \geq c_1\|u\|^2_{W_2^m(0,1)} - c_2\|u\|^2_{L_2(0,1)}, \qquad u \in D(L). \tag{3.1}$$

We consider the spectral problem (1.1). First, we state conditions on the function g which will be used in what follows. First of all, we assume that $g \in L_1(0,1)$. Moreover, there exist open subsets G^+, G^- of $G = (0,1)$ such that $\mu(\overline{G^\pm}\backslash G^\pm) = 0$,

$g > 0$ a.e. (almost everywhere) in G^+, $g < 0$ a.e. in G^- and $g = 0$ a.e. in $G^0 = G \setminus (\overline{G^+} \cup \overline{G^-})$. Here μ is the Lebesgue measure. Without loss of generality, we assume that the interiors of the sets $\overline{G^\pm}$ and $\overline{G^0}$ coincide with $G^\pm$ and G^0, respectively.

(C) The number of points $\{x_k\}_{k=1}^N$ in the set $\partial G^+ \cap \partial G^-$ is finite. For each point x_k there exists either a right neighborhood or a left neighborhood of this point (i.e., a set $(x_k, x_k + \delta)$ (or $(x_k - \delta, x_k)$)), where the function g is equivalent to a function ω with the properties: ω is continuous and positive on $(x_k, x_k + \delta]$ (or $[x_k - \delta, x_k)$) and satisfies condition (A) (or condition (B)). The interval (a, b) in this particular case is the interval $(x_k, x_k + \delta)$ (or $(x_k - \delta, x_k)$).

(D) Let $\{I_k\}_{k=1}^M$ be components of connectedness of the set G^0 with the following properties: $\overline{I_k} \cap \partial G^+ \neq \emptyset$, $\overline{I_k} \cap \partial G^- \neq \emptyset$. Assume that $M < \infty$ and if the Dirichlet problem

$$Lu = 0, \qquad u^{(l)}\Big|_{\partial I_k} = 0, \quad l = \overline{0, m-1},$$

has a nontrivial solution, then there exists either a right neighborhood or a left neighborhood of this interval (i.e., if $I_k = (y_k^1, y_k^2)$, then it is the set $(y_k^2, y_k^2 + \delta)$ (or $(y_k^1 - \delta, y_k^1)$)) on which the function g is equivalent to a function ω with the properties: ω is continuous and positive on $(y_k^2, y_k^2 + \delta]$ (or $[y_k^1 - \delta, y_k^1)$) and satisfies condition (A) (or condition (B)). If the above Dirichlet problem has a nontrivial solution, then this interval is called *degenerate.*

This condition (condition (D)) is probably excessive and arises in the method of the proof. We shall use the results of the article [19].

We now introduce some definitions.

Put $H_1 = D(|L|^{1/2})$. From (3.1) we infer that $D(|L|^{1/2}) = \{u \in W_2^m(0,1) : B_k u = 0$ for all boundary operators B_k containing only the derivatives $u^{(i)}$ with $i < m\}$ (see [21]). Let H' be the antidual space to H_1. H' is the completion of $L_2(0,1)$ with respect to the norm $\|u\|_{H_1'} = \sup |(u,v)|$, where the supremum is taken over the set $\{v \in H_1 : \|v\|_{H_1} = 1\}$. A function $u \in H_1$, $u \neq 0$, is an eigenfunction of the problem (1.1) if, for some $\lambda \in \mathbb{C}$, equality (1.1) holds true in the space H_1'. A set $\{u_k\}_{k=0}^N$ is a chain of eigenvectors and associated vectors of the problem (1.1) corresponding to some eigenvalue λ if

$$Lu_k - \lambda g(x) u_k - g(x) u_{k-1} = 0, \qquad u_k \in H_1, \quad k = \overline{0, N}, \quad u_{-1} := 0.$$

Actually, eigenfunctions and associated functions are generalized solutions to the corresponding problems. But if $g \in L_2(0,1)$, for example, then these functions belong to $D(L)$ and equation (1.1) is satisfied in the usual sense.

Next, we define the class C_1,

$$C_1 = \{u \in H_1 : (Lu, v) = 0, \ v \in V_1 \cap H_1\},$$

where $V_1 = \{v \in L_2(0,1) : \operatorname{supp} v \in G^0\}$, and we define the spaces F_1, F_0 and F_{-1} as follows: We put $F_0 = L_{2,g}((0,1) \setminus G^0)$ and

$$F_1 = \{u \in F_0 : \exists\, v \in C_1 : v|_{G^+ \cup G^-} = u\}.$$

The norm in F_1 is defined by the equality

$$\|u\|_{F_1} = \inf_{v \in C_1,\, v|_{G^+ \cup G^-} = u} \|v\|_{H_1}.$$

The space F_{-1} is defined as the completion of the space F_0 with respect to the norm

$$\|u\|_{F_{-1}} = \frac{\sup_{v \in F_1} |[u,v]_0|}{\|v\|_{F_1}}, \qquad [u,v]_0 = \int_{(0,1)\setminus G^0} g(x)u(x)\overline{v(x)}\,dx.$$

Note that the uniqueness of the Cauchy problem implies that $V_1 \cap \ker L = \{0\}$.

We can state the main results of the article [19] in application to our particular case. The following theorem is a consequence of Theorem 2.1 in [19].

Theorem 3.1. *Under the above conditions except for conditions* (C) *and* (D), *the eigenfunctions and associated functions of the problem* (1.1) *are dense in* F_0 *and form an unconditional basis in the space* F_1.

Theorem 3.2. *Let the above conditions (including condition* (C) *and* (D)) *hold. Then from the eigenfunctions and associated functions we can construct an unconditional basis of the space* F_0 *with the following properties:*
Each function $f \in F_0 = L_{2,g}(G^+ \cup G^-)$ *can be represented in a unique way in the form*

$$f = \sum_{i=1}^{\infty} u_i^+ c_i^+ + \sum_{i=1}^{\infty} u_i^- c_i^- + \sum_{i=1}^{M} u_i c_i \quad (M < \infty), \tag{3.2}$$

where $u_i^{\pm}$ *are the eigenfunctions corresponding to positive (negative) eigenvalues* $\lambda_i^{\pm}$ *except for a finite number of them,*

$$[u_i^{\pm}, u_j^{\pm}]_0 = \pm\delta_{ij}, \quad c_i^{\pm} = \pm[f, u_i^{\pm}]_0, \quad [u_i^{\pm}, u_j]_0 = 0, \quad [u_i^+, u_j^-]_0 = 0,$$

and $\{u_j\}_{j=1}^M$ *is a basis in some finite-dimensional subspace which is the span of some eigenfunctions and associated functions of the problem* (1.1). *The norm in the space* F_0 *is equivalent to the norm*

$$\|f\|_{F_0}^2 = \sum_{i=1}^{\infty} (|c_i^+|^2 + |c_i^-|^2) + \sum_{i=1}^{M} |c_i|^2.$$

The last assertion means that this basis is a Riesz basis (see [9]). *If* $f \in F_1$, *then the function* f *is also representable in the form* (3.2) *and the norm of* F_1 *is equivalent to the norm*

$$\|f\|_{F_1}^2 = \sum_{i=1}^{\infty} (|\lambda_i^+||c_i^+|^2 + |\lambda_i^-||c_i^-|^2) + \sum_{i=1}^{M} |c_i|^2.$$

This basis of the space F_0 can be divided into two parts so that the corresponding "halves" are Riesz bases in the spaces $L_{2,g}(G^+)$ and $L_{2,g}(G^-)$. If $f \in L_{2,g}(G^+)$ or $f \in L_{2,g}(G^-)$, respectively, then the function f is representable as

$$f = \sum_{i=1}^{\infty} u_i^+ c_i^+ + \sum_{i=1}^{M^+} v_i^+ a_i^+ \quad (M^+ < \infty),$$

or

$$f = \sum_{i=1}^{\infty} u_i^- c_i^- + \sum_{i=1}^{M^-} v_i^- a_i^- \quad (M^- < \infty),$$

respectively, where $\{v_i^\pm\}$ are some finite sets of eigenfunctions and associated functions. The norm in $L_{2,g}(G^+)$ or in $L_{2,g}(G^-)$ is equivalent to the norm

$$\|f\|^2_{L_{2,g}(G^\pm)} = \sum_{i=1}^{\infty} |c_i^\pm|^2 + \sum_{i=1}^{M^\pm} |a_i^\pm|^2.$$

Proof. To prove the theorem, we need to check the condition $(F_1, F_{-1})_{1/2,2} = F_0$ and then use Theorems 2.2–2.4 from the article [19].
Let $x_k \in \partial G^+ \cap \partial G^-$. Then on the interval $O_k^- = (x_k - \varepsilon, x_k)$ (or on the interval $(x_k, x_k + \varepsilon)$) the function g is equivalent to a function ω satisfying condition (B) (or (A)). Let it be the interval O_k^-. In the case of the latter interval the arguments are the same. Reducing ε if necessary, we can assume that $(x_k - \varepsilon, x_k) \subset G^+$ or $(x_k - \varepsilon, x_k) \subset G^-$. There exists a right neighborhood $O_k^+ = (x_k, x_k + \varepsilon_1)$ of the point x_k such that either $O_k^+ \cap \overline{G^+} = \emptyset$ or $O_k^+ \cap \overline{G^-} = \emptyset$ and, moreover, $x_k + \varepsilon_1 \in G^-$ or $x_k + \varepsilon_1 \in G^+$, respectively. Put $O_k = O_k^+ \cup O_k^- \cup \{x_k\}$,

$$W_1 = \{u \in W_0 = L_{2,g}(O_k \cap (G^+ \cup G^-)) : \exists v \in W_2^m(O_k) : v|_{O_k \cap (G^+ \cup G^-)} = u,$$
$$\int_{O_k \cap G_0} v\overline{L\varphi}\, dx = 0,\ \varphi \in W_2^{2m}(G^0),\ \operatorname{supp}\varphi \in O_k \cap G^0\}.$$

Endow the space W_1 with the norm

$$\|u\|_{W_1} = \inf \|v\|_{W_2^m(O_k)},$$

where the infimum is taken over the set of functions v used in the definition of W_1. Denote $W_s = (W_1, W_0)_{1-s,2}$. Now we show that there exists $s_0 > 0$ such that the operators

$$S_k^\pm u = \begin{cases} u, & x \in G^\pm \cap O_k, \\ 0, & x \in G^\mp \cap O_k, \end{cases}$$

are continuous as operators from W_s into W_s for all $s \in [0, s_0)$. Define the auxiliary spaces $A_1 = W_2^m(O_k^-)$, $A_0 = L_{2,g}(O_k^-)$, and $A_1^0 = \{u \in A_1 : u^{(l)}(x_k) = 0,\ l = \overline{0, m-1}\}$. Theorem 2.4 ensures that there exists $s_0 > 0$ such that

$$A_s = (A_1, A_0)_{1-s,2} = A_s^0 = (A_1^0, A_0)_{1-s,2}, \qquad s < s_0.$$

Next, we consider the operator $P_0 : W_s \to A_s$, $P_0 u = u|_{O_k^-}$. Obviously, $P_0 \in L(W_s, A_s)$ for all s. Define also the operator

$$P_1 : A_s^0 \to W_s, \qquad P_1 u = \begin{cases} u, & x \in O_k^-, \\ 0, & x \in O_k^+. \end{cases}$$

Obviously, $P_1 \in L(A_s^0, W_s)$ for all $s \in [0,1]$. Hence, for $s < s_0$, we obtain that $P_1 P_0 \in L(W_s)$. By construction, we have $P_1 P_0 u = S_k^- u$ or $P_1 P_0 u = S_k^+ u$. Thus, the operators S_k^- and S_k^+ belong to the class $L(W_s)$ for all $s < s_0$.

Let I_k be an interval in G^0 such that $\overline{I_k} \cap \partial G^+ \neq \emptyset$, $\overline{I_k} \cap \partial G^- \neq \emptyset$, and the interval I_k is degenerate. We now repeat the previous arguments, where we take the interval $I_k = (y_1, y_2)$ rather than the point x_k. Either on the interval $O_k^- = (y_1 - \varepsilon, y_1)$ or on the interval $(y_2, y_2 + \varepsilon)$ the function g is equivalent to a function ω satisfying condition (B) (or (A)). Let it be the first interval. Reducing ε if necessary, we can assume that either $(y_1 - \varepsilon, y_1) \subset G^+$ or $(y_1 - \varepsilon, y_1) \subset G^-$. There exists a right neighborhood of the interval $O_k^+ = (y_2, y_2 + \varepsilon_1)$ such that $O_k^+ \cap \overline{G^+} = \emptyset$ ($O_k^+ \cap \overline{G^-} = \emptyset$); moreover, $x_k + \varepsilon_1 \in G^-$,($x_k + \varepsilon_1 \in G^+$). Put $O_k = O_k^+ \cup O_k^- \cup \overline{I_k}$,

$$\begin{aligned} W_1 &= \{u \in W_0 = L_{2,g}(O_k \cap (G^+ \cup G^-)) : \exists\, v \in W_2^m(O_k) : v|_{O_k \cap (G^+ \cup G^-)} = u, \\ &\quad \int_{O_k \cap G_0} v \overline{L\varphi}\, dx = 0,\ \varphi \in W_2^{2m}(G^0), \operatorname{supp} \varphi \in O_k \cap G^0\}. \end{aligned}$$

Endow the space W_1 with the norm

$$\|u\|_{W_1} = \inf \|v\|_{W_2^m(O_k)},$$

where the infimum is taken over the set of function v from the definition of W_1. We denote $W_s = (W_1, W_0)_{1-s,2}$ and show that there exists $s_0 > 0$ such that the operators

$$R_k^\pm u = \begin{cases} u, & x \in G^\pm \cap O_k, \\ 0, & x \in G^\mp \cap O_k, \end{cases}$$

are continuous as operators from W_s into W_s for all $s \in [0, s_0)$. We now define the auxiliary spaces $A_1 = W_2^m(O_k^-)$, $A_0 = L_{2,g}(O_k^-)$, and $A_1^0 = \{u \in A_1 : u^{(l)}(x_k) = 0,\ l = \overline{0, m-1}\}$. By Theorem 2.4, there exists $s_0 > 0$ such that

$$A_s = (A_1, A_0)_{1-s,2} = A_s^0 = (A_1^0, A_0)_{1-s,2}, \qquad s < s_0.$$

Define the operators $P_0 : W_s \to A_s$, $P_0 u = u|_{O_k^-}$. Obviously, $P_0 \in L(W_s, A_s)$ for all s. We define also the operator

$$P_1 : A_s^0 \to W_s, \qquad P_1 u = \begin{cases} u, & x \in O_k^-, \\ 0, & x \in O_k^+. \end{cases}$$

Obviously, $P_1 \in L(A_s^0, W_s)$ for all $s \in [0,1]$. Hence, we obtain that $P_1 P_0 \in L(W_s)$ for all $s < s_0$. By construction, $P_1 P_0 u = R_k^- u$ ($P_1 P_0 u = R_k^+ u$). Thus, the operators R_k^- and R_k^+ belong to the class $L(W_s)$ for all $s < s_0$.

We now take an interval $I_k = (y_1, y_2)$ in G^0 and assume that $\overline{I_k} \cap \partial G^+ \neq \emptyset$, $\overline{I_k} \cap \partial G^- \neq \emptyset$, and the interval I_k is nondegenerate. By our conditions, there exists a neighborhood O_k of the interval I_k such that $O_k = O_k^+ \cup O_k^- \cup \overline{I_k}$, $O_k^+ = (y_2, y_2 + \varepsilon_1)$, $O_k^- = (y_1 - \varepsilon, y_1)$, $O_k^+ \cap \overline{G^+} = \emptyset$ ($O_k^+ \cap \overline{G^-} = \emptyset$), and $O_k^- \cap \overline{G^-} = \emptyset$ ($O_k^- \cap \overline{G^+} = \emptyset$). We can assume that $y_1 - \varepsilon$, $y_2 + \varepsilon_1 \in G^+ \cup G^-$. As before, we can define the spaces W_s and the operators $R_k^{\pm}$. In this case the operators $R_k^{\pm}$ are continuous as operators from W_s into W_s for all $s \in [0,1]$. Indeed, it suffices to check it for $s = 0$ and $s = 1$. For $s = 0$, this fact is evident. Let $u \in W_1$. There exists $v \in W_2^m(O_k)$ such that $v|_{O_k \cap (G^+ \cup G^-)} = u$. We extend this function to a function $\widetilde{v}$ putting $\widetilde{v}(x) = u(x)$ for $x \in O_k^-$, $\widetilde{v}(x) = 0$ for $x \in O_k^+$, and the function $\widetilde{v}$ on the interval I_k is equal to a solution of the problem

$$L\widetilde{v} = 0, \qquad \widetilde{v}^{(l)}(y_2) = 0, \qquad \widetilde{v}^{(l)}(y_1) = v^{(l)}(y_1), \qquad l = \overline{0, m-1}.$$

The function $\widetilde{v}$ belongs to $W_2^m(O_k)$ and $\widetilde{v}|_{O_k \cap (G^+ \cup G^-)} = u$. Also either $\widetilde{v} = 0$ on $G^+ \cap O_k$ or $\widetilde{v} = 0$ on $G^- \cap O_k$. This implies that $R_k^{\pm} \in L(W_1)$.

For each point $x_k \in \partial G^+ \cap \partial G^-$ and for each interval I_k with the properties $\overline{I_k} \cap \partial G^+ \neq \emptyset$ and $\overline{I_k} \cap \partial G^- \neq \emptyset$, we can construct neighborhoods O_k with the above properties. Also we can define the functions $\varphi_k \in C_0^\infty(O_k)$ and $\psi_k \in C_0^\infty(O_k)$ such that $\varphi_k = 1$ and $\psi_k = 1$ in some neighborhoods of x_k and I_k, respectively, and such that $\operatorname{supp} \varphi_k' \in G^+ \cup G^-$ and $\operatorname{supp} \psi_k' \in G^+ \cup G^-$. Clearly, it is possible.

We show that there exists $s_0 > 0$ such that the operator

$$S : u \to \begin{cases} u, & x \in G^+, \\ 0, & x \in G^-, \end{cases}$$

is continuous as an operator from F_s into F_s for all $s < s_0$. As s_0 we take the minimum of the constants s_0 which were defined in the proof. Fix $s < s_0$ and consider the operators $S_k u = \varphi_k u$ and $R_k u = \psi_k u$. By definition, if $u \in F_1$, then there exists a function $v \in C_1$ such that $v|_{G^+ \cup G^-} = u$. By construction, $R_k v \in C_1$, $S_k v \in C_1$. We can easily check the estimates

$$\|R_k u\|_{F_1} \leq c\|u\|_{F_1}, \qquad \|S_k u\|_{F_1} \leq c\|u\|_{F_1},$$

where c is a positive constant. Thus, we have $R_k, S_k \in L(F_1)$ and hence $R_k, S_k \in L(F_s, F_s)$ for all $s \in [0,1]$. Moreover, the functions $R_k u$ and $S_k u$ vanish outside O_k. This implies that $R_k, S_k \in L(F_s, W_s)$ (the spaces W_s are distinct in different neighborhoods). $R_k^+ R_k, S_k^+ S_k \in L(F_s, W_s)$. We can construct functions $\widetilde{\varphi}_k \in C_0^\infty(O_k)$ and $\widetilde{\psi}_k \in C_0^\infty(O_k)$ such that $\operatorname{supp} \widetilde{\varphi_k}' \subset G^+ \cup G^-$, $\operatorname{supp} \widetilde{\psi_k}' \subset G^+ \cup G^-$, $\widetilde{\varphi}_k = 1$ and $\widetilde{\psi}_k = 1$ in some neighborhoods of $\operatorname{supp} \varphi_k$ and $\operatorname{supp} \psi_k$, respectively. It is easy to see that the operators $\widetilde{S}_k : u \to \widetilde{\varphi}_k u$ and $\widetilde{R}_k : u \to \widetilde{\psi}_k u$, where we have extended the functions $\widetilde{\varphi}_k u$, $\widetilde{\psi}_k u$ by zero on the whole interval $(0,1)$, possess the following properties: $\widetilde{R}_k, \widetilde{S}_k \in L(W_s, F_s)$. Hence, $\widetilde{S}_k S_k^+ S_k \in L(F_s, F_s)$ and

$\widetilde{R}_k R_k^+ R_k \in L(F_s, F_s)$. We consider the operator

$$P : u \to \left(1 - \sum_{k=1}^{N} \varphi_k - \sum_{k=1}^{M} \psi_k\right) u.$$

The definitions imply that $P \in L(F_s, F_s)$ for all $s \in [0,1]$. By construction, $SP \in L(F_s, F_s)$ for all $s \in [0,1]$. Then the operator

$$S = SP + \sum_{k=1}^{N} \widetilde{S}_k S_k^+ S_k + \sum_{k=1}^{N} \widetilde{R}_k R_k^+ R_k$$

possesses the property $S \in L(F_s, F_s)$. Next, we refer to Lemma 1.2 in [19] which ensures the equality $(F_1, F_{-1})_{1/2,2} = F_0$. □

References

[1] Beals, R.: Indefinite Sturm-Liouville problem and half-range completeness; J. Differential Equations 56 (1985), 391–407.

[2] Birman, M.S., Solomjak, M.Z.: Asymptotic behaviour of the spectrum of differential equations; J. Soviet Math. 12 (1979), 247–282.

[3] Birman, M.S., Solomjak, M.Z.: Quantitative analysis in Sobolev imbedding theorems and application to spectral theory; Amer. Math. Soc. Transl. Ser. 2 114 (1980).

[4] Ćurgus, B.: On the regularity of the critical point infinity of definitizable operators; Integral Equations Operator Theory 8 (1985), 462–488.

[5] Ćurgus, B., Langer, H.: A Krein space approach to symmetric ordinary differential operators with an indefinite weight function; J. Differential Equations 79 (1989), 31–62.

[6] Faierman, M., Roach, G.F.: Full and half-range eigenfunction expansions for an elliptic boundary value problem involving an indefinite weight; Lect. Notes in Pure and Appl. Math. 118 (1989), 231–236.

[7] Fleckinger, J., Lapidus, M.L.: Eigenvalues of elliptic boundary value problems with an indefinite weight function; Trans. Amer. Math. Soc. 295 (1986), 305–324.

[8] Fleckinger, J., Lapidus, M.L.: Remainder estimates for the asymptotics of elliptic eigenvalue problems with indefinite weights; Arch. Rational Mech. Anal. 98 (1987), 329–356.

[9] Gohberg, I.C., Krein, M.G.: Introduction to the Theory of Linear Nonselfadjoint Operators in Hilbert Spaces; Amer. Math. Soc. Transl. 18 (1969).

[10] Haupt, O.: Untersuchungen über Oszillationstheoreme; Teubner Verlag, Leipzig 1911.

[11] Haupt, O.: Über eine Methode zum Beweise von Oszillationstheoremen; Math. Ann. 76 (1915), 67–104.

[12] HILB, H.: Eine Erweiterung der Kleinschen Oszillationstheoreme; Jahresber. Deutsch. Math.-Verein. 16 (1907), 279–285.

[13] HILBERT, D.: Grundzüge einer allgemeinen Theorie der linearen Integralgleichungen; Chelsea, New York 1953.

[14] HOLMGREN, E.: Über Randwertaufgaben bei einer linearen Integralgleichung zweiter Ordnung; Ark. Mat. Astronom. Fysik 1 (1904), 401–417.

[15] PLEIJEL, A.: Sur la distribution des valeurs propres de problèmes régis par l'équation $\Delta u + k(x,y)u = 0$; Ark. Mat. Astronom. Fysik 29B (1942), 1–8.

[16] PYATKOV, S.G.: On the solvability of a boundary value problems for a parabolic equation with changing time direction; Soviet Math.Dokl. 32 (1985), 895–897.

[17] PYATKOV, S.G.: Properties of eigenfunctions of linear sheaves; Siberian Math. J. 30 (1989), 587–597.

[18] PYATKOV, S.G.: Properties of eigenfunctions of linear pencils; Mat. Zametki 51 (1992), 141–148.

[19] PYATKOV, S.G.: Elliptic eigenvalue problems with an indefinite weight function; Siberian Adv. in Math. 4 (1994), 87–121.

[20] RICHARDSON, R.G.D.: Theorems of oscillation for two linear differential equations of the second order with two parameters; Trans. Amer. Math. Soc. 13 (1912), 22–34.

[21] TRIEBEL, H.: Interpolation Theory, Function Spaces, Differential Operators; VEB Deutscher Verlag Wissenschaft, Berlin 1978.

[22] VOLKMER, H.: Sturm-Liouville problems with indefinite weights and Everitt's inequality; Technical Report No. 7 – 1994-1995 Academic Year, Technical Report Series of the Department of Mathematical Sciences, University of Wisconsin-Milwaukee.

[23] FLEIGE, A.: Spectral theory of indefinite Krein-Feller differential operators; Mathematical Research 98, Akadademie Verlag, Berlin 1996.

Institute of Mathematics
Siberian Department
Russian Academy of Sciences
Universitetskii pr.4
630090, Novosibirsk
Russia
sychev@math.nsc.ru

1991 Mathematics Subject Classification: Primary 34B24; Secondary 46E35

Submitted: May 7, 1996

Operator Theory:
Advances and Applications, Vol. 102

Mellin pseudodifferential operators techniques in the theory of singular integral operators on some Carleson curves

V.S. Rabinovich

We consider an algebra $\mathcal{A}_p(\Gamma, \omega)$ of singular integral operators with slowly oscillating bounded coefficients acting in $L_p(\Gamma, \omega)$, $1 < p < \infty$, where Γ is a composed Carleson curve with logarithmic whirl points and ω is a power weight. The local analysis of operators $A \in \mathcal{A}_p(\Gamma, \omega)$ at singular points of the contours is based on the Mellin pseudodifferential operators method. This method gives effective formulas for the local symbols. These formulas describe the influence on the local symbol of both the curve and the weight in an explicit form.

1. Introduction

Fredholm criteria for operators in the algebra generated by singular integral operators S_Γ on a composed Lyapunov curve and operators of multiplication by piecewise continuous functions, acting in $L_p(\Gamma, \omega)$ where $1 < p < \infty$ and ω is a power weight were first established by Gohberg and Krupnik (see [GK] and references given there).

The Gohberg/Krupnik theory was essentially extended in the papers of Simonenko, Duduchava, Plamenevskii, Senichkin, Roch, Silbermann and others (see [BS], [ROS], [HRS] and references contained therein).

The case of composed Lyapunov curves Γ and arbitrary Muckenhoupt weights was considered by Gohberg, Krupnik, and Spitkovsky [GKS], [S].

Recently Böttcher and Karlovich considered the algebra generated by the operator S_Γ and operators of multiplication by piecewise continuous functions on general closed Carleson curves with arbitrary Muckenhoupt weights. Several new phenomena were discovered by them: the circular arcs and horns which typically arise in the spectral theory of Gohberg, Krupnik, and Spitkovsky are converted into logarithmic double spirals, spiralic horns and so-called leaves [BK1], [BK2], [BK3].

There are two approaches now to singular integral operators on Carleson curves: the Wiener-Hopf factorization method and Mellin pseudodifferential operators techniques (see [BKR] for a comparing discussion).

On the whole, the approach of Böttcher/Karlovich is based on the Wiener-Hopf method. We, however, develop the Mellin pseudodifferential operators approach which was presented first in our papers [R1], [R2] devoted to C^*-algebras of singular integral operators with coefficients which have second kind discontinuities on curves with slowly oscillating or rotating tangents. Here we consider an algebra $\mathcal{A}_p(\Gamma,\omega)$ of singular integral operators with slowly oscillating bounded coefficients acting in $L_p(\Gamma,\omega)$, $1<p<\infty$, where Γ is a Carleson curve in a wider class than in [R1], [R2] and ω is a power weight.

For the investigation of the Fredholm property of operators in $\mathcal{A}_p(\Gamma,\omega)$ we apply the Allan-Douglas local principle. The local analysis of operators $A\in\mathcal{A}_p(\Gamma,\omega)$ is based on the Mellin pseudodifferential operators method, which gives effective formulas for the local symbols. These formulas describe the influence of both the curve and the weight on the local symbol in an explicit form.

The Mellin pseudodifferential operators method can be applied to operators on wide classes of Carleson curves acting on L_p-spaces with Muckenhoupt weight. It is no problem to consider algebras of singular integral operators with matrix coefficients by means of this method.

This method can also be applied to the justification of numerical methods for solving singular integral equations on Carleson curves [HRS].

These results will be given in forthcoming papers.

The paper is divided in two sections. Section 2 is devoted to some results of the Mellin pseudodifferential operators theory which we need in Section 3, where we formulate and prove the results on singular integral operators.

2. Banach algebras of Mellin pseudodifferential operators

2.1. Definitions

Let B be a complex Banach space. $\mathcal{L}(B)$ denotes the space of all bounded linear operators acting in B, $\mathcal{K}(B)$ denotes the ideal of compact operators in $\mathcal{L}(B)$. Let X be a topological space, $\mathcal{B}$ be a Banach algebra. Then $C_b(X,\mathcal{B})$ denotes the Banach algebra of bounded continuous functions on X with values in $\mathcal{B}$ and the norm

$$\|a\|_{C_b(X,\mathcal{B})}=\sup_{x\in X}\|a(x)\|_{\mathcal{B}}.$$

Definition 2.1. We will say that a matrix-function $(a_{ij}(r,\lambda))_{i,j=1}^{N}$ defined on $\mathbb{R}_+\times\mathbb{R}$ is in $\mathcal{E}^m(N)$ if the functions $a_{ij}(r,\lambda)$ belong to $C^\infty(\mathbb{R}_+\times\mathbb{R})$ and satisfy the estimate

$$\sup_{\mathbb{R}_+\times\mathbb{R}}|(r\partial_r)^\alpha\partial_\lambda^\beta a_{ij}(r,\lambda)|\langle\lambda\rangle^{-m+\beta}<\infty$$

for all α, $\beta\in\mathbb{Z}_+=\mathbb{N}\cup 0$, $\langle\lambda\rangle=(1+\lambda^2)^{1/2}$.
We will say that a matrix-function $a(r,\lambda)$ $(\in\mathcal{E}^0(N))$ slowly varies at the point 0 and $+\infty$, respectively, if

$$\lim_{r\to+0}\sup_{\lambda\in\mathbb{R}}|(r\partial_r)^\alpha\partial_\lambda^\beta a_{ij}(r,\lambda)|\langle\lambda\rangle^{-\beta}=0, \tag{2.1}$$

$$\lim_{r\to+\infty}\sup_{\lambda\in\mathbb{R}}|(r\partial_r)^\alpha\partial_\lambda^\beta a_{ij}(r,\lambda)|\langle\lambda\rangle^{-\beta}=0, \tag{2.2}$$

respectively, for all $\alpha\in\mathbb{N}$ and $\beta\in\mathbb{Z}_+$. We denote by $\widetilde{\mathcal{E}}(N)$ the class of matrix-functions in $\mathcal{E}^0(N)$ which slowly vary both at the origin and at infinity.
Let $J_0(N)$ ($J_\infty(N)$) be the set of matrix-functions for which the condition (2.1) ((2.2)) holds for all $\alpha,\beta\in\mathbb{Z}_+$. Let

$$J(N)=J_0(N)\cap J_\infty(N)\cap\mathcal{E}^{-1}(N).$$

Let $a(r,\lambda)\in\mathcal{E}^m(N)$. The operator

$$(Au)(r)=a(r,\mathcal{D}_r)u=\int_{\mathbb{R}}d'\lambda\int_{\mathbb{R}_+}a(r,\lambda)\,(r\rho^{-1})^{i\lambda}u(\rho)\rho^{-1}d\rho, \tag{2.3}$$

where $d'\lambda=(2\pi)^{-1}d\lambda$, $u\in C_0^\infty(\mathbb{R}_+,\mathbb{C}^N)$, is called the Mellin pseudodifferential operator with symbol $a(r,\lambda)$. The class of all such operators is denoted by $OPS^m(N)$. The notations $OP\widetilde{\mathcal{E}}(N)$, $OPJ_0(N)$, $OPJ_\infty(N)$, $OPJ(N)$ have the obvious meaning.

Definition 2.2. Let a matrix-function $a(r,\rho,\lambda)$ ($\in C^\infty(\mathbb{R}_+\times\mathbb{R}_+\times\mathbb{R})$) be such that

$$\sup_{\mathbb{R}_+^2\times\mathbb{R}}|(r\partial_r)^\beta(\rho\partial_\rho)^\gamma\partial^\alpha a_{ij}(r,\rho,\lambda)|\langle\lambda\rangle^\alpha<\infty \tag{2.4}$$

for all $\alpha,\beta,\gamma\in\mathbb{Z}_+$ and $i,j=1,\ldots,N$. We will denote by $\mathcal{E}_d^0(N)$ the class of matrix-functions satisfying (2.4).
An operator which is defined by (2.3) where $a(r,\lambda)$ is replaced by $a(r,\rho,\lambda)$ is called a Mellin pseudodifferential operator with double symbol. Let $OP\mathcal{E}_d^0(N)$ stand for the class of such operators.
We will say that a double symbol $a(r,\rho,\lambda)$ slowly varies if

$$\lim_{r\to+0}\sup_{\rho\in K,\lambda\in\mathbb{R}}|(r\partial_r)^\beta(\rho\partial_\rho)^\gamma\partial^\alpha a_{ij}(r,r\rho,\lambda)|\langle\lambda\rangle^\alpha=0, \tag{2.5}$$

$$\lim_{r\to+\infty}\sup_{\rho\in K,\lambda\in\mathbb{R}}|(r\partial_r)^\beta(\rho\partial_\rho)^\gamma\partial^\alpha a_{ij}(r,r\rho,\lambda)|\langle\lambda\rangle^\alpha=0, \tag{2.6}$$

for all $\alpha,\beta,\gamma\in\mathbb{Z}_+$ such that $\beta+\gamma\neq0$ and for every compact $K\subset\mathbb{R}_+$.
We denote by $\widetilde{\mathcal{E}}_d(N)$ the class of slowly varying double symbols and by $OP\widetilde{\mathcal{E}}_d(N)$ the corresponding class of operators.

It should be noted that the change of variables $r=e^{-x}$, $\rho=e^{-y}$, $x,y\in\mathbb{R}$, transfers a Mellin pseudodifferential operator on $\mathbb{R}_+$ to a standard pseudodifferential operator on $\mathbb{R}$. Thus the class $OP\mathcal{E}^m(N)$ is converted into the L. Hörmander class $OP\mathcal{E}_{1,0}^m(N)$ (see for instance [T]) and the class $OP\widetilde{\mathcal{E}}(N)$ is converted into the class of pseudodifferential operators with slowly varying symbols (see [G], [RF]).

Below we will state some propositions concerning the theory of Mellin pseudodifferential operators without proofs. These propositions are either reformulations of corresponding facts in the theory of pseudodifferential operators on $\mathbb{R}$ or their proofs are based on a standard pseudodifferential operator technique (see [R2] for more details in the case $p = 2$).

Let $L_p^N(\mathbb{R}_+, d\mu)$, $p \in (1, \infty)$, be the Banach space of all measurable complex vector-functions $u(r) = (u_1(r), \dots, u_N(r))$, $r \in \mathbb{R}_+$, with the norm

$$\|u\|_{L_p^N(\mathbb{R}_+, d\mu)} = \left(\sum_{j=1}^{N} \int_0^\infty |u_j(r)|^p d\mu \right)^{1/p},$$

where $d\mu = r^{-1}dr$ is an invariant measure on the multiplicative group $\mathbb{R}_+$.

Proposition 2.3. [T]
(a) *An operator $A \in OP\mathcal{E}^0(N)$ is bounded in $L_p^N(\mathbb{R}_+, d\mu)$, $p \in (1, \infty)$, and*

$$\|A\|_{\mathcal{L}(L_p^N)} \leq C_p M,$$

where

$$M = \max_{1 \leq i,j \leq N} \max_{\alpha \leq 2, \beta \leq 1} \sup_{\mathbb{R}_+ \times \mathbb{R}} \left| (r\partial_r)^\beta \partial_\lambda^\alpha a_{ij}(r, \lambda) \right| \langle \lambda \rangle^\alpha.$$

(b) *If $A \in OP\mathcal{E}^0(N)$ is invertible in $L_p^N(\mathbb{R}_+, d\mu), p \in (1, \infty)$, then $A^{-1} \in OP\mathcal{E}^0(N)$.*

Proposition 2.4. [R2]
(a) *Let A, $B \in OP\widetilde{\mathcal{E}}(N)$. Then $AB \in OP\widetilde{\mathcal{E}}(N)$ and the symbol $\sigma_{AB}(x, \xi)$ of AB is given by the formula*

$$\sigma_{AB}(r, \lambda) = a(r, \lambda) b(r, \lambda) + t_1(r, \lambda). \tag{2.7}$$

(b) *Let A ($\in OP\widetilde{\mathcal{E}}(N)$) act on $L_p^N(\mathbb{R}_+, d\mu)$, $p \in (1, \infty)$. Then $A^* \in OP\mathcal{E}(N)$ and $\sigma_{A^*}(r, \lambda)$ is given by the formula*

$$\sigma_{A^*}(r, \lambda) = a^*(r, \lambda) + t_2(r, \lambda). \tag{2.8}$$

(c) *Let A be a Mellin pseudodifferential operator with double symbol $a(r, \rho, \lambda) \in \widetilde{S}_d(N)$. Then $A \in OP\widetilde{\mathcal{E}}(N)$ and*

$$\sigma_A(r, \lambda) = a(r, r, \lambda) + t_3(r, \lambda). \tag{2.9}$$

The symbols $t_j(r, \lambda)$, $j = 1, 2, 3$, are in $J(N)$.
(d) *$OPJ(N) \subset \mathcal{K}(L_p^N(\mathbb{R}_+, d\mu))$, $p \in (1, \infty)$.*

2.2. Local invertibility

Let $A \in \mathcal{L}(L_p^N(\mathbb{R}_+, d\mu))$. We say that A is a locally invertible operator at the point 0 if there exist $R > 0$ and operators B', $B'' \in \mathcal{L}(L_p^N(\mathbb{R}_+, d\mu))$ such that

$$B'A\chi_R = \chi_R, \quad \chi_R AB'' = \chi_R,$$

where χ_R is the operator of multiplication by the characteristic function of the segment $[0, R]$. In the same way we define the local invertibility at infinity.

Let $a(r, \mathcal{D}_r) \in OP\widetilde{\mathcal{E}}(N)$. Let us introduce the quantities

$$\begin{aligned} d_0(a) &= \lim_{\varepsilon \to 0} \inf_{(0,\varepsilon)\times\mathbb{R}} |\det a(r, \lambda)|, \\ d_\infty(a) &= \lim_{R \to +\infty} \inf_{(R,\infty)\times\mathbb{R}} |\det a(r, \lambda)|. \end{aligned}$$

Theorem 2.5. *Let $A = a(r, \mathcal{D}_r) \in OP\widetilde{\mathcal{E}}(N)$. Then the following assertions are equivalent:*

(a) *An operator $A : L_p^N(\mathbb{R}_+, d\mu) \to L_p^N(\mathbb{R}_+, d\mu)$ is locally invertible at the origin (at infinity).*

(b) $d_0(a) > 0$ ($d_\infty(a) > 0$).

(c) *There exist locally inverse operators B', $B'' \in OP\widetilde{\mathcal{E}}(N)$.*

In the case $p = 2$ this theorem has been proved in [R2]. The proof for $p \neq 2$ is similar.

2.3. Fredholmness and index

Theorem 2.6. *Let $A = a(r, \mathcal{D}_r) \in OP\widetilde{\mathcal{E}}(N)$. Then the following assertions are equivalent:*

(a) *$A : L_p^N(\mathbb{R}_+, d\mu) \to L_p^N(\mathbb{R}_+, d\mu)$ is a Fredholm operator.*

(b) *$d_0(a) > 0$, $d_\infty(a) > 0$ and*

$$d(a) = \lim_{t \to +\infty} \inf_{\mathbb{R}_+ \times \{\lambda \in \mathbb{R} : |\lambda| > t\}} |\det a(r, \lambda)| > 0.$$

(c) *There exists a regularizator $R \in OP\widetilde{\mathcal{E}}(N)$.*

If $A \in OP\widetilde{\mathcal{E}}(N)$ is a Fredholm operator, then

$$\operatorname{ind} A = -(2\pi)^{-1}[\arg\det a(r, \lambda)]_{\Gamma(R', R'')} \tag{2.10}$$

where $\Gamma(R', R'')$ is the boundary of the rectangle $\Pi(R', R'') = \{(r, \lambda) \in \mathbb{R}_+ \times \mathbb{R} : 1/R' < r < R', |\lambda| < R''\}$, which is positively oriented. Here the numbers R', R'' are such that $a^{-1}(r, \lambda)$ exists for all points of the domain $\mathbb{R}_+ \times \mathbb{R} \backslash \Pi(R', R'')$.

Proposition 2.7. *Let $A = a(r, \mathcal{D}_r) \in OP\widetilde{\mathcal{E}}(N)$, and let the symbol $a(r,\lambda)$ be equal to 1 if $r \geq a - \varepsilon$ ($\varepsilon > 0$). Then $\chi_a A \chi_a$ is a Fredholm operator in $L_p^N([0,a], d\mu)$ and*

$$\operatorname{ind} \chi_a A \chi_a = \operatorname{ind} A.$$

2.4. The Banach algebras $\mathcal{M}_p(N)$, $\mathcal{R}_p(N)$, $r_p(N)$, $\widetilde{r}_p(N)$

We will say that a matrix-function $a(\lambda)$ ($L_\infty(\mathbb{R}) \otimes \mathcal{L}(\mathbb{C}^N)$) is a Mellin L_p-multiplicator if the operator

$$M^{-1} a M \,:\, L_2^N(\mathbb{R}_+, d\mu) \to L_2^N(\mathbb{R}_+, d\mu),$$

where M is the Mellin transform, extends from $L_2^N(\mathbb{R}_+, d\mu) \cap L_p^N(\mathbb{R}_+, d\mu)$ to $L_p^N(\mathbb{R}_+, d\mu)$, $1 < p < \infty$.

We denote by $\mathcal{M}_p(N)$ the set of all Mellin L_p-multiplicators. $\mathcal{M}_p(N)$ is a Banach algebra with pointwise operations and the norm

$$\|a\|_{\mathcal{M}_p(N)} = \|M^{-1} a M\|_{\mathcal{L}(L_p^N(\mathbb{R}_+, d\mu))}.$$

We denote by $\mathcal{R}_p(N)$ the Banach algebra $C_b(\mathbb{R}_+, \mathcal{M}_p(N))$ and by $r_p(N)$ the closure of the class $\widetilde{\mathcal{E}}(N)$ in $\mathcal{R}_p(N)$.

It is easily seen that $r_p(N) \in C_b(\mathbb{R}_+ \times \mathbb{R}, \mathcal{L}(\mathbb{C}_N))$. We will use the following two-sided ideals in the algebra $r_p(N)$:

$$\begin{aligned}
\mathcal{J}_p^0(N) &= \{a \in r_p(N) : \lim_{r \to +0} \|a(r,\cdot)\|_{\mathcal{M}_p(N)} = 0\},\\
\mathcal{J}_p^\infty(N) &= \{a \in r_p(N) : \lim_{r \to +\infty} \|a(r,\cdot)\|_{\mathcal{M}_p(N)} = 0\},\\
\mathcal{J}_p'(N) &= \{a \in r_p(N) : \lim_{\lambda \to \infty} \sup_{r \in \mathbb{R}_+} \|a(r,\lambda)\|_{\mathcal{L}(\mathbb{C}^N)} = 0\},\\
\mathcal{J}_p(N) &= \mathcal{J}_p^0(N) \cap \mathcal{J}_p^\infty(N) \cap \mathcal{J}_p'(N).
\end{aligned}$$

Set

$$\widetilde{r}_p^0(N) = r_p(N)/\mathcal{J}_p^0(N), \quad \widetilde{r}_p^\infty(N) = r_p(N)/\mathcal{J}_p^\infty(N), \quad \widetilde{r}_p(N) = r_p(N)/\mathcal{J}_p(N).$$

Proposition 2.8.

(a) *A quotient class $a + \mathcal{J}_p^0(N)$ ($a + \mathcal{J}_p^\infty(N)$) is invertible in $\widetilde{r}_p^0(N)$ ($\widetilde{r}_p^\infty(N)$) if and only if it is invertible in $\widetilde{r}_2^0(N)$ ($\widetilde{r}_2^\infty(N)$). The latter is equivalent to the condition*

$$d_0(a) > 0 \qquad (d_\infty(a) > 0).$$

(b) *A quotient class $a + \mathcal{J}_p(N)$ is invertible in $\widetilde{r}_p(N)$ if and only if it is invertible in $\widetilde{r}_2(N)$, which in turn is equivalent to the conditions*

$$d_0(a) > 0, \qquad d_\infty(a) > 0, \qquad d(a) > 0.$$

The proof of this proposition is based on the fact that there exists an involution in the algebra $r_p(N)$. This involution is defined in the following way (see [DO]):

$$r_p(N) \ni a(r,\lambda) \to a^*(r,\lambda) \in r_p(N).$$

Proposition 2.9. *Let $A = a(r, \mathcal{D}_r) \in OP\widetilde{\mathcal{E}}(N)$. Then*

$$\begin{aligned}(2.11)\qquad \|a + \mathcal{J}_p^0(N)\|_{\widetilde{r}_p^0(N)} &\le \inf_{T\in OPJ_0(N)} \|A - T\|_{\mathcal{L}(L_p^N(\mathbb{R}_+,d\mu))}\\ &\le C_p\|a + \mathcal{J}_p^0(N)\|_{\widetilde{r}_p^0(N)},\end{aligned}$$

$$\begin{aligned}(2.12)\qquad \|a + \mathcal{J}_p(N)\|_{\widetilde{r}_p(N)} &\le \inf_{T\in OPJ(N)} \|A - T\|_{\mathcal{L}(L_p^N(\mathbb{R}_+,d\mu))}\\ &\le C_p\|a + \mathcal{J}_p(N)\|_{\widetilde{r}_p(N)},\end{aligned}$$

where $C_2 = 1$ and $C_p = 4$ in the other cases.

In the case $p = 2$ this proposition has been proved in [R2].

Definition 2.10. We will denote by $V_p(N)$ the closure of the class $OP\widetilde{\mathcal{E}}(N)$ in the Banach algebra $\mathcal{L}(L_p^N(\mathbb{R}_+, d\mu))$.
We denote by $W_p^0(N)$ the smallest closed two-sided ideal in $V_p(N)$ containing $OPJ_0(N)$ (one can show that $V_p(N) \supset \mathcal{K}(L_p^N(\mathbb{R}_+, d\mu))$. Set

$$\widetilde{V}_p(N) = V_p(N)/\mathcal{K}(L_p^N(\mathbb{R}_+, d\mu)), \qquad \widetilde{V}_p^0(N) = V_p(N)/W_p^0(N).$$

Definition 2.11. Let $A = a(r, \mathcal{D}_r) \in OP\widetilde{\mathcal{E}}(N)$. Then we define the local symbol $\widetilde{\sigma}_0(A)$ at the origin to be the quotient class $a + \mathcal{J}_p^0(N)$ in $\widetilde{r}_p^0(N)$, and we define the essential symbol $\widetilde{\sigma}(A)$ to be the quotient class $a + \mathcal{J}_p(N)$ in $\widetilde{r}_p(N)$.

Passing to the limit in (2.11), (2.12) we define the local symbol $\widetilde{\sigma}_0(A) \in \widetilde{r}_p^0(N)$ and the essential symbol $\widetilde{\sigma}(A) \in \widetilde{r}_p(N)$ for an arbitrary operator $A \in V_p(N)$.
The estimates (2.11), (2.12) are valid for an arbitrary operator $A \in V_p(N)$ if we replace $OPJ_0(N)$ by $W_p^0(N)$ in (2.11), and $OPJ(N)$ by $\mathcal{K}(L_p^N(\mathbb{R}_+, d\mu))$ in (2.12).
Let us introduce the mappings

$$\begin{aligned}\widetilde{\Sigma}_p^0 &: \widetilde{V}_p^0(N) \to \widetilde{r}_p^0(N),\\ \widetilde{\Sigma}_p &: \widetilde{V}_p(N) \to \widetilde{r}_p(N),\end{aligned}$$

defined by the formulas

$$\begin{aligned}&\widetilde{\Sigma}_p^0(A + W_p^0(N)) = \widetilde{\sigma}_0(A),\\ &\widetilde{\Sigma}_p(A + \mathcal{K}(L_p^N(\mathbb{R}_+, d\mu))) = \widetilde{\sigma}(A).\end{aligned}$$

It is easily seen that $\widetilde{\Sigma}_p^0$, $\widetilde{\Sigma}_p$ are Banach algebras isomorphisms which are isometric if $p = 2$.

Theorem 2.12. *Let $A \in V_p(N)$, $p \in (1, \infty)$. Then the following assertions are equivalent:*

(a) $A : L_p^N(\mathbb{R}_+, d\mu) \to L_p^N(\mathbb{R}_+, d\mu)$ *is a locally invertible operator at the point zero.*

(b) *The local symbol $\widetilde{\sigma}_0(A)$ is invertible in $\widetilde{r}_p^0(N)$.*

(c) $\lim_{\varepsilon \to 0} \inf_{(0,\varepsilon)\times\mathbb{R}} |\det a_0(r, \lambda)| > 0$,
where $a_0(r, \lambda)$ is an element in the quotient class $\widetilde{\sigma}_0(A)$.

(d) *There exist left and right locally inverses of A in the algebra $V_p(N)$.*

Theorem 2.13. *Let $A \in V_p(N)$, $p \in (1, \infty)$. Then the following assertions are equivalent:*

(a) $A : L_p^N(\mathbb{R}_+, d\mu) \to L_p^N(\mathbb{R}_+, d\mu)$ *is a Fredholm operator.*

(b) *The essential symbol $\widetilde{\sigma}(A)$ is invertible in $\widetilde{r}_p(N)$.*

(c) $\lim_{R\to\infty} \inf_{(r,\lambda)\in(1/R,R)\times\mathbb{R}} |\det a(r, \lambda)| > 0$,
$\lim_{R\to\infty} \inf_{(r,\lambda)\in\mathbb{R}\times\{\lambda\in\mathbb{R}:|\lambda|>R\}} |\det a(r, \lambda)| > 0$,
where $a(r, \lambda)$ is an element of the quotient class $\widetilde{\sigma}(A)$.

(d) *There exists a regularizator of A in the algebra $V_p(N)$.*

(e) *The index of a Fredholm operator $A \in V_p(N)$ is given by the formula* (2.10) *where $a(r, \lambda)$ is a representative of the quotient class $\widetilde{\sigma}(A)$.*

3. Singular integral operators on contours composed of perturbed logarithmic spirals

3.1. The algebra $\mathcal{A}_p(\Gamma, \omega)$

We say that a simple rectifiable nonclosed arc $\gamma \subset \mathbb{C}$ with endpoints x_1, x_2 is a curve in the class $\mathcal{R}$ if:

(a) $\gamma\backslash\{x_1, x_2\}$ is locally a Lyapunov arc.

(b) If for $j = 1, 2$

$$x - x_j = \varphi_j(t) = t\,e^{i\omega_j(t)}, \quad \omega_j(t) = \delta_j \ln t + \theta_j(t), \quad t \in (0, s),$$

is a parametrization of γ in a neighborhood of the endpoint x_j, then $\delta_j \in \mathbb{R}$ and $\theta_j(t)$ is a real function in $C^\infty(0, s)$ satisfying the following conditions:

$$\left|(t\frac{d}{dt})^k\,\theta_j(t)\right| \le c_{jk}, \qquad k \in \mathbb{N}, \tag{3.1}$$

$$\lim_{t\to+0}(t\frac{d}{dt})\,\theta_j(t) = 0. \tag{3.2}$$

The union of finitely many simple arcs each pair of which has at most endpoints in common is called a composed curve.

If Γ is a composed curve and $x \in \Gamma$, then in a small neighborhood of x the curve is locally comprised of a finite number of simple arcs. This number is denoted by $N(x)$.

We will consider composed curves Γ which are the union of finitely many simple arcs in the class $\mathcal{R}$. Denote by F the set of all the endpoints of curves forming the contour Γ. We suppose that for any point $z \in F$ there exists a neighborhood U_z where the following conditions are satisfied:

(a) $\Gamma \cap U_z = \{z\} \cup \bigcup_{j=1}^{N(z)}(\Gamma_j^z \cap U_z)$ with curves Γ_j^z admitting parametrizations:

$$\Gamma_j^z = \{x \in \mathbb{C} \,:\, x = z + \varphi_j^z(t),\ \varphi_j^z(t) = t\,\exp i(\delta_z \ln t + \theta_j^z(t)),\ t \in (0,s)\},$$

where $\delta_z \in \mathbb{R}$, $\theta_j^z(t)$ satisfies (3.1), (3.2).

(b) There are numbers β_j^z, γ_j^z such that

$$0 \le \beta_1^z < \gamma_1^z < \beta_2^z < \gamma_2^z < \cdots < \beta_N^z < \gamma_N^z < 2\pi, \beta_j^z < \theta_j^z(t) < \gamma_j^z, j=1,..,N. \tag{3.3}$$

The class of composed curves described above will be denoted by $C\mathcal{R}$. It should be noted that a curve in the class $C\mathcal{R}$ is a composed Carleson curve [DO].

Definition 3.14. Let $a(x)$ be a bounded complex function on a curve $\Gamma(\subset \mathbb{C})$ with finitely many discontinuities. Let x_0 be a point of discontinuity of $a(x)$. We say that $a(x)$ slowly varies at the point x_0 if there exists a neighborhood U_{x_0} such that $a(x) \in C^\infty(\Gamma \cap U_{x_0}\backslash\{x_0\})$ and

$$\sup_{x\in\Gamma\cap(U_{x_0}\backslash\{x_0\})}\left|\left((x-x_0)\frac{d}{dx}\right)^k a(x)\right| < \infty, \tag{3.4}$$

$$\lim_{\Gamma\ni x\to x_0}|x-x_0|\,a'(x) = 0. \tag{3.5}$$

We denote by $Q(\Gamma)$ the set of all bounded functions $a(x)$ with finitely many discontinuities at which $a(x)$ slowly varies. We denote by $PSV(\Gamma)$ the C^*-algebra which is the closure of $Q(\Gamma)$ in the sup-norm on Γ.

It is evident that $PSV(\Gamma)$ contains the C^*-algebra of all piecewise continuous functions on Γ. $PSV(\Gamma)$ also contains functions with second kind discontinuities. For example, if

$$a(x) = \theta_+(x)\exp(i|\ln x|^\alpha) + \theta_-(x)\exp(i\sin|\ln x|^\beta), \qquad x \in \mathbb{R}\backslash\{0\},$$

where $\theta_+(x)$ is the characteristic function of $\mathbb{R}_+$, $\theta_-(x) = 1 - \theta_+(x)$, α, $\beta \in (0,1)$, then $a(x)$ is a function in $Q(\Gamma)$.

Let

$$\omega(x) = \prod_{m=1}^{L} |x - x_m|^{\beta_m}, \quad x_m \in F, \quad \beta_m \in (-1/p, 1 - 1/p), \quad x \in \Gamma,$$

be a power weight on the contour Γ. Set $\beta_x := \beta_m$ if $x = x_m$ and $\beta_x := 0$ if $x \in \Gamma \setminus F$.

We denote by $L_p(\Gamma, \omega)$ the Banach space of all measurable complex functions on Γ with the norm

$$\|u\|_{L_p(\Gamma,\omega)} = \left(\int_\Gamma (\omega(x)\, |u(x)|\,)^p |dx| \right)^{1/p}, \qquad p \in (1, \infty).$$

Let

$$(S_\Gamma u)(x) = \frac{1}{\pi i} \int_\Gamma \frac{u(y)dy}{y - x}, \qquad x \in \Gamma,$$

be the singular integral operator on Γ where $\Gamma \in \mathcal{CR}$. Since Γ is a composed Carleson curve the operator S_Γ is bounded in $L_p(\Gamma, \omega)$, $1 < p < \infty$ [DO].

Definition 3.15. Let $\Gamma \in \mathcal{CR}$, and let ω be a power weight. We denote by $\mathcal{A}_p(\Gamma, \omega)$ the smallest subalgebra of $\mathcal{L}(L_p(\Gamma, \omega))$ containing the operator S_Γ and all operators of multiplication by functions in $PSV(\Gamma)$.

It is evident that $\mathcal{A}_p(\Gamma, \omega)$ is the closure in $\mathcal{L}(L_p(\Gamma, \omega))$ of the set of all operators of the form

$$A = \sum_k \prod_l A^{kl}, \tag{3.6}$$

where A^{kl} is either the operator S_Γ or an operator of multiplication by a function $a \in PSV(\Gamma)$.

Proposition 3.16. *The algebra $\mathcal{A}_p(\Gamma, \omega)$ contains the ideal of compact operators $\mathcal{K}(L_p(\Gamma, \omega))$.*

We denote by $\mathcal{A}_p^\pi = \mathcal{A}_p^\pi(\Gamma, \omega)$ the quotient algebra $\mathcal{A}_p(\Gamma, \omega)/\mathcal{K}(L_p(\Gamma, \omega))$ and by $\pi : \mathcal{A}_p(\Gamma, \omega) \to \mathcal{A}_p^\pi$ the canonical homomorphism.

Proposition 3.17. *Let $c(x) \in C(\Gamma)$, $A \in \mathcal{A}_p(\Gamma, \omega)$. Then the operator $cS_\Gamma - S_\Gamma cI$ is compact.*

Therefore $C^\pi = \{\pi(cI),\, c \in C(\Gamma)\}$ is a subalgebra of the center of $\mathcal{A}_p^\pi$.

These propositions are well-known for $\mathcal{A}_p(\Gamma, \omega)$ where Γ is a piecewise Lyapunov curve and $c(x) \in PC(\Gamma)$ [GK, BS]. For curves $\Gamma \in \mathcal{CR}$ we refer to [BK1].

For the investigation of the property of $A \in \mathcal{A}_p(\Gamma, \omega)$ to be a Fredholm operator we will use the local principle of Allan and Douglas (see [BS, HRS]).

For $x \in \Gamma$, denote by $J_x^\pi := J_x^\pi(p,\Gamma,\omega)$ the smallest closed two-sided ideal of $\mathcal{A}_p^\pi$ containing the set

$$\{\pi(cI) : c \in C(\Gamma),\, c(x) = 0\}.$$

Set $\mathcal{A}_x^\pi = \mathcal{A}_x^\pi(p,\Gamma,\omega) = \mathcal{A}_p^\pi / J_x^\pi$ and abbreviate the coset $\pi(A) + J_x^\pi$ by $\pi_x(A)$.

The local principle of Allan and Douglas implies that if $A \in \mathcal{A}_p(\Gamma,\omega)$, then $\pi(A)$ is invertible in $\mathcal{A}_p^\pi(\Gamma,\omega)$ if and only if $\pi_x(A)$ is invertible in $\mathcal{A}_x^\pi(p,\Gamma,\omega)$ for every $x \in \Gamma$.

Assume that for each $x \in \Gamma$ we can find a Banach space X_x, a Banach algebra $B_x \subset \mathcal{L}(X_x)$ and a homomorphism $s_x : \mathcal{A}_x^\pi \to B_x$ such that $\pi_x(A) \in \mathcal{A}_x^\pi$ is invertible if and only if $s_x(\pi_x(A))$ is invertible in B_x. Let

$$\operatorname{Sym}_x := s_x \circ \pi_x : \mathcal{A}_p^\pi \to B_x.$$

It follows from the Allan-Douglas principle that $\pi(A)$ is invertible in $\mathcal{A}_p^\pi(\Gamma,\omega)$ if and only if $\operatorname{Sym}_x(A)$ is invertible in B_x for all $x \in \Gamma$.

For a local analysis of the algebra $\mathcal{A}_p^\pi(\Gamma,\omega)$ we will use the Banach spaces $X_x = L_p^{N(x)}(\mathbb{R}_+, d\mu)$, $x \in \Gamma$, and the Banach algebras $B_x = \widetilde{V}_p^0(N(x))$, $x \in \Gamma$.

Let $x_0 \in \Gamma$ and $\widetilde{\Gamma}_{x_0} = \Gamma \cap U_{x_0} = \bigcup_{j=1}^{N(x_0)} \Gamma_j^{x_0} \cap U_{x_0}$, $\omega_\beta(x_0) = |x - x_0|^{\beta(x_0)}$, $-1/p < \beta(x_0) < 1 - 1/p$, $U_{x_0} = \{x \in \mathbb{C} : |x - x_0| < s\}$. As usual let $\varepsilon_k := 1$ if $\Gamma_k^{x_0}$ is oriented away from x_0 and let $\varepsilon_k := -1$ in the opposite case.

The map

$$\Phi_{x_0} : L_p(\widetilde{\Gamma}_{x_0}, \omega_\beta(x_0)) \to L_p^N((0,s), d\mu),$$

defined by

$$(\Phi_{x_0} f)(t) = \operatorname{column}\left(t^{\beta(x_0)+1/p} f(x_0 + \varphi_1^{x_0}(t)), \ldots, t^{\beta(x_0)+1/p} f(x_0 + \varphi_N^{x_0}(t))\right)$$

is a Banach space isomorphism.

Below we will omit x_0 in the notations of weights and curves.

Choose any function $\chi \in C^\infty(\mathbb{R}^2)$ such that $\chi(x_0) = 1$ and $\operatorname{supp}\chi \subset U_{x_0}$. Then

$$S^{x_0} := \Phi_{x_0}\, \chi\, S_{\Gamma \cap U_{x_0}}\, \chi\, \Phi_{x_0}^{-1} : L_p^N(\mathbb{R}_+, d\mu) \to L_p^N(\mathbb{R}_+, d\mu)$$

is a well-defined and bounded operator. Let

$$\nu_{jk}(\alpha,\beta,\lambda) = \begin{cases} \dfrac{\exp[(\beta - \alpha - \pi)\lambda]}{\sinh \pi\lambda}, & k > j, \\[2ex] \dfrac{\exp[(\beta - \alpha + \pi)\lambda]}{\sinh \pi\lambda}, & k < j, \\[2ex] \coth \pi\lambda, & k = j, \end{cases} \qquad \lambda \in \mathbb{C}. \tag{3.7}$$

The following theorem was already stated in [BKR], but we omitted the proof there.

Theorem 3.18. *The operator S^{x_0} is a Mellin pseudodifferential operator with double symbol*

$$S^{x_0}(t,\tau,\lambda) = (S^{x_0}_{jk}(t,\tau,\lambda))_{j,k=1}^N \in \widetilde{S}_d(N).$$

Its usual symbol $\sigma(S^{x_0})(t,\lambda) = (\sigma^{jk}(S^{x_0})(t,\lambda))_{j,k=1}^N$ is given by the formulas

$$\sigma^{jk}(S^{x_0})(t,\lambda) = \varepsilon_k \nu_{jk}\left(\theta_j(t),\theta_k(t),\frac{\lambda+i(\beta+1/p)}{1+i\delta}\right) + r_{jk}(t,\lambda), \tag{3.8}$$

$j,k=1,\dots,N$, where $r_{jk}\in J(1)$.

Note that the formulas (3.8) are similar to the formulas for the local symbol of S_Γ on a piecewise Lyapunov contour [ROS]. But there are the following differences:

1) The functions $\theta_j(t)$, $j=1,\dots,N$, depend on the variable $t\in(0,s)$.
2) The parameter $\mu = \frac{\lambda+i(\beta+1/p)}{1+i\delta}$ runs over the straight line $\{\mu=\eta+i(-\delta\eta+\beta+1/p),\ \eta\in\mathbb{R}\}$ with the slope $-\delta$ passing through the point $i(\beta+1/p)$.

We now give a proof of Theorem 3.18. It is based on the following

Proposition 3.19. *Let $0<\Im(\lambda)<1$, $\Re(\xi)\ge 1$, and*

$$J(\lambda,a,\xi) = \frac{a\xi}{\pi i}\int_0^\infty (a-t^\xi)^{-1}\, t^{-i\lambda}\,\frac{dt}{t}. \tag{3.9}$$

Then

$$J(\lambda,a,\xi) = \begin{cases} \dfrac{2a\exp(-i\lambda/\xi-1)\ln a}{\exp(2\pi\lambda/\xi)-1}, & a\in\mathbb{C}\backslash\bar{\mathbb{R}}_+,\\ \coth(\pi\lambda/\xi), & a=1.\end{cases} \tag{3.10}$$

The integral (3.9) is understood in the sense of the principal value if $a=1$.

Proof. Let $a\in\mathbb{C}\backslash\bar{\mathbb{R}}_+$. Then $J(\lambda,a,\xi)$ is an analytic function with respect to ξ in the domain $\Re(\xi)>\Im(\lambda)$. If $\xi\in\mathbb{R}$, then

$$J(\lambda,a,\xi) = \frac{a}{\pi i}\int_0^\infty \frac{\tau^{-i\lambda/\xi}}{a-\tau}\frac{d\tau}{\tau}. \tag{3.11}$$

The integral (3.11) is well-known and equal to

$$\frac{2a\exp(-i\lambda/\xi-1)\ln a}{\exp(2\pi\lambda/\xi)-1} = J_1(\lambda,a,\xi).$$

$J_1(\lambda,a,\xi)$ is an analytic function in the domain $\Re(\xi)>0$, therefore

$$J(\lambda,a,\xi) = J_1(\lambda,a,\xi), \qquad \Re(\xi)>\Im(\lambda).$$

The equality $J(\lambda,1,\xi)=\coth(\pi\lambda/\xi)$ follows from the limit property of Cauchy type integrals. □

Proof (of Theorem 3.18). It is easily seen that

$$(S^{x_0}_{jk}v_k)(t)=\varepsilon_k\int_0^\infty k_{jk}(t,\tau)\,v_k(\tau)\,\frac{d\tau}{\tau},\quad t\in\mathbb{R}_+,\tag{3.12}$$

where $v_k(\tau)=u(x_0+\varphi_k(\tau))$,

$$k_{jk}(t,\tau)=\frac{1}{\pi i}\chi_j(t)\chi_k(\tau)\frac{(1+i\tau\omega_k'(\tau))\,(t\tau^{-1})^{\beta+1/p}}{1-(t\tau^{-1})\exp i(\omega_j(t)-\omega_k(\tau))},\tag{3.13}$$

$\chi_j(t)=\chi(x_0+\varphi_j(t))$.
If $j=k$ then we can write (3.13) in the form

$$k_{jj}(t,\tau)=\frac{1}{\pi i}\chi_j(t)\chi_j(\tau)\frac{(1+i\tau\omega_j'(\tau))\,(t\tau^{-1})^{\beta+1/p}}{1-(t\tau^{-1})^{1+ih_j(t,\tau)}},\tag{3.14}$$

where

$$h_j(t,\tau)=\frac{\omega_j(t)-\omega_j(\tau)}{\ln t-\ln\tau}=((1-q)t+q\tau)\,\omega_j'((1-q)t+q\tau),\qquad q\in(0,1).$$

Let us note that

$$\lim_{\substack{t\to+0\\ \tau\to+0}}h_j(t,\tau)=\delta,\tag{3.15}$$

by virtue of the condition (3.2). If $j\neq k$, then

$$k_{jk}(t,\tau)=\frac{\varepsilon_k}{\pi i}\chi_j(t)\chi_k(\tau)\,\frac{\mathrm{e}^{\,i(\theta_k(\tau)-\theta_j(t))}(1+i\tau\omega_j'(\tau))\,(t\tau^{-1})^{\beta+1/p}}{\mathrm{e}^{\,i(\theta_k(\tau)-\theta_j(t))}-(t\tau^{-1})^{1+i\delta}}.\tag{3.16}$$

Applying Proposition 3.19 we obtain

$$\begin{aligned}
k_{jj}(t,\tau)&=\varepsilon_k\chi_j(t)\chi_j(\tau)\frac{(1+i\tau\omega_j'(\tau))}{1+ih_j(t,\tau)}\\
&\quad\cdot\frac{1}{2\pi}\int_{\mathbb{R}}\coth\left(\frac{\pi(\mu+i(\beta+1/p))}{1+ih(t,\tau)}\right)(t\tau^{-1})^{i\mu}\,d\mu,\\
k_{jk}(t,\tau)&=\varepsilon_k\chi_j(t)\chi_k(\tau)\frac{(1+i\tau\omega_k'(\tau))}{1+i\delta}\\
&\quad\cdot\frac{1}{2\pi}\int_{\mathbb{R}}\frac{\exp\frac{\mu+i(\beta+1/p)}{1+i\delta}\,(\theta_k(\tau)-\theta_j(t)\mp\pi)}{\sinh\frac{\pi(\mu+i(\beta+1/p))}{1+i\delta}}(t\tau^{-1})^{i\mu}\,d\mu.
\end{aligned}\tag{3.17}$$

The sign $(-)$ before π is taken if $k > j$ and the sign $(+)$ is taken if $k < j$. It should be noted that

$$\begin{aligned} 0 < \theta_k(\tau) - \theta_j(t) < 2\pi, \qquad & k > j, \quad \text{for all } (t,\tau) \in (0,s), \\ -2\pi < \theta_k(\tau) - \theta_j(t) < 0, \qquad & k < j, \quad \text{for all } (t,\tau) \in (0,s), \end{aligned}$$

by virtue of the conditions (3.3). Let

$$a^{x_0}(t,\tau,\mu) = (a^{x_0}_{jk}(t,\tau,\mu))^N_{j,k=1}$$

where for $\mu \in \mathbb{R}$,

$$\begin{aligned} a^{x_0}_{jj}(t,\tau,\mu) &= \varepsilon_k \chi_j(t)\chi_j(\tau)\frac{(1+i\tau\omega'_j(\tau))}{1+ih_j(t,\tau)} \\ &\quad \cdot \coth\frac{\pi(\mu+i(\beta+1/p))}{1+ih(t,\tau)}, \\ a^{x_0}_{jk}(t,\tau,\mu) &= \varepsilon_k \chi_j(t)\chi_k(\tau)\frac{(1+i\tau\omega'_k(\tau))}{1+i\delta} \\ &\quad \cdot \frac{\exp\frac{\mu+i(\beta+1/p)}{1+i\delta}(\theta_k(\tau)-\theta_j(t)\mp\pi)}{\sinh\frac{\pi(\mu+i(\beta+1/p))}{1+i\delta}}. \end{aligned} \tag{3.18}$$

It is easy to check that $a^x(t,\tau,\mu) \in \widetilde{\mathcal{E}}_d(N)$ and hence S^{x_0} is a Mellin pseudodifferential operator in the class $OP\widetilde{\mathcal{E}}_d(N)$. If we take into account that

$$\lim_{\tau\to+0} \tau\omega'_j(\tau) = \delta, \qquad \lim_{t\to+0} h_j(t,t) = \delta,$$

and Proposition 2.4 (c) we obtain the formulas (3.8) for the symbol $\sigma_{S^{x_0}}(t,\mu)$. □

It follows from Theorem 3.18 that $\pi_x(S_\Gamma)$ is invertible in $\mathcal{A}^\pi_x$ if and only if the operator $S^{x_0} : L_p((0,s),d\mu) \to L_p((0,s),d\mu)$ is locally invertible at the point 0. It follows from Section 2 that the local invertibility of S^{x_0} at the point 0 is equivalent to the invertibility of the quotient class $S^{x_0} + W^0_p(N)$.

Therefore we can define

$$\operatorname{Sym}_x(S_\Gamma) = S^{x_0} + W^0_p(N(x)) \in \widetilde{V}^0_p(N(x))$$

and

$$\sigma_x(S_\Gamma) = \left(\varepsilon_k \nu_{jk}(\theta_j(t),\theta_k(t),\frac{\lambda+i(\beta+1/p)}{1+i\delta})\right)^{N(x)}_{j,k=1} + \mathcal{J}^0_p(N(x)).$$

Let $a(x) \in PSV(\Gamma)$. Set $a^{x_0}_j(t) = a(x_0+\varphi^{x_0}_j(t))$, $j = 1,\dots,N(x_0)$. With a point x_0 and a function $a(x)$ we will associate the operator of multiplication by the diagonal matrix

$$a^{x_0}(t)I = \operatorname{diag}(a^{x_0}_1(t),\dots,a^{x_0}_{N(x_0)}(t))I$$

acting on $L^{N(x_0)}_p(\mathbb{R}_+,d\mu)$.

It follows from the definition of the class $PSV(\Gamma)$ that $a^x(t)I \in V_p(N(x))$. Define

$$\begin{aligned}\mathrm{Sym}_x(aI) &= a^x(t)I + W_p^0(N(x)) \in \widetilde{V}_p^0(N(x)),\\ \sigma_x(aI) &= a^x(t) + \mathcal{J}_p^0(N(x)) \in \widetilde{r}_p(N(x)).\end{aligned}$$

Let $A \in \mathcal{A}_p(\Gamma, \omega)$ be an operator of the form (3.6). Then we can define

$$\begin{aligned}\mathrm{Sym}_x(A) &= \sum_k \prod_l \mathrm{Sym}_x(A^{kl}),\\ \sigma_x(A) &= \sum_k \prod_l \sigma_x(A^{kl}).\end{aligned}$$

It should be noted that if x is a point of continuity of a and there exists a neighborhood U of x such that $\Gamma \cap U$ is a Lyapunov curve then

$$\sigma_x(aI) = a(x), \qquad \sigma_x(aS_\Gamma) = a(x)\mathrm{sgn}\,\xi, \quad \xi \in \mathbb{R}.$$

Proposition 3.20. *Let A be an operator of the form* (3.6). *Then*

$$\|\sigma_x(A)\|_{\widetilde{r}_p^0(N(x))} \le \|\pi_x(A)\|_{\mathcal{A}_x^\pi} \le C_p \|\sigma_x(A)\|_{\widetilde{r}_p^0(N(x))}, \tag{3.19}$$

where $C_p = 1$ if $p = 2$ and $C_p = 4$ in the other cases.

This proposition follows from Proposition 2.9.

Passing to the limit in (3.19) we define a local matrix symbol $\sigma_x(A) \in \widetilde{r}_p^0(N(x))$ for an arbitrary operator $A \in \mathcal{A}_p(\Gamma, \omega)$.

Theorem 3.21. *The element $\pi_x(A)$ is invertible in $\mathcal{A}_x^\pi$ if and only if $\sigma_x(A)$ is invertible in $\widetilde{r}_p^0(N(x))$. The latter condition is fulfilled if and only if*

$$\lim_{\varepsilon \to 0} \inf_{\substack{t \in (0,\varepsilon)\\ \mu \in \mathbb{R}}} |\det \widetilde{\sigma}_x(A)(t, \mu)| > 0, \tag{3.20}$$

where $\widetilde{\sigma}_x(A)$ is a representative of the quotient class $\sigma_x(A)$.

This theorem follows from Theorem 2.12.

Theorem 3.22. *$A \in \mathcal{A}_p(\Gamma, \omega)$ is a Fredholm operator in $L_p(\Gamma, \omega)$, $p \in (1, \infty)$, if and only if $\sigma_x(A)$ is invertible in $\widetilde{r}_p^0(N(x))$ for every point $x \in \Gamma$, that is, if and only if for each point $x \in \Gamma$ the condition* (3.20) *holds.*

Theorem 3.22 follows from the Allan-Douglas local principle and Theorem 3.21.

Theorem 3.23. *Let $A \in \mathcal{A}_p(\Gamma, \omega)$. Then*

$$\sup_{x \in \Gamma} \|\sigma_x(A)\|_{\widetilde{r}_p^0(N(x))} \le \|A\|_{\mathcal{A}_p^\pi(\Gamma,\omega)} \le C_p \sup_{x \in \Gamma} \|\sigma_x(A)\|_{\widetilde{r}_p^0(N(x))}, \tag{3.21}$$

where $C_p = 1$ if $p = 2$ and $C_p = 4$ in the other cases.

Proof. It is easily seen that $\mathcal{A}_p(\Gamma,\omega)$ is KMS with respect to the C^*-algebra $C^0(\Gamma)$ (see [ROS, p. 31]). Hence the right hand side of the estimate (3.21) follows from Proposition 3.20 (see [ROS, Theorem 5.2]). The estimate from the left is evident. □

We will denote by $\sigma(A) = \{\sigma_x(A)\}_{x\in\Gamma}$ the matrix symbol of an operator $A \in \mathcal{A}_p(\Gamma,\omega)$. We denote by $\mathcal{B}_p(\Gamma,\omega)$ the Banach algebra of all matrix symbols of operators $A \in \mathcal{A}_p(\Gamma,\omega)$ with pointwise operations and the norm

$$\|\sigma(A)\| = \sup_{x\in\Gamma} \|\sigma_x(A)\|_{\widetilde{r}_p(N(x))}.$$

Theorem 3.24. *The mapping*

$$\Sigma_p^\pi : \mathcal{A}_p^\pi(\Gamma,\omega) \to \mathcal{B}_p(\Gamma,\omega)$$

is a Banach algebra isomorphism.

3.2. Index formula

Let $E \subset \Gamma$ be a finite set, $PSV(\Gamma,E)$ be a subalgebra of $PSV(\Gamma)$ consisting of functions with points of discontinuities in E. We will suppose that E is the set of all nodes in Γ, i.e., $\Gamma\backslash E$ is a local Lyapunov curve.

We denote by $\mathcal{A}_p(\Gamma,E,\omega)$ the smallest subalgebra of $\mathcal{A}_p(\Gamma,\omega)$ containing S_Γ and operators of multiplication by functions $a \in PSV(\Gamma,E)$.

Theorem 3.25. *Let $A \in \mathcal{A}_p(\Gamma,E,\omega)$ be a Fredholm operator. Then*

$$\operatorname{ind} A = -\sum_{j=1}^{L}(2\pi)^{-1}\left[\arg\frac{\sigma_x^+(A)}{\sigma_x^-(A)}\right]_{x\in\Gamma_j} - \sum_{j=1}^{L}(2\pi)^{-1}\left[\arg\det\widetilde{\sigma}_{x_j}(A)(t_j,\mu)\right]_{\mu=-\infty}^{\infty}. \tag{3.22}$$

In the formula (3.22), $\sigma_x(A) = (\sigma_x^+(A),\sigma_x^-(A))$ is a local symbol of A at the point $x \in \Gamma\backslash E$, Γ_j, $j = 1,\dots,L$, are arcs composing the contour Γ with induced orientation, $x_j \in E$ and t_j is a point in a small neighborhood of the origin such that

$$\inf_{\mu\in\mathbb{R}} |\det\widetilde{\sigma}_{x_j}(A)(t,\mu)| > 0.$$

Theorem 3.25 is proved by the standard method of separation of singularities (see [GK, BS, GKS]). The essential role in this proof is played by Proposition 2.7 and the formula (2.10).

Acknowledgements

The paper is supported by Russian Found of Foundamental Investigations (RFFI), Grant: 96-01-01195 RFFI.

References

[BK1] BÖTTCHER, A., KARLOVICH, YU.I.: Toeplitz and singular integral operators on Carleson curves with logarithmic whirl points; Integral Equations Operator Theory 22 (1995), 127–161.

[BK2] BÖTTCHER, A., KARLOVICH, YU.I.: Toeplitz and singular integral operators on general Carleson curves; Operator Theory: Adv. Appl. 90 (1996), 119–152.

[BK3] BÖTTCHER, A., KARLOVICH, YU.I.: Toeplitz operators with PC symbols on general Carleson curves with arbitrary Muckenhoupt weights; Preprint, TU Chemnitz, September 1995.

[BKR] BÖTTCHER, A., KARLOVICH, YU.I., RABINOVICH, V.S.: Emergence, persistence, and disappearance of logarithmic spirals in the spectra of singular integral operators; Integral Equations Operator Theory 25 (1996), 406–444.

[BS] BÖTTCHER, A., SILBERMANN, B.: Analysis of Toeplitz operators; Akademie Verlag, Berlin 1990.

[DO] DYNKIN, E.M., OSILENKER, B.P.: Weight norm estimates of singular integral operators and their applications; J. Sov. Math. 30 (1985), 2094–2154.

[GK] GOHBERG I.C., KRUPNIK, N.YA.: One-dimensional linear singular integral equations, Vol. 1, 2; Birkhäuser Verlag, Basel, Boston, Berlin 1992 (Russian original: Shtiintsa, Kishinev 1973).

[GKS] GOHBERG, I.C., KRUPNIK, N.YA., SPITKOVSKY, I.M.: Banach algebra of singular integral operators with piecewise continuos coefficients, general contour and weight; Integral Equations Operator Theory 17 (1993), 322–337.

[G] GRUSHIN, V.V.: Pseudodifferential operators on $\mathbb{R}_n$ with bounded symbols; Funkts. Anal. Prilozh. 4 (1970), 202–212 (in Russian).

[HRS] HAGEN, R., ROCH, S., SILBERMANN, B.: Spectral theory of approximation methiods for convolution equations; Operator Theory: Adv. App. Vol. 74; Birkhäuser Verlag, Basel, Boston, Berlin 1994.

[R1] RABINOVICH, V.S.: Singular integral operators on a complex contour with oscillating tangent and pseudodifferential Mellin operators; Soviet Math. Dokl. 44 (1992), 791–796.

[R2] RABINOVICH, V.S.: Algebras of singular integral operators on composed contour and pseudodifferential operators; Mat. Zametki 58 (1995), 68–85 (in Russian).

[RF] RABINOVICH, V.S.: Fredholm property of pseudodifferential operators with symbols in the class $S^m_{\rho,\delta}$ $(0 \leq \delta = \rho < 1)$; Mat. Zametki 27 (1980), 226–231 (in Russian).

[ROS] ROCH, S., SILBERMANN, B.: Algebras of convolution operators and their image in the Calkin algebra; Report R-Math-05190, Karl-Weierstraß-Institut für Mathematik, Berlin 1990.

[SCNM] SIMONENKO, I.B., CHIN NGOK MINH: A local method in the theory of one-dimensional singular integral equations with piecewise continuous coefficients. Fredholmness; Izd. Rostov Univ., Rostov-na-Donu 1986 (in Russian).

[S] SPITKOVSKY, I.M.: Singular integral operators with PC symbols on spaces with general weight; J. Funct. Anal. 105 (1992), 129–149.

[T] TAYLOR, M.E.: Pseudodifferential operators; Princeton University Press, Princeton, New Jersey 1981.

Rostov State University
Dept. of Mathematics and Mechanics
5, ul. Zorge
Rostov-na-Donu, 344104
Russia
rabinov@ns.unird.ac.ru

1991 Mathematics Subject Classification: Primary 58G15, 47G30; Secondary 47A56

Submitted: July 4, 1996

Operator Theory:
Advances and Applications, Vol. 102

Wiener-Hopf factorization of singular matrix functions

M. Rakowski

We discuss properties of Wiener-Hopf factorization of singular matrix functions relative to a rectifiable contour.

1. Introduction

Wiener-Hopf factorization of nonsingular matrix functions on a rectifiable contour has been studied extensively. Fairly complete accounts can be found in the monographs [5], [9], and, in the canonical case, [4]. The understanding of the singular case is less satisfactory. A system-theoretic interpretation of factorization indices at infinity of a singular matrix polynomial has been given in [8]. A necessary and sufficient condition for existence of a canonical Wiener-Hopf factorization of a rational matrix function has been obtained in [11]. The necessary and sufficient condition for existence of a generalized Wiener-Hopf factorization of a measurable singular matrix function G has been indicated in [12]. The condition involves the defect numbers of the Riemann problem with coefficient G and the dual problem and the closure of the images of both problems with respect to rational vector functions. In [13], the normal solvability of the Riemann problem with coefficient G has been characterized in terms of continuity of certain operators induced by a generalized Wiener-Hopf factorization of G in the case where G takes injective values almost everywhere (a.e.) on the contour, or the right factor in the factorization and its multiplicative inverse are essentially bounded.

The definitions of a generalized Wiener-Hopf factorization of a singular matrix function which have been adopted in [12] and [13] differ. The objective of this note is to relate both definitions. Below, we make use of rational vector or matrix functions with no poles on the contour. Due to engineering applications, such functions and their properties have been widely investigated in the literature (see e.g. [1] and the references cited there and, in the singular case, [3]). We will use some of these techniques.

Throughout this note, we will assume that the contour Γ is rectifiable. A rectifiable contour is *regular* if there exists an upper bound (independent of a disk) for the ratio of the length of the part of the contour inside a disk D to the radius of D. A contour Γ is regular if and only if the operator of singular integration along Γ is bounded in the space $L_p := L_p(\Gamma)$, $1 < p < \infty$ [7]. Throughout the note, we will also assume that the contour Γ consists of a finite number of disjoint simple curves. All other assumptions regarding Γ, including regularity, will be stated explicitly.

The region to the left of Γ will be denoted by $\mathcal{D}_+$, and the complementary region $\mathbb{C}_\infty \backslash \{\mathcal{D}_+ \cup \Gamma\}$ by $\mathcal{D}_-$. $E_{\infty+}$ will denote the space of functions bounded and analytic in $\mathcal{D}_+$ and, if $1 \leq p < \infty$, E_{p+} will denote the space of functions f analytic in $\mathcal{D}_+$ for which there exists a sequence of expanding regions $\mathcal{D}_k$ with rectifiable boundaries Γ_k such that

i) $\Gamma_k \subset \mathcal{D}_+$ for $k = 1, 2, \dots$,
ii) $\bigcup_{k=1}^{\infty} \mathcal{D}_k = \mathcal{D}_+$,
iii) $\sup_{k\to\infty} \int_{\Gamma_k} |f|^p < \infty$.

We will denote by L_{p+} the subspace of L_p formed by functions f such that f is a.e. a nontangential boundary limit of an element of E_{p+}, $1 \leq p \leq \infty$. The spaces E_{p-} and L_{p-} are the $\mathcal{D}_-$ analogues of E_{p+} and L_{p+}. We will denote by $\dot{E}_{p-}$ (or $\dot{L}_{p-}$) the subspace of E_{p-} formed by functions vanishing at infinity. $\mathcal{R}$ will denote the space of rational functions without poles on Γ. If $X \in \{L_{p+}, L_{p-}, \dot{L}_{p-}, \mathcal{R}\}$, we will denote the space of $m \times n$ matrices over X by $X^{m\times n}$. If $m = 1$ or $n = 1$, we will usually write X^n or X^m. $\mathcal{S}$ will denote the operator of singular integration along Γ,

$$(1.1) \qquad (\mathcal{S}f)(t) = \frac{1}{\pi i} \int_\Gamma \frac{f(\tau)}{\tau - t}\, d\tau$$

with the integral understood in the sense of Cauchy principal value, and $\mathcal{P} = \frac{1}{2}(I + \mathcal{S})$ and $\mathcal{Q} = \frac{1}{2}(I - \mathcal{S})$ will be the usual analytic and antianalytic projections. If the function f in (1.1) takes values in $\mathbb{C}^n$, we will assume that $\mathcal{S}$, $\mathcal{P}$ and $\mathcal{Q}$ act componentwise.

The note is organized as follows. In Section 2 we discuss definitions of Wiener-Hopf factorization of singular matrix functions used in literature. Section 3 relates Wiener-Hopf factorization of an $m \times n$ singular matrix function G with the Riemann problem

$$(1.2) \qquad \phi_+ + G\phi_- = g$$

and the dual problem

$$(1.3) \qquad \psi_- + G^T\psi_+ = h.$$

Here $g \in L_p^m$ and $h \in L_q^n$ are given, and one looks for $\phi_+ \in L_{p+}^m$, $\phi_- \in \dot{L}_{p-}^n$, $\psi_+ \in L_{q+}^m$, $\psi_- \in \dot{L}_{q-}^n$. In Section 4 we give a necessary and sufficient condition for factorability of a function whose restriction to each component of Γ is rational.

2. Generalized Wiener-Hopf factorization

Let G be an $m \times n$ matrix function with entries in L_1 and let $p, q > 1$ be the conjugate exponents, $\frac{1}{p} + \frac{1}{q} = 1$. A *generalized* (*left*) *Wiener-Hopf factorization* of G is a factorization $G = G_+\Lambda G_-$ where

$$(2.1) \quad G_+ \in L_{p+}^{m\times m}, \qquad G_+^{-1} \in L_{q+}^{m\times m}, \qquad G_- \in L_{q-}^{n\times n}, \qquad G_-^{-1} \in L_{p-}^{n\times n},$$

and

$$(2.2)\quad \Lambda(t) = \begin{bmatrix} \tau(t)^{\kappa_1} & & & 0 \\ & \ddots & & \\ & & \tau(t)^{\kappa_k} & \\ 0 & & & 0 \end{bmatrix} = \begin{bmatrix} \mathrm{diag}\,(\tau(t)^{\kappa_1}, \ldots, \tau(t)^{\kappa_k}) & 0 \\ 0 & 0 \end{bmatrix}$$

with $\kappa_1 \geq \kappa_2 \geq \cdots \geq \kappa_k$ integers, and $\tau(t)$ a scalar rational function of McMillan degree 1 with the pole in $\mathcal{D}_-$ and the zero in $\mathcal{D}_+$. Typically, $\tau(t) = \frac{t-t_+}{t-t_-}$ where $t_+ \in \mathcal{D}_+$ and $t_- \in \mathcal{D}_-$ are fixed. If $0 \in \mathcal{D}_+$ and $\infty \in \mathcal{D}_-$, one usually takes $\tau(t) = t$. If G_+, G_- and their multiplicative inverses are continuous up to the boundary, the term *generalized* is omitted.

Let G be an $m \times n$ matrix function, and consider the linear space

$$(2.3)\quad \mathcal{Y}_p^m = \{g \in L_p^m : \langle g, \psi_+\rangle = 0 \ \text{ for all } \psi_+ \in L_{q+}^m \text{ such that } G^T\psi_+ = 0\}$$

where

$$(2.4)\quad \langle g, \psi\rangle = \int_\Gamma g^T(t)\psi(t)\,dt$$

for all $g \in L_p^m$ and $\psi \in L_q^m$. View L_p^m and L_q^m as real vector spaces, so that L_q^m is, via (2.4), the dual space of L_p^m. If g belongs to the image of the Riemann problem and $\psi_+ \in L_{q+}^m$ is such that $G^T\psi_+ = 0$, $g^T\psi_+ \in L_{1+}$ and so $\langle g,\, \psi_+\rangle = 0$. Thus, $\mathcal{Y}_p^m$ contains the image of the Riemann problem. Since $\mathcal{Y}_p^m$ is the intersection of closed subspaces of L_p^m, it is complete. In the special case where G takes nonsingular values a.e. on Γ, $G^T\psi_+ = 0$ implies $\psi_+ = 0$ and $\mathcal{Y}_p^m = L_p^m$.

Suppose $G_+\Lambda G_-$ is a Wiener-Hopf factorization of G and define operators

$$(2.5)\quad K_+ = G_+\mathcal{P}G_+^{-1} : \mathcal{Y}_p^m \to L_p^m,$$

$$(2.6)\quad K_- = G_-^{-1}\Lambda^\dagger\mathcal{Q}G_+^{-1} : \mathcal{Y}_p^m \to L_p^n,$$

where $\Lambda^\dagger$ denotes the pointwise Moore-Penrose inverse of Λ. If K_+ and K_- are bounded, the image of the Riemann problem is closed (cf. Proposition 4.2 in [13]). The generalized Wiener-Hopf factorization of G for which K_+ and K_- are bounded is said to be *Fredholm*.

If G takes nonsingular values a.e. on Γ, continuity of K_+ (or, equivalently, $G_+\mathcal{Q}G_+^{-1}$) and essential boundedness of G^{-1} imply boundedness of K_-. (In fact, boundedness of K_+ and G^{-1} are equivalent to boundedness of K_+ and K_-, or to normal solvability of the Riemann problem with coefficient G; cf. Theorem 3.14 in [9]). In [5] (cf. [2]), the boundedness of K_+ and G^{-1} is part of the definition of a generalized Wiener-Hopf factorization. That is, a generalized Wiener-Hopf factorization in [5] is what we call a Fredholm factorization.

A nonsingular matrix valued function is Fredholm factorable if and only if the Riemann problem with coefficient G is normally solvable. Hence, if a nonsingular G admits a Fredholm factorization in L_p, each generalized Wiener-Hopf factorization of G in L_p is Fredholm. If G takes singular values on Γ, a generalized Wiener-Hopf factorization of G in L_p need not be Fredholm although G may be Fredholm factorable. Indeed, let Γ be the unit circle, let $p = \frac{3}{2}$, let $q = 3$, and pick a branch of $\left(\frac{t-1}{t}\right)^{\frac{1}{4}}$ on $\mathbb{C}_\infty \backslash [0,1]$. Then

$$(2.7) \qquad G_+(t)\Lambda(t)G_-(t) := \begin{bmatrix} 1 & 0 \\ 0 & 1 \end{bmatrix} \begin{bmatrix} 1 & 0 \\ 0 & 0 \end{bmatrix} \begin{bmatrix} 1 & 0 \\ \left(\frac{t-1}{t}\right)^{-\frac{1}{4}} & \left(\frac{t-1}{t}\right)^{\frac{1}{4}} \end{bmatrix}$$

is a complete Wiener-Hopf factorization of a function $G(t) = \begin{bmatrix} e_1 & 0 \end{bmatrix}$. The function G is Fredholm factorable. In fact, one can find a Fredholm factorization of G with all factors constant. Since

$$(2.8) \qquad K_-\left(\begin{bmatrix} \left(\frac{t-1}{t}\right)^{-\frac{1}{2}} / t \\ 0 \end{bmatrix}\right) = \begin{bmatrix} \left(\frac{t}{t-1}\right)^{\frac{1}{2}} / t \\ -\left(\frac{t}{t-1}\right) / t \end{bmatrix} \notin L^2_{\frac{3}{2}},$$

the factorization (2.7) is not Fredholm.

Suppose the zero bottom rows and zero right columns in $\Lambda(t)$ are deleted, and G_+ and G_- in (2.1) are modified accordingly. The restricted G_+ and G_- have one-sided inverses in L_{p+} and L_{q-}. We will call the resulting factorization a *restricted* factorization. To emphasize the difference, we will call factorization (2.1) the *complete* generalized Wiener-Hopf factorization. The definition of restricted factorization was used in [6], [11] and [12], while the definition of a complete factorization was used in [13]. We note that the uniqueness of factors in the complete factorization is not as satisfactory as in the restricted factorization (cf. Theorem 2.4 in [12] and Theorem 2.2 in [13]).

Proposition 2.1. *Suppose G admits a restricted Wiener-Hopf factorization (with all factors and their multiplicative inverses continuous up to the boundary) and at least one of the following two conditions holds:*

(i) *The restriction of G to each component of Γ is a rational matrix function,*

(ii) *Γ is the unit circle.*

Then G admits a complete Wiener-Hopf factorization in L_p.

Proof. Since a function which admits a Wiener-Hopf factorization has neither zeros nor poles on Γ, the assertion in case (i) follows from Proposition 4.1 below. We verify the proposition in case (ii).

Denote by $\mathcal{D}$ ($\overline{\mathcal{D}}$, respectively) the open (closed, respectively) unit disk, and by $\mathbb{T}$ the unit circle $\overline{\mathcal{D}} \backslash \mathcal{D}$. Suppose (ii) holds and $G = G_+ \Lambda G_-$ is a Wiener-Hopf factorization such that G_+ has a multiplicative left inverse analytic in $\mathcal{D}$ and continuous in $\overline{\mathcal{D}}$. Let $\epsilon = \min\{\underline{\sigma}(G_+(z)) : z \in \mathbb{T}\}$ where $\underline{\sigma}(*)$ denotes the smallest singular value of a matrix $*$. By Mergelyan's Theorem, G_+ can be approximated by a rational matrix function R_+ with no poles in $\overline{\mathcal{D}}$ so that $\|G_+(z) - R_+(z)\| < \epsilon/2$ whenever $z \in \overline{\mathcal{D}}$. Then $\operatorname{im} R_+(z_0) \cap \operatorname{im} G_+(z_0)^\perp = \{0\}$ for all $z_0 \in \mathbb{C}$ such that

$|z_0| = 1$. Let $R_i R_o$ with R_i rational be an inner-outer factorization of R_+ so that $\operatorname{im} R_i(z_0) = \operatorname{im} R_+(z_0)$ whenever $|z_0| = 1$. Let $W(z) = I - R_i(z)R_i(\frac{1}{\bar{z}})^*$ so that the columns of $W(z_0)$ span the orthogonal complement of $\operatorname{im} R_+(z_0)$ for all $z_0 \in \mathbb{T}$ except possibly for a finite number of points where the rank of W drops (cf. Theorem 5.5 in [10]). Assume without loss of generality that W has a constant rank on $\mathbb{T}$, let $W_+\Lambda_W W_-$ be a Wiener-Hopf factorization of W relative to $\mathbb{T}$, and let $W_i W_o$ be an inner-outer factorization of W_+, so that W_i is inner co-outer and $\operatorname{im} W_i(\lambda) = \operatorname{im} W_+(\lambda)$ for all λ in the closed unit disk. Then $\operatorname{im} W_i(\lambda) \perp \operatorname{im} R_+(\lambda)$ whenever $|\lambda| = 1$, and so $\widetilde{G}_+ = [\,G_+ \ W_i\,]$ is analytic in $\mathcal{D}$, continuous in $\overline{\mathcal{D}}$, and $\widetilde{G}_+(\lambda)$ is nonsingular whenever $\lambda \in \mathcal{D}$ and $1 - |\lambda|$ is sufficiently small.
The last property implies that $\widetilde{G}_+$ has finitely many zeros in $\mathcal{D}$, all of which have finite multiplicity. Let $\lambda \in \mathcal{D}$ be a zero of $\widetilde{G}_+$. One can find a unimodular matrix polynomial P such that the leading coefficients of the Taylor expansions at λ of the columns of $\widetilde{G}_+P$ are linearly independent. Multiplying this matrix function by

$$\operatorname{diag}\left(I, \frac{1}{(z-\lambda)^{\kappa_1}}, \ldots, \frac{1}{(z-\lambda)^{\kappa_r}}\right), \tag{2.9}$$

where r is the multiplicity of the zero of the corresponding column of $\widetilde{G}_+P$ at λ and $\kappa_1, \kappa_2, \ldots, \kappa_r$ are positive integers, we obtain a matrix function $[\,G_+ \ \widetilde{W}_i\,]$ which is analytic and takes a nonsingular value at λ. Finitely many such operations yield $\widehat{W}_i$ such that $\widehat{G} = [\,G_+ \ \widehat{W}_i\,]$ has no zeros in $\mathcal{D}$. The multiplicative inverse of $\widehat{G}$ is analytic in $\mathcal{D}$ and continuous in $\overline{\mathcal{D}}$. In an analogous way, working with polynomials in $\frac{1}{z-z_+}$ for some $z_+ \in \mathcal{D}$, we can extend G_- to obtain a (complete) Wiener-Hopf factorization of G. □

In particular, if the components of G are members of the Wiener algebra on the disk, the two definitions of Wiener-Hopf factorization are equivalent.

3. Riemann problem with singular coefficient

Let G be an $m \times n$ matrix function on Γ. The *defect numbers* of the Riemann problem with coefficient G [13] are the dimension α_R of the kernel of the problem in the space

$$\mathcal{X}_p = L^m_{p+} \dotplus \frac{\{\phi_- \in \dot{L}^n_{p-} \,:\, G\phi_- \in L^m_p\}}{\{\phi_- \in \dot{L}^n_{p-} \,:\, G\phi_- = 0\}} \tag{3.1}$$

and the co-dimension β_R of the closure of the image of the problem in the Banach space $\mathcal{Y}^m_p$ defined in (2.3). If at least one of these two numbers is finite, the difference $\alpha_R - \beta_R$ is the *index* of the problem.
The number α_R can be equivalently defined [12] as the dimension of the linear space

$$L^m_{p+} \cap G\dot{L}^n_{p-}. \tag{3.2}$$

Moreover, suppose β_R is finite, i.e., there are finitely many linearly independent functions $g_1, g_2, \ldots, g_N$ in L_p^m modulo the closure of the image of the problem for which $\int_\Gamma \psi_+ g = 0$ whenever $\psi_+ \in L_{q+}^{1\times m}$ and $\psi_+ G = 0$. The closed subspace $\Phi \subset L_p^m$ generated by those functions and the image of the Riemann problem is the annihilator of the space

$$\Psi = \{\psi_+ \in L_{q+}^m : \psi_+ G = 0\}. \tag{3.3}$$

Equivalently, the closed subspace $\Psi \subset L_q^m$ is the annihilator of Φ. By the Hahn-Banach Theorem, there exist linearly independent $h_1, h_2, \ldots, h_N \in L_q^m$ such that the linear span of the h_i's and Ψ is the annihilator of the image of the Riemann problem. Thus, β_R (finite or infinite) equals the co-dimension of the space (3.3) in the annihilator of the image of the Riemann problem. The last quantity was the basis for the definition of β_R in [12].

The defect numbers of the dual problem are the dimension α_D of the kernel of the dual problem in

$$\widehat{\mathcal{X}}_q = \dot{L}_{q-}^n \dot{+} \frac{\{\psi_+ \in L_{q+}^m : G^T\psi_+ \in L_q^n\}}{\{\psi_+ \in L_{q+}^m : G^T\psi_+ = 0\}}, \tag{3.4}$$

and the co-dimension of the closure of the image of the dual problem in

$$\widehat{\mathcal{Y}}_q^n = \{h \in L_q^n : \langle \phi_-, h\rangle = 0 \ \text{ for all } \phi_- \in \dot{L}_{p-}^n \text{ such that } G\phi_- = 0\}. \tag{3.5}$$

If at least one of the two numbers is finite, the difference $\alpha_D - \beta_D$ is the *index* of the dual problem. Similarly as in the Riemann problem, the defect numbers of the dual problem are equal to the dimension of the space

$$\dim\left(\dot{L}_{q-}^n \cap G^T L_{p+}^m\right) \tag{3.6}$$

and the co-dimension of the closed space

$$\left\{\phi_- \in \dot{L}_{p-}^n : G\phi_- = 0\right\} \tag{3.7}$$

in the annihilator of the image of the dual problem. In (3.7), and similarly in the formulas above, $G\phi_- = 0$ means that $G(t)\phi(t) = 0$ for all $t \in \Gamma$ except, possibly, for a set of measure 0.

In this terminology, Theorem 4.5 in [12] can be restated as follows.

Theorem 3.1. *If $G \in L_1^{m\times n}$ and* $\operatorname{rank} G = k$ *a.e. on a contour Γ, then the following assertions are equivalent:*

(i) *The indices of the Riemann problem with coefficient G and its dual are finite and opposite, and the image of each of the problems contains all rational vector functions in its closure.*

(ii) *G admits a restricted generalized Wiener-Hopf factorization in L_p relative to Γ.*

By Proposition 2.1, if the restriction of G to each component of Γ is a rational matrix function, or if Γ is the unit circle, the conclusion of Theorem 3.1 holds with complete factorization in place of restricted.

If G is essentially bounded, the spaces (3.1) and (3.4) simplify to

$$\begin{aligned} (3.8) \qquad \mathcal{X}_p &= L^m_{p+} \dot{+} \dot{L}^n_{p-} \,/\, \{\phi_- \in \dot{L}^n_{p-} : \, G\phi_- = 0\}, \\ (3.9) \qquad \widehat{\mathcal{X}}_q &= \dot{L}^n_{q-} \dot{+} L^m_{q+} \,/\, \{\psi_+ \in L^m_{q+} : \, G^T\psi_+ = 0\}. \end{aligned}$$

We emphasize that (3.1), (3.4), and (3.8), (3.9) are internal direct sums. That is, if $n \geq m$ and $f \in \mathcal{X}_p$, the bottom $n-m$ components of f are in $\dot{L}^n_{p-}$ and

$$\begin{aligned} \|f\| &= \operatorname{dist}(f, \ker G \cap \dot{L}^n_{p-}) = \inf\left\{\|f-\phi_-\|_p : \, \phi_- \in \ker G \cap \dot{L}^n_{p-}\right\} \\ (3.10) \qquad &= \inf\left\{\left(\int_\Gamma \left(\sum_{j=1}^n |f_j(t) - \phi_{j-}(t)|^2\right)^{\frac{p}{2}} dt\right)^{\frac{1}{p}} : \, \phi_- \in \dot{L}^n_{p-}, \, G\phi_- = 0\right\}. \end{aligned}$$

If $n < m$ and $f \in \mathcal{X}_p$, $f_j \in L_{p+}$ for $j = n+1, n+2, \ldots, m$ and

$$\begin{aligned} \|f\| &= \inf\left\{\left(\int_\Gamma \left(\sum_{j=1}^n |f_j(t) - \phi_{j-}(t)|^2 + \sum_{j=n+1}^m |f_j(t)|^2\right)^{\frac{p}{2}} dt\right)^{\frac{1}{p}} : \right. \\ (3.11) \qquad & \qquad \left. \phi_- \in \dot{L}^n_{p-}, \, G\phi_- = 0\right\}. \end{aligned}$$

The norm in (3.4), or (3.9), is defined in a similar manner.

If the contour is not regular, $\mathcal{X}_p$ and $\widehat{\mathcal{X}}_q$ need not be complete. Indeed, suppose G is bounded and takes nonsingular values a.e. on Γ, so that $\mathcal{X}_p = L^m_{p+} \dot{+} \dot{L}^m_{p-}$. For a nonregular Γ, $L_{p+} \dot{+} \dot{L}_{p-}$ is a proper dense subset of L_p and so $\overline{\mathcal{X}}_p \backslash \mathcal{X}_p$ is nonempty.

Suppose Γ is regular and G is bounded. For the sake of definiteness, assume $m \leq n$. The subspace of L^m_p containing functions whose last $n-m$ components belong to $\dot{L}_{p-}$ is complete. Hence $\mathcal{X}_p$ is complete. Similarly, $\widehat{\mathcal{X}}_q$ is a Banach space. In this case, the defect numbers of the Riemann problem and its dual coincide with the defect numbers of certain Toeplitz operators.

Proposition 3.2. *Suppose G is a bounded $m \times n$ matrix function and the contour Γ is regular. Then:*

(i) *The image of the Riemann problem with coefficient G is closed if and only if the Toeplitz operator*

$$(3.12) \qquad T_Q = QG : \, \dot{L}^n_{p-} \,/\, \{\phi_- \in \dot{L}^n_{p-} : \, G\phi_- = 0\} \;\rightarrow\; Q\mathcal{Y}^m_p$$

is normally solvable; moreover, the defect numbers of the Riemann problem coincide with the defect numbers of $T_{\mathcal{Q}}$.
(ii) *The image of the dual problem is closed if and only if the Toeplitz operator*

$$(3.13)\qquad T_{\mathcal{P}} = \mathcal{P}G^T : L^m_{q+} \,/\, \{\psi_+ \in L^m_{q+} : G^T\psi_+ = 0\} \rightarrow \mathcal{P}\widehat{\mathcal{Y}}^n_q$$

is normally solvable; moreover, the defect numbers of the dual problem coincide with the defect numbers of $T_{\mathcal{P}}$.
(iii) *The operator* $T_{\mathcal{Q}}$ *is normally solvable if and only if* $T_{\mathcal{P}}$ *is normally solvable.*

Proof. Since the space $L_{p+} \dot{+} \dot{L}_{p-}$ is complete, the mapping $T: L_{p+} \times \dot{L}_{p-} \rightarrow L_p$ defined by $T(f,g) = f + g$ is open. Hence, if $T_{\mathcal{Q}}$ is normally solvable, the image of the Riemann problem is closed. Conversely, if the image of the Riemann problem is closed and $\phi \in \overline{\operatorname{im} T_{\mathcal{Q}}}$, ϕ belongs to the image of the Riemann problem and so $\phi \in \operatorname{im} T_{\mathcal{Q}}$. Thus, the image of the Riemann problem is closed if and only if $T_{\mathcal{Q}}$ is normally solvable. Since (ϕ_+, ϕ_-) belongs to the kernel of the Riemann problem if and only if $\phi_- \in \ker T_{\mathcal{Q}}$, $\dim\ker T_{\mathcal{Q}} = \alpha_R$. Plainly, the co-dimension of the closure of $\operatorname{im} T_{\mathcal{Q}}$ is infinite if and only if β_R is infinite and, if finite, both numbers are equal. Thus, (i) holds.
Statement (ii) can be proved in the same manner as (i). We verify (iii). The image of the operator $T_{\mathcal{Q}}$ coincides with the image of an operator $\widehat{T}_{\mathcal{Q}} = \mathcal{Q}G\mathcal{Q}$ acting on L^n_p, and the image of $T_{\mathcal{P}}$ coincides with the image of $\widehat{T}_{\mathcal{P}} = \mathcal{P}G^T\mathcal{P}$ acting on L^m_p. Since the operators $\widehat{T}_{\mathcal{Q}}$ and $\widehat{T}_{\mathcal{P}}$ are adjoint to each other with respect to our inner product, their images are closed simultaneously. □

Theorem 3.1 and Proposition 3.2 above, and Proposition 4.1 in [13], imply the following.

Theorem 3.3. *Suppose an* $m \times n$ *matrix function* G *is bounded and a contour* Γ *is regular. Then the following statements are equivalent:*
(i) G *admits a restricted generalized Wiener-Hopf factorization in* L_p *and the image of the Riemann problem with coefficient* G *is closed.*
(ii) *The operator* $T_R: \mathcal{X}_p \rightarrow \mathcal{Y}^m_p$ *defined by* $T_R(\phi_+, \phi_-) = \phi_+ + G\phi_-$ *is Fredholm.*
Moreover, if the equivalent conditions (i) *and* (ii) *are satisfied, the dimension of the kernel of* T_R *equals the sum of positive indices of the factorization and the co-dimension of the image of* T_R *equals the absolute value of the sum of negative indices of the factorization.*

We end this section with a straightforward proposition which justifies our terminology. Recall that a complete generalized Wiener-Hopf factorization $G = G_+\Lambda G_-$ is said to be Fredholm if the operators K_+ and K_- defined in (2.5) and (2.6) are continuous. By Proposition 4.2 in [13], the image of the Riemann problem with a Fredholm factorable coefficient is closed. That is, we have the following.

Proposition 3.4. *Suppose $G_+\Lambda G_-$ is a Fredholm factorization of a bounded matrix function G on a regular contour Γ. Then the operator T_R defined in Theorem 3.3 is Fredholm.*

4. Conditions for factorability

Each continuous, nonsingular matrix function admits a generalized Wiener-Hopf factorization relative to Γ (see [9], Theorem 5.3). A singular matrix function G may fail to have a Wiener-Hopf factorization even if the restriction of G to each component of Γ is constant. Indeed, suppose Γ has two components, Γ_1 and Γ_2,

$$G|_{\Gamma_1} = \begin{bmatrix} 1 \\ 0 \end{bmatrix} \quad \text{and} \quad G|_{\Gamma_2} = \begin{bmatrix} 0 \\ 1 \end{bmatrix}, \tag{4.1}$$

and $G = G_+\Lambda G_-$ is a Wiener-Hopf factorization relative to Γ. By the first equality in (4.1), the bottom component of G_+ equals 0 and, by the second equality in (4.1), the top component of G_+ equals 0. Thus, $G_+ = 0$ and so $G = 0$, a contradiction.

Theorem 3.1 gives a criterion for existence of a restricted generalized Wiener-Hopf factorization in L_p. Condition (i) in Theorem 3.1 is not satisfied by the function (4.1). Indeed, in this example $\alpha_R = 0$, $\mathcal{Y}_p^m = L_p^m$, and β_R is infinite. The function (4.1) falls into the following more general category.

Proposition 4.1. *Suppose $G \in \mathbb{C}^{m\times n}(\Gamma)$, Γ has N connected components $\Gamma_1, \Gamma_2, \ldots, \Gamma_N$, and $G_i = G|_{\Gamma_i}$ is a rational matrix function without poles or zeros on Γ_i, $i = 1, 2, \ldots, N$. Then each of the following three statements implies the other two:*

(i) *G admits a restricted Wiener-Hopf factorization.*

(ii) *G admits a complete Wiener-Hopf factorization with the left factor rational.*

(iii) $G_1\mathcal{R}^{n\times 1} = G_2\mathcal{R}^{n\times 1} = \cdots = G_N\mathcal{R}^{n\times 1}$.

Proof. Plainly, (ii) implies (i). Suppose $G = G_+\Lambda G_-$ is a restricted Wiener-Hopf factorization and $\psi \in \mathcal{R}^{1\times m}$ is such that $\psi G_i = 0$. Then $\psi(z)G_+(z) = 0$ for all $z \in \mathcal{D}$, $i = 1, 2, \ldots, N$. Since $G_i\mathcal{R}^{n\times 1} \subset \mathcal{R}^{m\times 1}$, $G_i\mathcal{R}^{n\times 1}$ does not depend on i and (iii) holds.
Suppose $G_1\mathcal{R}^{n\times 1} = \cdots = G_N\mathcal{R}^{n\times 1}$. Then $G_i = G_L G_{iR}$, $i = 1, 2, \ldots, N$, with G_L a left invertible rational matrix function and each G_{iR} right invertible. Since G_L admits a Wiener-Hopf factorization, we may assume that G_L is analytic in $\mathcal{D}_+$. Each G_{iR} admits a factorization $G_{iR_A}G_{iR_B}$ with G_{iR_A} square and G_{iR_B} right invertible and analytic outside Γ_i. By Lemma I.2.1 in [5], the function whose restriction to Γ_i is G_{iR_A} admits a Wiener-Hopf factorization $\widetilde{G}_+\Lambda\widetilde{G}_-$ with $\widetilde{G}_+$ a rational matrix function. If $G_+ = G_L\widetilde{G}_+$ and $G_-(z) = \widetilde{G}_-(z)G_{iR_B}(z)$ for z in the component of $\mathcal{D}_-$ bounded by Γ_i, $G_+\Lambda G_-$ is a restricted Wiener-Hopf factorization of G. It follows from the existence of a Smith-McMillan factorization

(or from Smith's Theorem) that a rational matrix function analytic in a domain $\mathcal{D} \subset \mathbb{C}$ which has a one sided inverse analytic in $\mathcal{D}$ can be extended to a rational matrix function which is analytic and takes nonsingular values in $\mathcal{D}$. Thus, (ii) holds. □

In the special case where the restriction of G to each component of Γ is a nonsingular rational matrix function without poles or zeros on Γ, condition $G_1\mathcal{R}^{n\times 1} = \cdots = G_N\mathcal{R}^{n\times 1}$ is always satisfied and Proposition 4.1 reduces to Theorem I.2.2 in [5].

The continuity of a singular matrix function G is not sufficient for the existence of a generalized Wiener-Hopf factorization even when Γ is connected. Indeed, suppose Γ is the unit circle,

$$G(e^{it}) = \begin{bmatrix} 1 \\ 2 \end{bmatrix} \quad \text{if } 0 \le t \le \frac{\pi}{2}, \qquad G(e^{it}) = \begin{bmatrix} 2 \\ 1 \end{bmatrix} \quad \text{if } \pi \le t \le \frac{3}{2}\pi, \tag{4.2}$$

and the value of G changes linearly with angle in the second and fourth quadrant so as to make the function continuous. If G admitted a generalized Wiener-Hopf factorization, there would exist $G_+ \in L_{p+}^{2\times 1}$ whose image would coincide a.e. on Γ with the image of G. Such G would have to be identically zero, a contradiction. Note that G has a constant rank on Γ. Also note that G can be completed to a continuous matrix function which takes nonsingular values at each point of Γ. The completed function admits a generalized Wiener-Hopf factorization.

Suppose G satisfies (4.2) and $\psi_+ \in L_{q+}^2$ is such that $\psi_+ G = 0$. Then the first component of ψ_+ equals -2 times the second component a.e. on the set $\{e^{it} : 0 \le t \le \frac{\pi}{2}\}$, and the second component equals -2 times the first component a.e. on $\{e^{it} : \pi \le t \le \frac{3}{2}\pi\}$. Thus, $\psi_+ = 0$ and $\mathcal{Y}_p^2 = L_p^2$. Since the set

$$\left\{ \begin{bmatrix} t^{-j} \\ 0 \end{bmatrix} : j = 1, 2, \ldots \right\} \tag{4.3}$$

does not belong to the closure of the image of the Riemann problem, $\alpha_R = \infty$. Since $\beta_R = 0$, the index of the problem is not finite. That is, condition (i) in Theorem 3.1 is not satisfied.

Suppose $G = G_+ \Lambda G_-$ is a restricted generalized Wiener-Hopf factorization in L_p and $\psi_+ \in L_{q+}^m$ is such that $\psi_+ G$ vanishes on a set of positive measure. Then $\psi_+ G_+ \in L_{1+}^k$ vanishes on a set of positive measure and, since $\mathcal{D}_+$ is connected, $\psi_+ G_+ = 0$. Considering $\psi_- \in \dot{L}_{p-}^n$ in the same way, we obtain the following criterion for factorability.

Proposition 4.2. *Suppose $G \in L_1^{m\times n}(\Gamma)$ admits a generalized Wiener-Hopf factorization in L_p.*

(i) If $\psi_+ \in L_{q+}^m$ is such that $\psi_+ G$ vanishes on a set of positive measure, then $\psi_+ G = 0$.

(ii) *If* Γ_i *is a component of* Γ *and* $\psi_- \in L^n_{p-}(\Gamma_i)$ *is such that* $G\psi_-$ *vanishes on a subset of* Γ_i *of positive measure, then* $G\psi_-$ *vanishes a.e. on* Γ_i.

If $r \neq 0$ is a scalar rational function and ψ is a measurable vector function on Γ, $r(t)\psi(t) = 0$ on a set of positive measure (a.e. on Γ, respectively) if and only if $\psi(t) = 0$ on a set of positive measure (a.e. on Γ, respectively). Thus, if G admits a generalized Wiener-Hopf factorization and $\psi \in \mathcal{R}^{1\times m}$ ($\phi \in \mathcal{R}^{n\times 1}$, respectively) is such that ψG ($G\phi$, respectively) vanishes on a subset of Γ (Γ_i, respectively) of positive measure, then $\psi G = 0$ ($G(t)\phi(t) = 0$ a.e. on Γ_i, respectively).

Acknowledgements

This research was partially supported by the National Science Foundation under Grant DMS-9302706.

References

[1] Ball, J.A., Gohberg, I.C., Rodman, L.: Interpolation of Rational Matrix Functions; Birkhäuser Verlag, Basel Boston Berlin 1990.

[2] Ball, J.A., Helton, J.W.: Beurling-Lax Representations Using Classical Lie Groups with Many Applications II: $GL(n, \mathbb{C})$ and Wiener-Hopf Factorization; Integral Equations Operator Theory 7 (1984), 291–309.

[3] Ball, J.A., Rakowski, M.: Interpolation by Rational Matrix Functions and Stability of Feedback Systems: The 2-Block Case; J. Math. Systems, Estimation, and Control 4 (1994), 261–318.

[4] Bart, H., Gohberg, I.C., Kaashoek, M.A.: Minimal Factorization of Matrix and Operator Functions; Operator Theory: Adv. Appl. 1, Birkhäuser Verlag, Basel Boston Stuttgart 1979.

[5] Clancey, K., Gohberg, I.C.: Factorization of Matrix Functions and Singular Integral Operators; Operator Theory: Adv. Appl. 3, Birkhäuser Verlag, Basel Boston Stuttgart 1981.

[6] Clancey, K., Rakowski, M.: Factorization of Rectangular Matrix Functions Relative to a Contour; Preprint, 1990.

[7] David, G.: Opérateurs intégraux singuliers sur certains courbes du plan complexe; Ann. Sci. École Norm. Sup. 17 (1984), 157–189.

[8] Fuhrmann, P.A., Willems, J.C.: Factorization Indices at Infinity for Rational Matrix Functions; Integral Equations Operator Theory 2/3 (1979), 287–301.

[9] Litvinchuk, G.S., Spitkovsky, I.M.: Factorization of Measurable Matrix Functions; Operator Theory: Adv. Appl. 25, Birkhäuser Verlag, Basel Boston 1987.

[10] Rakowski, M.: Generalized Pseudoinverses of Matrix Valued Functions; Integral Equations Operator Theory 14 (1991), 564–585.

[11] ———: Spectral Factorization of Rectangular Rational Matrix Functions with Application to Discrete Wiener-Hopf Equations; J. Funct. Anal. 10 (1992), 410–433.

[12] Rakowski, M., Spitkovsky, I.M.: Spectral Factorization of Measurable Rectangular Matrix Functions and the Vector Valued Riemann Problem; Revista Matemática Iberoamericana 12 (1996), 669–696.

[13] ______: On Normal Solvability of the Riemann Problem with Singular Coefficient; Proc. Amer. Math. Soc. 125 (1997), 815–826.

634 Sharon Mill Court
Worthington, OH 43085
USA

1991 Mathematics Subject Classification: Primary 47A68; Secondary 45F15

Submitted: May 31, 1996

Operator Theory:
Advances and Applications, Vol. 102

Elliptic boundary value problems for general elliptic systems in complete scales of Banach spaces

I. ROITBERG

Elliptic boundary value problems for general elliptic systems are studied in complete scales of Banach type in the case where the boundary conditions contain both the function from the systems, and the additional functions, defined at the boundary of the domain.

1. Introduction

Elliptic boundary value problems for general systems of differential equations were studied in the classes of sufficiently smooth functions by Agmon, Douglis, Nirenberg [1], Volevich [9], Solonnikov [8]. These problems were investigated in complete scales of Banach spaces by Ya.Roitberg [6], [7]; the theorems on complete collection of isomorphisms and their applications were established there.

In problems of elasticity theory and hydrodynamics, for example in the works of A. Aslanyan, D. Vassiliev, V. Lidskii [2], P.G. Garlet [4], S. Nazarov, K. Pileskas [5], there arise boundary value problems for general elliptic systems whose boundary conditions contain both the functions $u_1, \dots, u_N$ contained in the system and the additional functions $u'_{N+1}, \dots, u'_{N+k}$ defined at the boundary. The number of the boundary conditions respectively increases .

The present work is devoted to the investigation of these problems in complete scales of Banach spaces.

2. Statement of the problem

Let $G \subset \mathbb{R}^n$ be a bounded domain with boundary $\partial G \in C^\infty$. Consider the following boundary value problem:

$$l(x, D)u = f(x), \qquad x \in G, \tag{2.1}$$

$$b(x, D)u + b'(x', D')u' = \varphi(x'), \qquad x' \in \partial G. \tag{2.2}$$

The system

$$lu := (l_{rj}(x, D))_{r,j=1,\dots,N}$$

and the boundary differential expressions

$$b(x, D) := (b_{hj}(x, D))_{\substack{h=1,\dots,m+k'' \\ j=1,\dots,N}}, \quad b'(x', D') := (b'_{hj}(x', D'))_{\substack{h=1,\dots,m+k'' \\ j=N+1,\dots,N+k'}}$$

are described by the following relations:

$$\begin{array}{lll} \text{ord } l_{rj} \le s_r + t_j & \text{for } s_r + t_j \ge 0, & \multirow{2}{*}{$r, j = 1, \dots, N,$} \\ l_{rj} = 0 & \text{for } s_r + t_j < 0, & \\ \text{ord } b_{hj} \le \sigma_h + t_j & \text{for } \sigma_h + t_j \ge 0, & j = 1, \dots, N,\ h = 1, \dots, m + k'', \\ b_{hj} = 0 & \text{for } \sigma_h + t_j < 0, & \\ \text{ord } b'_{hj} \le \sigma_h + t_j & \text{for } \sigma_h + t_j \ge 0, & j = N+1, .., N+k', h = 1, .., m+k''. \\ b'_{hj} = 0 & \text{for } \sigma_h + t_j < 0, & \end{array}$$

Here $t_1, \dots, t_{N+k'}, s_1, \dots, s_N$ and $\sigma_1, \dots, \sigma_{m+k''}$ are given integers such that

$$s_1 + \dots + s_N + t_1 + \dots + t_N = 2m$$

and

$$t_1 \ge \dots \ge t_N \ge 0 = s_1 \ge \dots \ge s_N, \qquad \sigma_1 \ge \dots \ge \sigma_{m+k''}.$$

In addition,

$$\begin{aligned} u(x) &= (u_1(x), \dots, u_N(x)), \qquad x \in \overline{G}, \\ u'(x') &= \big(u'_{N+1}(x'), \dots, u'_{N+k'}(x')\big), \qquad x' \in \partial G, \end{aligned}$$

where

$$u'_j(x') = \big(u'_{j1}(x'), \dots, u_{j,\sigma_h+t_j+1}(x')\big).$$

Moreover,

$$b_{hj}(x, D)u_j(x) = \sum_{1 \le k \le \sigma_h + t_j + 1} \Lambda_{hjk}(x', D')u_{jk}, \qquad j = 1, \dots, N,$$

where $u_{jk} = D_\nu^{k-1} u_j|_{\partial G}$, $D_\nu = i\partial/\partial\nu$, ν is the normal to ∂G,

$$b'_{hj}(x', D')u'_j(x') = \sum_{1 \le k \le \sigma_h + t_j + 1} \Lambda_{hjk}(x', D')u'_{jk}(x'), \qquad j = N+1, \dots, N+k',$$

where $\Lambda_{hjk}(x', D')$ are tangential operators of order $\sigma_h + t_j - k + 1$ for h, j such that $\sigma_h + t_j \ge 0$.

In a natural way we introduce the notion of ellipticity of the problem (2.1), (2.2).

3. Definition of the ellipticity of the problem (2.1), (2.2)

Let $L_0(x, \xi) := \det(l_0(x, \xi))$, where $l_0(x, \xi) = (l_{rj0}(x, \xi))$ is the principal symbol of the matrix $l(x, D)$. The expression $l(x, D)$ is called *elliptic* in $\overline{G}$, if

$$L_0(x, \xi) \ne 0, \qquad x \in \overline{G},\ \xi \in \mathbb{R}^N \setminus \{0\}. \tag{3.1}$$

Let $x \in \partial G$, let $\tau \neq 0$ be an arbitrary real vector, tangential to ∂G at the point x, and let ν be a unit normal to ∂G at this point. If the system (2.1) is elliptic, then the polynomial $L_0(\xi) = L_0(x, \tau+\zeta\nu)$ of order $r = s_1+\cdots+s_N+t_1+\cdots+t_N$ does not have real roots. A system (2.1) elliptic in $\overline{G}$ is called *properly elliptic*, if for every point $x \in \partial G$ and any $\tau \neq 0$ the polynomial $L_0(\zeta)$ has even order $(r = 2m)$, and accurately m of its roots have positive imaginary parts.

For $n > 2$, any elliptic system is properly elliptic.

If the system (2.1) is properly elliptic, then $L_0(\zeta) = L_+(\zeta)L_-(\zeta)$, where L_+ (L_-) is the mth-order polynomial all roots of which lie in the upper (lower) half-plane. Therefore, if $l_0(\zeta) = l_0(x, \tau + \zeta\nu)$, then, for every point $x \in \partial G$ and any vector $\tau \neq 0$, tangential to ∂G at the point $x \in \partial G$, the space $\mathfrak{M}_+ = \mathfrak{M}_+(x, \tau)$ of the stable (i.e., decreasing as $t \longrightarrow \infty$) solutions of the equations $l_0(D_t) = 0$ $(D_t = i\partial/\partial t)$ is an m-dimensional space.

Let $b_0(x, D)$ and $b'_0(x', D')$ be the principal parts of the matrix $b(x, D)$ and $b'(x', D')$, respectively, and let

$$b_0(\zeta) = b_0(x, \tau + \zeta\nu), \qquad b'_0(\tau) = b'_0(x', \tau).$$

Definition 3.1. The problem (2.1), (2.2) is called *elliptic*, if the system (2.1) is properly elliptic, and the following condition (Lopatinskii's condition) is valid: for every point $x \in \partial G$, for any vector $\tau \neq 0$, tangential to ∂G at the point $x \in \partial G$, and for every $h = (h_1, \ldots, h_{m+k'}) \in \mathbb{C}^{m+k'}$, the problem

$$l_0(D_t)V = 0, \quad t > 0, \qquad b_0(D_t)V|_{t=0} + b'_0(\tau)V' = h,$$

has a solution (V, V'), $V \in \mathfrak{M}_+(x, \tau)$.

In what follows we assume that the problem (2.1), (2.2) is elliptic.

4. Functional spaces

For any real $s \geq 0$ and $p \in (1, \infty)$ we denote by $H^{s,p}(G)$ the space of Bessel potentials (Liouville classes). Let $H^{-s,p}(G) := (H^{s,p'}(G))^*$ $(1/p + 1/p' = 1)$ and let $\|\cdot\|$ denote the norm in $H^{s,p}(G)$. Let $B^{s,p}(\partial G)$ denote the Besov spaces with the norm $\langle\langle\cdot\rangle\rangle_{s,p}$, and let $B^{-s,p}(\partial G) = (B^{s,p'}(\partial G))^*$.

For a fixed $r \in \mathbb{N}$ and any $s, p \in \mathbb{R}$, $1 < p < \infty$, $s \neq k + 1/p$, $k = 0, 1, \ldots, r-1$, by $\widetilde{H}^{s,p,(r)}(G)$ (see [7]) we denote the completion of $C^\infty(\overline{G})$ with respect to the norm

$$|||u|||_{s,p,(r)} := \left(||u||^p_{s,p} + \sum_{1\leq j\leq r} \langle\langle D^{j-1}_\nu u\rangle\rangle^p_{s-j+1-1/p,p} \right)^{1/p}. \tag{4.1}$$

The closure $S = S_{s,p}$ of the mapping

$$u \longmapsto (u|_{\overline{G}}, u|_{\partial G}, \ldots, D^{r-1}_\nu u|_{\partial G}), \qquad u \in C^\infty(\overline{G}),$$

is an isometry between $\widetilde{H}^{s,p,(r)}(G)$ and the subspace of the direct product

$$\mathfrak{F}^{s,p} := H^{s,p}(G) \times \prod_{1\le j\le r} B^{s-j+1-1/p,p}(\partial G).$$

Therefore one can identify elements $u \in \widetilde{H}^{s,p,(r)}(G)$ with elements $Su \in \mathfrak{F}^{s,p}$. For any $u \in \widetilde{H}^{s,p,(r)}(G)$ we write $u = (u_0, \dots, u_r)$.

Using the complex interpolation method, the space $\widetilde{H}^{s,p,(r)}(G)$ and the norm $|||\cdot|||_{s,p,(r)}$ may be defined also for $s = k + 1/p$, $k = 0, \dots, r-1$.

Finally, if $r = 0$ then $\widetilde{H}^{s,p,(0)}(G) := H^{s,p}(G)$, and $|||\cdot|||_{s,p,(0)} = ||\cdot||_{s,p}$.

Let

$$M(x,D) = \sum_{0\le j\le q} M_j(x,D')D_\nu^j, \qquad x \in \overline{G}, \tag{4.2}$$

be a differential expression of order q with infinitely smooth coefficients in $\overline{G}$, where $M_j(x,D)$ is a tangential expression of order $\le q-j$. If $q \le r$, then the closure M of the mapping $u \mapsto Mu$, $u \in C^\infty(\overline{G})$, acts continuously from the whole space $\widetilde{H}^{s,p,(r)}(G)$ into $H^{s-q,p}(G)$. If $q \le r-1$, then the closure of the mapping $u \mapsto Mu|_{\partial G}$ acts continuously from the whole space $\widetilde{H}^{s,p,(r)}(G)$ into $B^{s-q-1/p,p}(\partial G)$, $s \in \mathbb{R}$.

By integration by parts it is easy to obtain that

$$(Mu,v) = (u, M^+v) - i\sum_{j=1}^{q}\sum_{k=1}^{j}\langle D_\nu^{k-1}u, D_\nu^{j-k}M_j^+v\rangle, \qquad u,v \in C^\infty(\overline{G}).$$

This implies that

$$(Mu)_+ = Mu_+ - i\sum_{j=1}^{q}\sum_{k=1}^{j} M_j D_\nu^{j-k}\big(D_\nu^{k-1}u|_{\partial G} \times \delta(\partial G)\big), \tag{4.3}$$

where $(Mu)_+$ and u_+ are the extensions by zero of the functions Mu and u to $\mathbb{R}^n$, respectively, and $\delta(\partial G)$ is the Dirac measure. On the set of the expressions (4.2) we introduce the operator J such that for $q = 0$ we have $JM(x,D') = 0$, and for $q \ge 1$ we have $JM(x,D) = \sum_{1\le j\le q} M_j(x,D')D_\nu^{j-1}$. Then one can rewrite (4.3) in the form of

$$(Mu)_+ = Mu_+ - i\sum_{1\le j\le q}(J^kM)\big(D_\nu^{k-1}u|_{\partial G} \times \delta(\partial G)\big).$$

If $u = (u_0, \dots, u_r) \in \widetilde{H}^{s,p,(r)}(G)$, $s \in \mathbb{R}$, then Mu belongs to $H^{s-q,p}(G)$ if and only if

$$(Mu)_+ = Mu_+ - i\sum_{1\le j\le q}(J^kM)\big(u_k \times \delta(\partial G)\big). \tag{4.4}$$

Similarly, if $q \le r-1$, then

$$Mu|_{\partial G} = \sum_{1\le j\le q} M_j(x, D')u_{j+1}. \tag{4.5}$$

We will consider these spaces also in the cases where

$$G = \mathbb{R}^n_+ := \{(x_1, \dots, x_n) \in \mathbb{R}^n,\, x_n > 0\}, \qquad \partial G = \{x \in \mathbb{R}^n : x_n = 0\} =: \mathbb{R}^{n-1}.$$

5. Theorem on complete collection of isomorphisms

Let

$$\kappa = \max\{0, \sigma_1 + 1, \dots, \sigma_m + 1\}, \qquad \tau_j = t_j + \kappa,\ j = 1, \dots, N.$$

Assume that $\widetilde{H}^{T+s,p,(\tau)} := \prod_{1\le j\le N} \widetilde{H}^{t_j+s,p,(\tau_j)}(G)$, and

$$\mathfrak{B}^{T'+s,p} := \prod_{N+1\le j\le N+k'} \ \prod_{1\le k\le \sigma_1+t_j+1} B^{t_j+s-k+1-1/p,p}(\partial G).$$

It is easy to see that the closure $A = A_{s,p}$ of the mapping

$$(u, u') \mapsto (lu, bu + b'u'), \quad (u, u') \in (C^\infty(\overline{G}))^N \times \prod_{N+1\le j\le N+k'} (C^\infty(\partial G))^{\sigma_h+t_j+1}$$

acts continuously from $\mathfrak{H}^{s,p}$ to $K^{s,p}$ where the spaces $\mathfrak{H}^{s,p}$ and $K^{s,p}$ are given by

$$\begin{aligned}
\mathfrak{H}^{s,p} &:= \widetilde{H}^{T+s,p,(\tau)} \times \mathfrak{B}^{T+s,p}, \\
K^{s,p} &:= \prod_{j=1}^{N} \widetilde{H}^{s-s_j,p,(\kappa-s_j)}(G) \times \prod_{h=1}^{m+k''} B^{s-\sigma_h-1/p,p}(\partial G).
\end{aligned}$$

Theorem 5.1. *Let the problem (2.1), (2.2) be elliptic. Then for any $s \in \mathbb{R}$ and $p \in (1, +\infty)$ the operator A is Noetherian, the kernel $\mathfrak{N}$ and the cokernel $\mathfrak{N}^\star$ are finite dimensional; $\mathfrak{N}$ and $\mathfrak{N}^\star$ do not depend on s and p and consist of infinitely smooth elements:*

$$\begin{aligned}
\mathfrak{N} &= \{(u, u') \in (C^\infty(\overline{G}))^N \times \prod_{N+1\le j\le N+k'} (C^\infty(\partial G))^{\sigma_1+t_j+1} : A(u, u') = 0\}, \\
\mathfrak{N}^\star &= (C^\infty(\overline{G}))^N \times \prod_{1\le j\le N} (C^\infty(\partial G))^{\kappa-s_j} \times (C^\infty(\partial G))^{m+k''}.
\end{aligned}$$

The equation

$$A(u, u') = F = (f, \phi) = (f_1, \dots, f_N, \phi_1, \dots, \phi_{m+k''}) \in K^{s,p}$$

is solvable in $\mathfrak{H}^{s,p}$ if and only if the relation

$$[F, V] := \sum_{1\le j\le N} (f_{j0}, v_{j0}) + \sum_{1\le j\le N} \ \sum_{1\le k\le \kappa-s_j} \langle f_{jk}, v_{jk}\rangle + \sum_{1\le h\le m+k''} \langle \phi_j, \psi_j\rangle = 0$$

is valid for any element $V \in \mathfrak{N}^\star$.

6. Some applications of the theorem on isomorphisms

Theorem 6.1. *Let $U = (u, u') \in \mathfrak{H}^{s,p}$ be a generalized solution of the problem (2.1), (2.2), i.e., $AU = F \in K^{s,p}$. Assume that $G_0 \subset G$ is a subdomain of G, adherent to a piece Γ of the surface ∂G. If F belongs to K^{s_1,p_1} locally in G_0 up to Γ, $s_1 \geq s$, $p_1 \geq p$, then U belongs to $\mathfrak{H}^{s_1,p_1}$ locally in G_0 up to Γ.*

Theorems 5.1 and 6.1 permit us to investigate the problem (1.1), (1.2) in the case where the right-hand sides have arbitrary power singularities along the manifolds of different dimensions: instead of the function $F(x)$ we consider the regularization F which is an element of the space $K^{s,p}$ for some $s < 0$, depending on the singularities of $F(x)$. These theorems give a possibility to prove the existence of the solution and to investigate its behaviour near the manifold of singularities of the function $F(x)$.

Theorems 5.1 and 6.1 further allow us to investigate a class of strongly degenerate elliptic problems of the type of (2.1), (2.2).

Theorems 5.1 and 6.1 also give a possibility to prove the existence and to investigate the properties of regularities of Green's matrix of the problem (2.1), (2.2).

7. Proof of Theorem 5.1

Consider a problem of the type (2.1), (2.2) in the half-space $\mathbb{R}^n_+$ in the case where the coefficients are complex constant, and the differential expressions (2.1) and (2.2) consist of their principal parts only. For $j = 1, \dots, n-1$, in place of each operator

$$D_j = F'^{-1}\xi_j F'$$

we substitute the operator

$$\widehat{D}_j = F'^{-1}(\xi_j/|\xi'|)(1+|\xi'|)F', \qquad |\xi'| = (\xi_1^2 + \dots, \xi_{n-1}^2)^{1/2}.$$

As a result, we obtain the model problem in the half-space

$$(7.1) \qquad \widehat{l}_0(D)u(x) = f(x), \qquad x \in \mathbb{R}^n_+,$$

$$(7.2) \qquad \widehat{b}_0(D)u(x)|_{x_n=0} + \widehat{b}'_0(D')u'(x') = \varphi(x'), \qquad x' \in \mathbb{R}^{n-1}.$$

The closure $\widehat{A}_0 = \widehat{A}_{0\,s,p}$ of the mapping

$$U = (u, u') \to (\widehat{l}_0 u, b_0 u + b'_0 u'),$$

where

$$U \in \left(C_0^\infty(\overline{\mathbb{R}^n_+})\right)^N \times \prod_{N+1\leq j\leq N+k'} \left(C_0^\infty(\mathbb{R}^{n-1})\right)^{\sigma_1+t_j+1},$$

acts continuously from $\mathfrak{H}^{s,p}(\mathbb{R}^n_+)$ to $K^{s,p}(\mathbb{R}^n_+)$ where the spaces $\mathfrak{H}^{s,p}(\mathbb{R}^n_+)$ and $K^{s,p}(\mathbb{R}^n_+)$ are given by

$$\begin{aligned}\mathfrak{H}^{s,p}(\mathbb{R}^n_+) &:= \widetilde{H}^{T+s,p,(\tau)}(\mathbb{R}^n_+)\times\mathfrak{B}^{T'+s,p}(\mathbb{R}^{n-1}),\\ K^{s,p}(\mathbb{R}^n_+) &:= \prod_{j=1}^{N}\widetilde{H}^{s-s_j,p,(\kappa-s_j)}(\mathbb{R}^n_+)\times\prod_{h=1}^{m+k''}B^{s-\sigma_h-1/p,p}(\mathbb{R}^{n-1}).\end{aligned}\tag{7.3}$$

Theorem 7.1. *Let the problem (2.1), (2.2) be elliptic. Then for any $s\in\mathbb{R}$ and any $p\in(1,\infty)$ the operator $\widehat{A}=\widehat{A}^{s,p}$ realizes the isomorphism between the spaces $\mathfrak{H}^{s,p}(\mathbb{R}^n_+)$ and $K^{s,p}(\mathbb{R}^n_+)$.*

Theorem 7.1 permits us, by using the standard procedure, to construct the right and left regularizers of the operator A and to prove the Theorem 5.1 (cf. [7]). We reduce the proof of Theorem 7.1 to a series of the lemmas.

Let $s,\sigma,p\in\mathbb{R}$, $1<p<\infty$. By $H^{s,\sigma,p}(\mathbb{R}^n)$ we denote the space of distributions f such that

$$||f,\mathbb{R}^n||_{s,\sigma,p}:=||F^{-1}(1+|\xi|^2)^{s/2}(1+|\xi'|^2)^{\sigma/2}Ff||_{L_p(\mathbb{R}^n)}<\infty.$$

Lemma 7.2. *For any element $\Phi=(\Phi_1,\dots,\Phi_N)$, such that $\Phi_r\in H^{s-s_r,\sigma,p}(\mathbb{R}^n)$, there exists one and only one solution $v=(v_1,\dots,v_N)$, $v_i\in H^{t_j+s,\sigma,p}(\mathbb{R}^n)$, of the problem $\widehat{l_0}v=\Phi$. In addition, the inequality*

$$\sum_{1\le j\le N}||v_j,\mathbb{R}_n||_{t_j+s,\sigma,p}\le c\sum_{1\le j\le N}||\Phi_j,\mathbb{R}^n||_{s-s_j,\sigma,p}$$

holds where the constant $c>0$ does not depend on Φ, v, s and σ.

Lemma 7.3. [7] *Under the assumption of Lemma 7.2, the inclusion* supp $\Phi\subset\overline{\mathbb{R}^n_+}$ *implies that* supp $v\subset\overline{\mathbb{R}^n_+}$ *if and only if the equalities*

$$\int_{-\infty}^{\infty}\widehat{L}_{-1}(\xi',\xi_n)\left(\xi_n+i\sqrt{1+|\xi'|^2}\right)^{s+t_k-m}\sum_{r=1}^{N}L_{rk}(\xi',\xi_n)\widetilde{\Phi}_r(\xi',\xi_n)\xi_n^j\,d\xi_n=0,$$
$$k=1,\dots,N,\ j=0,\dots,m-1.$$

are valid for almost all $\xi'=0$.

Here $\widehat{L}(\xi',\xi_n)=\det\widehat{l_0}(\xi)$, the expression $\widehat{L}_{rk}(\xi',\xi_n)$ is the cofactor of the element $\widehat{l}_{rk0}(\xi',\xi_n)$ of the determinant $L(\xi',\xi_n)$, and $\widetilde{\Phi}_r$ is the Fourier transform of the element Φ_r.

If

$$\begin{aligned}u &= (u_1,\dots,u_n)\in\widetilde{H}^{T+s,p,(\tau)}(\mathbb{R}^n_+),\quad u_j=(u_{j0},\dots,u_{j\tau_j})\in\widetilde{H}^{t_j+s,p,(\tau_j)}(\mathbb{R}^n_+),\\ F &= (f,\phi)=(f_1,\dots,f_N,\phi_1,\dots,\phi_{m+k''})\in K^{s,p}(\mathbb{R}^n_+),\quad f_j=(f_{j0},\dots,f_{j,\kappa-s_j}),\end{aligned}$$

then, by virtue of formulas (4.4) and (4.5), the equality (7.1) holds if and only if the relations

$$(7.4)\quad \left(\widehat{l}_{0r}u\right)_{+} := \sum_{j=1}^{N} \widehat{l}_{rj0}(D)u_{j0+} - \sum_{j:s_r+t_j\geq 1} \sum_{k=1}^{s_r+t_j} J^k \widehat{l}_{rj0}\big(u_{jk}\times\delta(x_n)\big) = f_{r0+},$$

$$r = 1,\dots,N,$$

$$(7.5)\quad D_n^{h-1}\big(\widehat{l}_{0r}u\big)|_{x_n=0} := \sum_{j=1}^{N}\sum_{k=1}^{s_r+t_j+1} \widehat{l}_k^{rj}(D')u_{j,k+h-1}(x') = f_{rh},$$

$$h = 1,\dots,\kappa - s_r,\ r : \kappa - s_r \geq 1.$$

are valid. By using Lemma 7.3, instead of the equalities (7.4) we substitute the equivalent equalities

$$(7.6)\quad \sum_{\alpha=1}^{N_1}\sum_{\beta=1}^{t_\alpha} \widehat{C}_{kj\alpha\beta}(\xi')\widehat{u}_{\alpha\beta}(\xi') = \widehat{g}_{kj}(\xi'), \qquad k = 1,\dots,N,\ j = 0,\dots,m-1.$$

Here $N_1 \leq N$ is equal to a number of t_j such that $t_j \geq 1$. It turns out ([7]) that for any $\xi' \neq 0$, in the set of Nm conditions (7.6) there exist m linearly independent conditions; the rest of them are expressed in terms of these m conditions. Therefore, in what follows we will assume that in every point $\xi' \neq 0$ we leave m linearly independent equations. It is clear, that these equations are linearly independent also in some neighborhood of the point ξ. If we add to (7.6) the Fourier transforms of equalities (7.5) and (7.2), then we obtain a system of $\tau_1 + \cdots + \tau_N + k''$ linear equations with respect to the $\tau_1 + \cdots + \tau_N$ variables $\widehat{u}_{jk}(\xi')$, and the $k'(\sigma_1 + 1) + t_{N+1} + t_{N+k'}$ variables $\widehat{u}'_{jk}(\xi')$.

Lemma 7.4. *The ellipticity of the problem* (2.1), (2.2) *is equivalent to the fact that the obtained linear system is quadratic, and its determinant* $\widehat{\Delta}(\xi')$ *is not equal to zero.*

Therefore, a specific relation between k' and k'' can be obtained. Lemma 7.4 permits us to express and to estimate the expressions $\widehat{u}_{jk}(\xi')$ and $u'_{jk}(\xi')$ in the terms of right-hand parts, and to prove Theorem 7.1.

Now we complete the proof of Theorem 5.1. Represent the expression

$$A(x,D) = \big(l(x,D), b(x,D), b'(x,D)\big)$$

in the form

$$A(x,D) = a\ A_0(x,D) + A'(x,D),$$

where A_0 is the principal part of the expression $A(x,D)$, and $A'(x,D)$ is an operator whose order, evidently, is less than the order of A_0.

Let $x_0 \in \partial G$. We write the expression $A(x, D)$ in the form

$$A(x, D) = A_0(x_0, D) + Q_1(x, D) + A'(x, D),$$

where $A_0(x_0, D)$ is an expression with constant coefficients (fixed at the point x_0), and the coefficients of $Q_1(x, D)$ are small in a sufficiently small neighborhood of the point x_0.

Let $U(x_0)$ be a neighborhood of the point x_0 in $\mathbb{R}^n$, and let $\chi, \chi_1 \in C_0^\infty(U)$, $\chi_1 = 1$ on the support of the function χ. Then

$$\chi A(x, D) = \chi\big(A_0(x_0, D) + Q_1(x, D)\chi_1 + A'(x, D)\chi_1\big).$$

Therefore the next lemma is true.

Lemma 7.5. *Let $x_0 \in \partial G$, and let $p \in (1, \infty)$. For any $\varepsilon > 0$ and any bounded domain $E \subset \mathbb{R}$ there exists a neighborhood $U(x_0) = U(x_0, \delta) = \{x \in \overline{G} : |x - x_0| < \delta\}$ of the point x_0 such that for any function $\chi \in C_0^\infty(U)$ and any $s \in E$ the representation*

$$\chi A(x, D) = \chi\big(A_0(x_0, D) + Q(x, D) + A'(x, D)\big) \tag{7.7}$$

is true. Here $Q(x, D) = Q_1(x, D)\chi_1$ is an operator whose norm is small , i.e.,

$$||Q||_{\mathfrak{H}^{s,p}(\mathbb{R}^n_+) \to K^{s,p}(\mathbb{R}^n_+)} \leq \varepsilon, \tag{7.8}$$

and the operator $A'(x, D)$ acts continuously from $\mathfrak{H}^{s,p}(\mathbb{R}^n_+)$ into $K^{s+1,p}(\mathbb{R}^n_+)$.

Since $A_0(x_0, D) - \widehat{A}_0(x_0, D)$ is an operator whose order is less than the order of A_0 by 1, expression (7.7) implies the representation

$$\chi A(x, D) = \chi\big(\widehat{A}_0(x_0, D) + Q(x, D) + A''(x, D)\big). \tag{7.9}$$

Here, by virtue of Theorem 7.1, the operator $\widehat{A}_0(x_0, D)$ establishes an isomorphism between the spaces in (7.3), $Q(X, D)$ is the operator with small norm (7.8), and the operator $A''(x, D)$ acts continuously from $\mathfrak{H}^{s,p}(\mathbb{R}^n_+)$ into $K^{s+1,p}(\mathbb{R}^n_+)$.

Since Q is an operator whose norm is small, by choosing sufficiently small $\varepsilon > 0$ we obtain that the operator

$$R(x_0) = R_{s,p}(x_0) = \big(\widehat{A}_0(x_0, D) + Q(x, D)\big)^{-1}$$

is an isomorphism from the space $K^{s,p}(\mathbb{R}^n_+)$ onto $\mathfrak{H}^{s,p}(\mathbb{R}^n_+)$. In addition, the following relations are true:

$$\begin{aligned} R(x_0)(\widehat{A}_0 + Q + A'') &= I_1 + T_0', \\ (\widehat{A}_0 + Q + A'')R(x_0) &= I_2 + T_0'', \end{aligned} \tag{7.10}$$

where I_1 and I_2 are the identity operators in $\mathfrak{H}^{s,p}(\mathbb{R}^n_+)$ and $K^{s,p}(\mathbb{R}^n_+)$, respectively, and

$$\begin{aligned} T_0' &= R(x_0)A'' : \mathfrak{H}^{s,p}(\mathbb{R}^n_+) \to \mathfrak{H}^{s+1,p}(\mathbb{R}^n_+), \\ T_0'' &= A''R(x_0) : K^{s,p}(\mathbb{R}^n_+) \to K^{s+1,p}(\mathbb{R}^n_+), \end{aligned}$$

are smoothing operators.

Thus, for any point $x_0 \in \partial G$, there exist a sufficiently small neighborhood $U(x_0)$ and an operator $R(x_0)$ such that the relations (7.10) hold. In the case $x_0 \in G$, a similar statement is true. As a result, we obtain a covering of the compact set $\overline{G} = G \cup \partial G$. Let us select a finite subcovering

$$\{U^j : j = 1, \dots, \nu\}, \qquad U^j = U^j(x_j),$$

of this covering. Denote by R_j the operator $R(x_j)$. Let $\{\chi_j : j = 1, \dots, \nu\}$ be the decomposition of unity subordinate to the covering $\{U^j\}$,

$$\chi_j \in C_0^\infty(U^j), \qquad \sum_{j=1}^{\nu} \chi_j(x) \equiv 1.$$

Let $\widetilde{\chi}_j \in C_0^\infty(U^j)$, and let $\widetilde{\chi}_j = 1$ in some neighborhood of the support of the function χ_j, $0 \le \widetilde{\chi}_j \le 1$. It is clear that $\widetilde{\chi}_j\chi_j = \chi_j$, $j = 1, \dots, \nu$. We set

$$RF = R_{s,p}F = \sum_{j=1}^{\nu} \widetilde{\chi}_j R_j \chi_j F, \qquad F \in K^{s,p}(G).$$

By commuting A with $\widetilde{\chi}_j$, we obtain

$$ARF = \sum_{j=1}^{\nu} A\widetilde{\chi}_j R_j \chi_j F = \sum_{j=1}^{\nu} (\widetilde{\chi}_j A R_j \chi_j F + A''' R_j \chi_j F), \tag{7.11}$$

where the order of the operator A''' is lower than the order of A. It follows from (7.10) that

$$ARF = I_2F + T_2F, \tag{7.12}$$

where I_2 is the identity operator in $K^{s,p}(G)$, and the operator T_2 acts continuously from $K^{s,p}(G)$ into $K^{s+1,p}(G)$. One can prove by analogy that

$$RA = I_1 + T_1, \tag{7.13}$$

where I_1 is the identity operator in $\mathfrak{H}^{s,p}(G)$, and the operator T_1 acts continuously from $\mathfrak{H}^{s,p}(G)$ into $\mathfrak{H}^{s+1,p}(G)$. Equalities (7.12) and (7.13) show that the operator $R = R_{s,p}$ is a regularizer of the operator $A_{s,p}$, and the statement of the Theorem 5.1 follows (cf., for example, [2], [5]).

Acknowledgements

The author expresses deep gratitude to Professor S. Nazarov for the statement of the problem and to Professors D. Vassiliev and Ya. Roitberg for useful discussions and valuable remarks.

The author would like to thank the organizers of IWOTA 95 for hospitality and financial support of her participation.

References

[1] AGMON, S., DOUGLIS, A., NIRENBERG, L.: Estimates near the boundary for solutions of elliptic partial differential equations satisfying general boundary conditions, I, II; Comm. Pure Appl. Math. 12 (1959), 623-727; 17 (1964), 35–92.

[2] ASLANYAN, A.G., VASSILIEV, D.G., LIDSKII, V.B.: Frequencies of free oscillations of thin shell interacting with fluid; Functional Anal. Appl. 15:3 (1981), 1–9.

[3] BEREZANSKII, YU.M.: Expansions in eigenfunctions of selfadjoint operators; Naukova Dumka, Kiev 1965.

[4] GARLET, P.G.: Plates and junctions in ellastic multistructures. An asymptotic analysis; Masson: Paris Milan, Barcelona Mexico 1990.

[5] NAZAROV, S., PILECKAS, K.: On noncompact free boundary problems for the plane stationary Navier-Stokes equations; J. Reine Angew. Math. 438 (1993), 103–141.

[6] ROITBERG, YA.A.: Theorem on complete collection of isomorphisms for Douglis–Nirenberg elliptic systems; Ukrain. Mat. Zh. 27:4 (1975), 554–548.

[7] ROITBERG,YA.A.: Elliptic Boundary Value Problems in the Spaces of Distributions; Kluwer Academic Publishers, Dordrecht Boston London 1996.

[8] SOLONNIKOV, V.A.: On general boundary value problems elliptic according to Douglis–Nirenberg, I, II; Izv. Akad. Nauk SSSR Ser. Mat. 29:3 (1964), 665–706; Trudy Mat. Inst. Steklov. 92 (1966), 233–297.

[9] VOLEVICH, L.R.: Solvability of boundary-value problems for general elliptic systems; Mat. Sb. 68:3 (1965), 373–416.

Chernigov Pedagogical Institute
Sverdlova str., 53
Chernigov 250038
Ukraine
alex@elit.chernigov.ua

1991 Mathematics Subject Classification: Primary 35J55; Secondary 35J40

Submitted: May 31, 1996

Operator Theory:
Advances and Applications, Vol. 102

Classic spectral problems

L.A. Sakhnovich

This article is dedicated to the detailed investigation of classic spectral problems. We show that the string matrix equation and the Sturm-Liouville equation belong to a class of canonical systems.

1. Generalized string equation (direct spectral problem)

1.1. Let us consider the canonical system of the equations

$$\frac{d\mathcal{W}}{dx} = izJ\mathcal{H}(x)\mathcal{W}, \qquad \mathcal{W}(0,z) = E_{2m}, \qquad J = \begin{bmatrix} 0 & E_m \\ E_m & 0 \end{bmatrix}. \tag{1.1}$$

We shall suppose that the Hamiltonian $\mathcal{H}(x)$ is continuous and belongs to the class $N(M)$, i.e., the matrix $J\mathcal{H}(x)$ is linearly similar to the matrix

$$M = \begin{bmatrix} 0 & 0 \\ E_m & 0 \end{bmatrix}. \tag{1.2}$$

This means that the matrix $\mathcal{H}(x)$ has the following block structure

$$J\mathcal{H}(x) = \begin{bmatrix} p(x) \\ q(x) \end{bmatrix} [q^*(x),\, p^*(x)] \tag{1.3}$$

where $p(x)$ and $q(x)$ are continuous matrices of size $m \times m$ and the relations

$$q^*(x)p(x) + p^*(x)q(x) = 0, \qquad \text{rank } \mathcal{H}(x) = m \tag{1.4}$$

are true.

In what follows we assume that the normalization condition

$$p(x) > 0 \tag{1.5}$$

is satisfied. In case of (1.3), (1.4) both the canonical system (1.1) and the corresponding isometric operator V (see [3]) can be written in a simpler form. Indeed, we consider the matrix functions

$$\varphi_1(x,z) = p(x)\mathcal{W}_{21}(x,z) + q^*(x)\mathcal{W}_{11}(x,z), \tag{1.6}$$

$$\varphi_2(x,z) = p(x)\mathcal{W}_{22}(x,z) + q^*(x)\mathcal{W}_{12}(x,z). \tag{1.7}$$

It follows from (1.1) that

$$\frac{dW_{1k}}{dx} = izp(x)\varphi_k(x,z), \qquad \frac{dW_{2k}}{dx} = izq(x)\varphi_k(x,z), \qquad k = 1,2. \tag{1.8}$$

Integrating (1.8) and bearing (1.6), (1.7) in mind we obtain

$$\varphi_1(x,z) = iz\int_0^x [p(x)q(t) + q^*(x)p(t)]\,\varphi_1(t,z)\,dt + q^*(x), \tag{1.9}$$

$$\varphi_2(x,z) = iz\int_0^x [p(x)q(t) + q^*(x)p(t)]\,\varphi_2(t,z)\,dt + p^*(x). \tag{1.10}$$

Let us note that the functions $\varphi_1(x,z)$ and $\varphi_2(x,z)$ are uniquely defined by the equations (1.9) and (1.10), respectively.

The transformation operator can be written in the form (see [3])

$$F(u) = V_1 f(x) = \int_0^l Y^*(x,u)f(x)\,dx \tag{1.11}$$

where

$$Y(x,u) = \varphi_1(x,u)D_1^* + \varphi_2(x,u)D_2^*, \tag{1.12}$$

$$f(x) = q^*(x)g_1(x) + p(x)g_2(x). \tag{1.13}$$

It is supposed here that the matrices D_1, D_2 of size $m \times m$ satisfy the relations

$$D_1D_2^* + D_2D_1^* = 0, \qquad D_1D_1^* + D_2D_2^* = E_m.$$

According to the general definition (see [3]), the spectral matrix function $\tau(u)$ of the system (1.1) can be characterized by the relation

$$\int_{-\infty}^{\infty} F^*(u)[d\tau(u)]F(u) = \int_0^l f^*(x)f(x)\,dx$$

as the following identity

$$\int_0^l g^*(x)\mathcal{H}(x)g(x)\,dx = \int_0^l f^*(x)f(x)\,dx$$

holds.

Thus the operator V_1 maps $\mathrm{L}_m^2(0,\infty)$ into $\mathrm{L}_m^2(\tau)$ isometrically. Formulas (1.9), (1.10), (1.12) imply that $Y(x,z)$ satisfies the equation

$$Y(x,z) = iz\int_0^x [p(x)q(t) + q^*(x)p(t)]\,Y(t,z)\,dt + q^*(x)D_1^* + p(x)D_2^*. \tag{1.14}$$

Equation (1.14) is the generalized string equation.

1.2. Under additional requirements the integral system (1.14) can be reduced to a differential system of second order (matrix string equation).

Theorem 1.1. *Let the following conditions be fulfilled:*
1) *The matrices* $p(x)$ *and* $q(x)$ *are continuous and satisfy relations* (1.4), (1.5).
2) *The matrix* $p(x)$ *is invertible.*
3) *The matrix* $p^{-1}(x)q^*(x)$ *is continuously differentiable.*
4) *The matrix*

$$r(x) = \frac{d}{dx}\left[p^{-1}(x)q^*(x)\right] \tag{1.15}$$

is invertible.

Then the integral system (1.14) *is equivalent to the differential system*

$$p^{-1}(x)\frac{d}{dx}\left\{A(x)\frac{d}{dx}\left[p^{-1}(x)Y(x,z)\right]\right\} = zY(x,z) \tag{1.16}$$

and to the conditions

$$Y(0,z) = q^*(0)D_1^* + p(0)D_2^*, \tag{1.17}$$

$$\frac{d}{dx}\left[p^{-1}(x)Y(x,z)\right]\bigg|_{x=0} = r(0)D_1^*, \tag{1.18}$$

where

$$A(x) = -ir^{-1}(x) \tag{1.19}$$

Proof. By (1.4), (1.6)–(1.8) we have

$$\frac{d}{dx}\left[p^{-1}(x)\varphi_k(x,z)\right] = r(x)\mathcal{W}_{1k}(x,z). \tag{1.20}$$

Suppose in addition that the matrix function $r(x)$ is invertible. We deduce from (1.8) and (1.20) that

$$p^{-1}(x)\frac{d}{dx}\left\{A(x)\frac{d}{dx}\left[p^{-1}(x)\varphi_k(x,z)\right]\right\} = z\varphi_k(x,z). \tag{1.21}$$

From (1.12) and (1.22) we obtain equation (1.16). Conditions (1.17), (1.18) follow from (1.12), (1.20). The theorem is proved. □

Remark 1.2. The differential operator on the right-hand side of (1.16) is formally self-adjoint since by (1.4), (1.15), (1.19) we have

$$A(x) = A^*(x). \tag{1.22}$$

By strengthening the conditions of Theorem 1.1 we can write them in a simpler form.

Theorem 1.3. *Let the following conditions be fulfilled:*

1) *The matrices* $p(x)$ *and* $q(x)$ *are continuously differentiable and satisfy the relations* (1.4), (1.5).

2) *The matrices* $p(x)$ *and*

$$r(x) = \frac{d}{dx}\left[p^{-1}(x)q^*(x)\right]$$

are invertible.

Then the integral system (1.14) *is equivalent to the differential system* (1.16) *and to the conditions* (1.17), (1.18).

1.3. System (1.16) is itself as is known of great interest. Therefore it is important to formulate the results in terms of this system, i.e., with the help of the matrices $A(x)$ and $p(x)$.

Theorem 1.4. *Let the following conditions be fulfilled:*

1) *The matrices* $p(x)$ *and* $A(x)$ *are continuous and invertible.*

2) *Relations* (1.5) *and* (1.22) *are valid.*

Then the differential system (1.16)–(1.18) *is equivalent to the integral system* (1.14) *where*

$$q^*(x) = p(x)\left[\int_0^x r(t)\,dt + r(0)\right], \tag{1.23}$$

$$r(x) = -iA^{-1}(x). \tag{1.24}$$

Proof. It is easy to see that $p(x)$ and $q(x)$ satisfy all the conditions of Theorem 1.1. Putting

$$Y_1(x,z) = p^{-1}(x)Y(x,z), \tag{1.25}$$

we can rewrite (1.16) in the form

$$\frac{d}{dx}\left\{A(x)\left[\frac{d}{dx}Y_1(x,z)\right]\right\} = zp^2(x)Y_1(x,z). \tag{1.26}$$

□

Equation (1.26) is a matrix generalization of the classical string equation.

Example 1.5. Let the relations

$$p(x) = E_m, \qquad q(x) = -ixE_m, \qquad D_1 = 0, \qquad D_2 = E_m \tag{1.27}$$

be true. Then the equalities

$$A(x) = -E_m, \qquad Y(x,z) = \cos\sqrt{z}xE_m \tag{1.28}$$

are valid.

Setting

$$D_1 = 0, \qquad D_2 = E_m, \tag{1.29}$$

let us apply the general spectral theorems (see [3]) to equality (1.16). The following assertion comes from the results in [3].

Theorem 1.6. *Let the matrices $p(x)$, $q(x)$ of size $m \times m$ be continuous and satisfy the relations* (1.4), (1.5), *and let $\lambda = 0$ not be an eigenvalue of the operator*

$$A_M f = i \int_0^x [p(x)q(t) + q^*(x)p(t)]\, f(t)\, dt \tag{1.30}$$

acting in $\mathrm{L}^2_m(0, l)$.
Then the set of the Weyl-Titchmarsh matrix functions $v(z)$ corresponding to the system (1.14), (1.29) *is described with the help of the formula*

$$v(z) = i\,[a(z)\mathcal{P}(z) + b(z)\mathcal{Q}(z)]\,[c(z)\mathcal{P}(z) + d(z)\mathcal{Q}(z)]^{-1} \tag{1.31}$$

where $\mathcal{P}(z)$, $\mathcal{Q}(z)$ is a non-special pair with the J-property and the matrix of the coefficients has the form

$$\mathfrak{A}(l, z) = \begin{bmatrix} a(z) & b(z) \\ c(z) & d(z) \end{bmatrix} = \mathcal{W}^*(l, \bar{z}). \tag{1.32}$$

Let us remind that the spectral data $\tau(u)$ and $\alpha = \alpha^*$ of equality (1.14) are connected with its Weyl-Titchmarsh matrix function by the equality (see [3])

$$v(z) = \alpha + \int_{-\infty}^{\infty} \left(\frac{1}{u - z} - \frac{u}{1 + u^2} \right) d\tau(u). \tag{1.33}$$

Remark 1.7. In case (1.29) the conditions (1.17), (1.18) have the form

$$Y(0, z) = p(0), \qquad \left.\frac{d}{dx}\left[p^{-1}(x)\, Y(x, z)\right]\right|_{x=0} = 0. \tag{1.34}$$

1.4 If the conditions of Theorem 1.3 are fulfilled, then the conditions of Theorem 1.6 are also fulfilled, i.e., the following assertion is true.

Theorem 1.8. *Let the the conditions of Theorem* 1.3 *be fulfilled. Then the set of Weyl-Titchmarsh matrix functions $v(z)$ corresponding to the system* (1.16), (1.34) *is described with the help of the formulas* (1.31), (1.32).

Proof. Let us suppose that $\lambda = 0$ is an eigenvalue of the operator A_M. Then the equality

$$\int_0^x [p(x)q(t) + q^*(x)p(t)]\, f_0(t)\, dt = 0 \tag{1.35}$$

where $\|f_0\| \neq 0$ holds.

We denote by $H_{1,x}$ the space the basis of which are the rows of the matrix $[p(x), q^*(x)]$ and we denote by $H_{2,x}$ the space with the basis formed by the rows of the matrix $[q^*(x), p(x)]$. Due to (1.4) these spaces are mutually orthogonal and have dimensions equal to m. It follows from (1.35) that the vector

$$g_0(x) = \left[\int_0^x f_0^*(t)q^*(t)\,dt,\ \int_0^x f_0^*(t)p(t)\,dt\right]$$

belongs to $H_{2,x}$. Hence there exists a matrix $h_0(x)$ of size $1 \times m$ such that

$$h_0(x)\,[q^*(x),\ p(x)] = g_0(x),$$

i.e.,

$$h_0(x)q^*(x) = \int_0^x f_0^*(t)q^*(t)\,dt, \tag{1.36}$$

$$h_0(x)p(x) = \int_0^x f_0^*(t)p(t)\,dt. \tag{1.37}$$

Since the matrices $p(x)$ and $q^*(x)$ are invertible and differentiable the matrix $h_0(x)$ is absolutely continuous due to (1.36), (1.37). Hence the equalities

$$h'_0(x)q^*(x) + h_0(x)q^{*\prime}(x) = f_0^*(x)q^*(x), \tag{1.38}$$

$$h'_0(x)p(x) + h_0(x)p'(x) = f_0^*(x)p(x) \tag{1.39}$$

are true for almost all x. Multiplying from the left both sides of (1. 38) by $p(x)$ and both sides of (1.39) by $q(x)$ and adding the obtained expressions we have

$$h_0(x)r_1(x) = 0 \tag{1.40}$$

where

$$r_1(x) = q^{*\prime}(x)p(x) + p'(x)q(x). \tag{1.41}$$

Comparing (1.15) and (1.41) we have

$$r(x) = p^{-1}(x)r_1(x)p^{-1}(x). \tag{1.42}$$

It follows from the invertibility of the matrix $r(x)$ (condition 2) of Theorem 1.3) that the matrix $r_1(x)$ is invertible too. Then in view of (1.40) the equality $h_0(x) = 0$ is valid, i.e., $f_0(x) = 0$. Hence $\lambda = 0$ is not an eigenvalue of the operator A_M.
Thus the conditions of Theorem 1.3 are fulfilled. This proves the theorem. □

Corollary 1.9. *Let the conditions of Theorem* 1.1 *be fulfilled, and let the matrix function $p(x)$ be continuously differentiable. Then the assertion of Theorem* 1.8 *is true.*

A survey of the results of the spectral problem for the scalar string equation is contained in the paper by I.S. Kac and M.G. Krein (see [2], and also [1]).

2. Matrix Sturm-Liouville equation (direct spectral problem)

2.1. Let $U(x)$ be a matrix function of size $m \times m$, and let

$$U(x) = U^*(x). \tag{2.1}$$

We introduce another matrix function

$$G(x,z) = \begin{bmatrix} 0 & E_m \\ U(x) - zE_m & 0 \end{bmatrix} \tag{2.2}$$

and consider the following system of equations:

$$\frac{dW}{dx} = G(x,z)W, \qquad 0 \leqslant x \leqslant l \leqslant \infty, \qquad W(0,z) = E_{2m}. \tag{2.3}$$

Further, we put

$$\mathcal{P} = \begin{bmatrix} 0 & E_m \\ U & 0 \end{bmatrix}, \quad j_1 = i\begin{bmatrix} 0 & -E_m \\ E_m & 0 \end{bmatrix}, \quad T = \frac{1}{\sqrt{2}}\begin{bmatrix} iE_m & E_m \\ iE_m & -E_m \end{bmatrix}. \tag{2.4}$$

Let us define the matrix $B(x)$ by the equations

$$\frac{dB}{dx} = \mathcal{P}B, \qquad B(0) = T. \tag{2.5}$$

We suppose here that the entries of $U(x)$ are locally integrable. Then the matrix function

$$\mathcal{W}(x,z) = B^{-1}(x)W(x,z)T \tag{2.6}$$

is a solution of the system

$$\frac{d\mathcal{W}}{dx} = izJ\mathcal{H}(x)\mathcal{W}, \qquad \mathcal{W}(0,z) = E_{2m} \tag{2.7}$$

where

$$\mathcal{H}(x) = B^*(x)\begin{bmatrix} E_m & 0 \\ 0 & 0 \end{bmatrix}B(x). \tag{2.8}$$

Here we have taken into account the relations

$$B^*(x)j_1B(x) = J, \qquad T^*j_1T = J \tag{2.9}$$

which follow from (1.5),(1.6). From (2.8) and (2.9) we deduce the equality

$$J\mathcal{H}(x) = B^{-1}(x)\begin{bmatrix} 0 & 0 \\ iE_m & 0 \end{bmatrix}B(x). \tag{2.10}$$

It means that $\mathcal{H}(x)$ belongs to $N(M)$ (see [4]) where

$$M = \begin{bmatrix} 0 & 0 \\ iE_m & 0 \end{bmatrix}. \tag{2.11}$$

The operator V has the form (see [3])

$$F(u) = Vg(x) = \int_0^l [D_1, D_2]T^* W^*(x,u)\mathrm{col}\,[f_1(x), 0]\ dx \tag{2.12}$$

where

$$f(x) = \mathrm{col}\,[f_1(x), f_2(x)] = A(x)g(x). \tag{2.13}$$

Let us introduce $Y_1(x,z)$, $Y_2(x,z)$, $Y(x,z)$ by the relations

$$Y_1(x,z) = \frac{i}{\sqrt{2}}\left[W_{11}(x,z) + W_{12}(x,z)\right], \tag{2.14}$$

$$Y_2(x,z) = \frac{1}{\sqrt{2}}\left[W_{11}(x,z) - W_{12}(x,z)\right], \tag{2.15}$$

$$Y(x,z) = Y_1(x,z)D_1^* + Y_2(x,z)D_2^*, \tag{2.16}$$

where $W_{ij}(x,z)$ are $m \times m$ blocks of the matrix $W(x,z)$. Formula (2.12) can be written in the form

$$F(u) = V_1 f_1(x) = \int_0^l Y^*(x,u) f_1(x)\, dx. \tag{2.17}$$

Thus the operator V_1 maps $\mathrm{L}_m^2(0,\infty)$ into $\mathrm{L}_m^2(\tau)$ isometrically. Formulas (2.2), (2.3) and (2.14)–(2.16) imply that $Y(x,z)$ satisfies the Sturm-Liouville matrix equation

$$-\frac{d^2Y}{dx^2} + U(x)Y(x,z) = zY(x,z) \tag{2.18}$$

and the conditions

$$Y(0,z) = (iD_1^* + D_2^*)\frac{1}{\sqrt{2}}, \qquad Y'(0) = (iD_1^* - D_2^*)\frac{1}{\sqrt{2}}. \tag{2.19}$$

The boundary conditions (2.19) can be written in the form

$$C_1 Y(0) = C_2 Y'(0) \tag{2.20}$$

where

$$C_1 = D_1 - iD_2, \qquad C_2 = D_1 + iD_2. \tag{2.21}$$

Here we have taken into account the relations

$$D_1 D_2^* + D_2 D_1^* = 0, \qquad D_1 D_1^* + D_2 D_2^* = E_m. \tag{2.22}$$

Therefore the following assertion is true.

Theorem 2.1. *The spectral matrix function* $\tau(u)$ *introduced above coincides with the classical spectral matrix function of the Sturm-Liouville problem* (2.18), (2.20).

Remark 2.2. Due to (2.22) the equality

$$C_1 C_2^* = C_2 C_1^* \tag{2.23}$$

is fulfilled.

Hence the operator defined by the differential expression (2.18) and the boundary conditions (2.20) admits a self-adjoint extension.

2.2. As in Section 1 we set

$$D_1 = 0, \qquad D_2 = E_m. \tag{2.24}$$

Then condition (2.20) takes the form

$$Y(0) = -Y'(0). \tag{2.25}$$

We apply the general spectral theorems (see [3]) to system (2.18), (2.25). For this purpose let us write the block representation of the matrix

$$B(x) = \begin{bmatrix} B_{11}(x) & B_{12}(x) \\ B_{21}(x) & B_{22}(x) \end{bmatrix}$$

where all the blocks are of size $m \times m$.

Theorem 2.3. *Let the matrix* $U(x)$ *of size* $m \times m$ *satisfy condition* (2.1) *and let*

$$\int_0^l \|U(x)\|^2 \, dx < \infty. \tag{2.26}$$

Then the set of Weyl-Titchmarsh matrix functions $v(z)$ *corresponding to system* (2.18), (2.25) *is described with the help of formulas* (1.31), (1.32), *where* $\mathcal{W}(l,z)$ *is defined by relation* (2.6).

3. Inverse spectral problem

3.1. Let us apply the general procedure of solving the inverse spectral problem (see [3]) to the case when $\mathcal{H}(x) \in N(M)$.

Let us introduce the operator

$$A\vec{f} = \int_0^x (t-x)\vec{f}(t)\, dt, \qquad 0 \leqslant x \leqslant l, \tag{3.1}$$

where $\vec{f}(x) \in \mathrm{L}^2_m(0,l)$.

The operators Φ_2 and P_ζ are defined by the equalities

$$\Phi_2 g = g, \qquad P_\zeta f = \begin{cases} f(x), & 0 \leqslant x \leqslant \zeta, \\ 0, & x > \zeta. \end{cases} \tag{3.2}$$

Here g is a constant vector of size $m \times 1$. The unperturbed problem is of the form

$$J\mathcal{H}_0(x) = \begin{bmatrix} ixE_m & E_m \\ x^2 E_m & -ixE_m \end{bmatrix}. \tag{3.3}$$

In this case we have

$$\tau_0(\lambda) = \begin{cases} 0, & \lambda < 0, \\ \dfrac{2}{\pi}\sqrt{\lambda}, & \lambda \geqslant 0, \end{cases} \qquad S_0 = E. \tag{3.4}$$

The matrix $J\mathcal{H}_0(x)$ is linearly similar to the matrix

$$M = \begin{bmatrix} 0 & 0 \\ E_m & 0 \end{bmatrix}.$$

Applying the general theory (see [3]) to (3.1), (3.2) we obtain the following result.

Theorem 3.1. *Let the operators A, Φ_2 and P_ζ be defined by the formulas (1.2). Suppose that the monotonically increasing matrix function $\tau(\lambda)$ satisfies the following conditions:*

1) *The matrix function*

$$K(x,u) = \int_{-\infty}^{\infty} \cos\sqrt{\lambda}x\,\{d[\tau(\lambda) - \tau_0(\lambda)]\}\cos\sqrt{\lambda}u, \qquad 0 \leqslant x,\ u < \infty, \tag{3.5}$$

is continuous.

2) *If*

$$\int_{-\infty}^{\infty} F^*(\lambda)[d\tau(\lambda)]F(\lambda) = 0,$$

then $F(\lambda) = 0$ where

$$F(\lambda) = \int_0^l \cos\sqrt{\lambda}x f(x)\,dx. \tag{3.6}$$

Then $\tau(\lambda)$ is a spectral matrix function of system (1.1). The corresponding Hamiltonian $\mathcal{H}(x)$ is continuous, belongs to the class $N(M)$ and can be found by formulas

$$S_\zeta f = f(x) + \int_0^\zeta K(x,u)f(u)\,du, \tag{3.7}$$

$$\mathcal{H}(\zeta) = \frac{d}{d\zeta}\left(\Pi^* S_\zeta^{-1} P_\zeta \Pi\right), \qquad \Pi = [\Phi_1, \Phi_2], \tag{3.8}$$

$$\Phi_1 = -i\int_{-\infty}^{\infty} \frac{\cos\sqrt{\lambda}x - 1}{\lambda(1+\lambda^2)}\Phi_2\,d\tau(\lambda). \tag{3.9}$$

Proof. The general formula for the operator S has the form (see [3])

$$S = \int_{-\infty}^{\infty} (E - Au)^{-1}\Phi_2[d\tau(u)]\Phi_2^*(E - A^*u)^{-1}. \tag{3.10}$$

Since in the considered case the equality

$$(E - Au)^{-1}\Phi_2 = \left(\cos\sqrt{u}x\right)\Phi_2 \tag{3.11}$$

holds, from the relations (3.2), (3.5) we deduce equality (3.7) where $S_\zeta = P_\zeta S P_\zeta$. The general formula for Φ_1 has the form (see [3])

$$\Phi_1 = -i\int_{-\infty}^{\infty}\left[A(E - Au)^{-1} + \frac{u}{1+u^2}E\right]\Phi_2\,d\tau(u). \tag{3.12}$$

In case (3.1), (3.2) formula (3.12) coincides with (3.9). It follows from the representation (3.10) that $S \geqslant 0$. Due to condition 2) of the theorem, the strict inequality

$$S > 0 \tag{3.13}$$

is true. From (3.7), (3.13) we obtain the existence and boundedness of the operator S_ζ^{-1}. So formula (3.8) is valid.

According to results in [4] the Hamiltonian $\mathcal{H}(x)$ is continuous, belongs to the class $N(M)$, i.e., the matrix $J\mathcal{H}(x)$ is linearly similar to the matrix M. This proves the theorem. □

Remark 3.2. Condition 1) of Theorem 3.1 is fulfilled if

$$\int_0^{\infty} \|d\left[\tau(\lambda) - \tau_0(\lambda)\right]\| + \int_{-\infty}^{0} e^{\sqrt{|\lambda|}x}\|d\tau(\lambda)\| < \infty.$$

References

[1] Dym, H., Kravitsky, N.: On the Inverse Problem for the String Equation; Integral Equations Operator Theory 1:2 (1978), 270–277.

[2] Kac, I.S., Krein, M.G.: On the Spectral functions of the String; Amer. Math. Soc. Transl. Ser. 2 103 (1974), 19–102.

[3] Sakhnovich, L.A.: Factorization Problems and Operator Identities; Russian Math. Surveys 41:1 (1986), 1–64.

[4] Sakhnovich, L.A.: On one Hypothesis Concerning Hamiltonians of Canonical Systems; Ukrainian Math. J. 46:10 (1994), 1428–1431.

Department of Mathematics
Ukrainian National Academy of Communication
Odessa
Ukraine
alex@alex.intes.odessa.ua

1991 Mathematics Subject Classification: Primary 30B99; Secondary 31C45, 42A16

Submitted: June 7, 1996

Operator Theory:
Advances and Applications, Vol. 102

Mellin operators in a pseudodifferential calculus for boundary value problems on manifolds with edges

E. SCHROHE and B.-W. SCHULZE

As an integral part of a pseudodifferential calculus for boundary value problems on manifolds with edges we introduce the algebra of Mellin operators. They represent the typical operators near the edge. In fact we show how to associate an operator-valued Mellin symbol to an arbitrary edge-degenerate pseudodifferential boundary value problem, the so-called 'Mellin quantization' procedure. Furthermore, we introduce a class of adequate Sobolev spaces based on the Mellin transform on which these operators act continuously.

1. Introduction

The analysis of partial differential operators on manifolds with piecewise smooth geometry, in particular, on manifolds with polyhedral singularities, is of central interest in models in mathematical physics, engineering, and applied sciences.

An essential aspect is the understanding of the solvability of elliptic differential equations in terms of a Fredholm theory. It is very desirable, for example, to have an appropriate notion of ellipticity implying the Fredholm property and the possibility of constructing parametrices to elliptic elements within a specified calculus, for this allows a precise analysis of the solutions to elliptic equations.

We shall deal with these questions in the context of boundary value problems on a manifold with edges by constructing an algebra of pseudodifferential operators adapted particularly to this situation.

The present paper is an important step in this direction. It focusses on the Mellin type operators, their properties, and the (Mellin) Sobolev spaces they naturally act on. It follows the general strategy of an iterative construction of operator algebras for situations of increasing complexity: Our local model of a manifold with an edge is the wedge $C \times \mathbb{R}^q$, where C is a manifold with boundary and conical singularities. We can therefore rely on the analysis of boundary value problems on manifolds with conical singularities given in [15], [16]. Technically, we regard the operators on the wedge as pseudodifferential operators along the edge of the wedge, taking values in the algebra of boundary value problems on the cone, and we employ the concept of operator-valued symbols on Banach spaces with group actions as presented, e.g., in [20].

The operators we are considering in this article correspond to boundary value problems on a manifold with edges localized to a neighborhood of the edge. They show a typical edge-degeneracy: Denoting the variable in the direction of the cone by t and the variables along the edge by y, derivatives ∂_t or ∂_y will only appear with an additional factor t. This suggests the use of the Mellin transform and associated Mellin Sobolev spaces.

There are two crucial constructions in this context. The first is the Mellin quantization procedure which shows how to pass from an edge-degenerate boundary symbol to a Mellin symbol which induces the same operator up to smoothing errors and vice versa. The second is the so-called kernel cut-off, an analytical procedure that allows to switch to holomorphic Mellin symbols (up to regularizing symbols). While the first step shows that the Mellin calculus is indeed the appropriate tool for this situation, the second one is indispensable for a Fredholm theory within the calculus, for it enables us to work on Sobolev spaces with different weights.

Historically, this paper has several roots. One is Kondrat'ev's article [10], where he analyzed boundary value problems on domains with conical points, another Agranovich, Vishik [1], who employed parameter-dependent operators, furthermore Vishik, Eskin [23], who analyzed boundary value problems without the transmission property, and Boutet de Monvel [3], who constructed a pseudodifferential calculus for boundary value problems. Primarily, however, there is the Mellin calculus for manifolds with conical singularities in the boundaryless case, see, e.g., Schulze [20], as well as the corresponding calculus for manifolds with edges in [7].

2. Basic constructions for pseudodifferential boundary value problems

Operator-valued symbols and wedge Sobolev spaces

2.1. Operator-valued symbols. A strongly continuous group action on a Banach space E is a family $\kappa = \{\kappa_\lambda : \lambda \in \mathbb{R}_+\} \subseteq \mathcal{L}(E)$ such that, for $e \in E$, the mapping $\lambda \mapsto \kappa_\lambda e$ is continuous and $\kappa_\lambda \kappa_\mu = \kappa_{\lambda\mu}$. In particular, each κ_λ is an isomorphism.

It will be useful to know that there are constants c and M with

$$\|\kappa_\lambda\|_{\mathcal{L}(E)} \leq c \max\{\lambda, \lambda^{-1}\}^M. \tag{2.1}$$

This can be easily deduced from the corresponding well-known result on the growth of (additive) strongly continuous semi-groups.

We let $H^s(\mathbb{R})$ be the usual Sobolev space on $\mathbb{R}$, while $H^s(\mathbb{R}_+) = \{u|_{\mathbb{R}_+} : u \in H^s(\mathbb{R})\}$ and $H_0^s(\mathbb{R}_+)$ is the set of all $u \in H^s(\mathbb{R})$ whose support is contained in $\overline{\mathbb{R}}_+$. Furthermore, $H^{s,t}(\mathbb{R}_+) = \{\langle r\rangle^{-t} u : u \in H^s(\mathbb{R}_+)\}$ and $H_0^{s,t}(\mathbb{R}_+) = \{\langle r\rangle^{-t} u : u \in H_0^s(\mathbb{R}_+)\}$. Finally, $\mathcal{S}(\mathbb{R}^q_+) = \{u|_{\mathbb{R}^q_+} : u \in \mathcal{S}(\mathbb{R}^q)\}$.

For all Sobolev spaces on $\mathbb{R}$ and $\mathbb{R}_+$, we will use the group action

$$(\kappa_\lambda f)(r) = \lambda^{\frac{1}{2}} f(\lambda r). \tag{2.2}$$

This action extends to distributions by $\kappa_\lambda u(\varphi) = u(\kappa_{\lambda^{-1}}\varphi)$. On $E = \mathbb{C}^l$ use the trivial group action $\kappa_\lambda = id$.

In the above definition, $\langle r\rangle = (1+|r|^2)^{1/2}$ is the function used frequently for estimates in connection with pseudodifferential operators. The definition extends $\langle\eta\rangle$ to $\eta\in\mathbb{R}^q$. It is equivalent, but sometimes more convenient, to estimate in terms of a function $[\eta]$, where $[\eta]$ is strictly positive, and $[\eta]=|\eta|$ for large $|\eta|$. We then have *Peetre's inequality*: For each $s\in\mathbb{R}$ there is a constant C_s with

$$[\eta+\xi]^s \le C_s[\eta]^s[\xi]^{|s|}.$$

Let E, F be Banach spaces with strongly continuous group actions $\kappa, \widetilde{\kappa}$, let $\Omega\subseteq\mathbb{R}^k$, $a\in C^\infty(\Omega\times\mathbb{R}^n, \mathcal{L}(E,F))$, and $\mu\in\mathbb{R}$. We shall write

$$a\in S^\mu(\Omega,\mathbb{R}^q;E,F),$$

provided that, for every $K\subset\subset\Omega$ and all multi-indices α,β, there is a constant $C=C(K,\alpha,\beta)$ with

$$\|\widetilde{\kappa}_{\langle\eta\rangle^{-1}} D_\eta^\alpha D_y^\beta a(y,\eta)\kappa_{\langle\eta\rangle}\|_{\mathcal{L}(E,F)} \le C\,\langle\eta\rangle^{\mu-|\alpha|}. \tag{2.3}$$

The space $S^\mu(\Omega,\mathbb{R}^q;E,F)$ is Fréchet topologized by the choice of the best constants C.

The space $S^\mu(\Omega,\mathbb{R}^q;\mathbb{C}^k,\mathbb{C}^l)$ coincides with the ($l\times k$ matrix-valued) elements of Hörmander's class $S^\mu(\Omega,\mathbb{R}^q)$.

Just like in the standard case one has asymptotic summation: Given a sequence $\{a_j\}$ with $a_j\in S^{\mu_j}(\Omega,\mathbb{R}^q;E,F)$ and $\mu_j\to-\infty$, there is an $a\in S^\mu(\Omega,\mathbb{R}^q;E,F)$, $\mu=\max\{\mu_j\}$ such that $a\sim\sum a_j$; a is unique modulo $S^{-\infty}(\Omega,\mathbb{R}^q;E,F)$. Note that $S^{-\infty}(\Omega,\mathbb{R}^q;E,F)$ is independent of the choice of κ and $\widetilde{\kappa}$.

A symbol $a\in S^\mu(\Omega,\mathbb{R}^q;E,F)$ is said to be *classical,* if it has an asymptotic expansion $a\sim\sum_{j=0}^\infty a_j$ with $a_j\in S^{\mu-j}(\Omega,\mathbb{R}^q;E,F)$ satisfying the homogeneity relation

$$a_j(y,\lambda\eta) = \lambda^{\mu-j}\widetilde{\kappa}_\lambda\, a_j(y,\eta)\,\kappa_{\lambda^{-1}} \tag{2.4}$$

for all $\lambda\ge1, |\eta|\ge R$, for a suitable constant R. We write $a\in S^\mu_{cl}(\Omega,\mathbb{R}^q;E,F)$. For $E=\mathbb{C}^k$, $F=\mathbb{C}^l$ we recover the standard notion.

There is an extension to projective and inductive limits: Let $\widetilde{E},\widetilde{F}$ be Banach spaces with group actions. If $F_1\hookleftarrow F_2\hookleftarrow\dots$ and $E_1\hookrightarrow E_2\hookrightarrow\dots$ are sequences of Banach spaces with the same group action, and $F=\text{proj}-\lim F_k$, $E=\text{ind}-\lim E_k$, then let

$$\begin{aligned}
S^\mu(\Omega,\mathbb{R}^q;\widetilde{E},F) &= \text{proj}-\lim\nolimits_k S^\mu(\Omega,\mathbb{R}^q;\widetilde{E},F_k),\\
S^\mu(\Omega,\mathbb{R}^q;E,\widetilde{F}) &= \text{proj}-\lim\nolimits_k S^\mu(\Omega,\mathbb{R}^q;E_k,\widetilde{F}),\\
S^\mu(\Omega,\mathbb{R}^q;E,F) &= \text{proj}-\lim\nolimits_{k,l} S^\mu(\Omega,\mathbb{R}^q;E_k,F_l).
\end{aligned}$$

Example 2.2. Let $\gamma_j:\mathcal{S}(\mathbb{R}_+)\to\mathbb{C}$ be defined by

$$\gamma_j f = \lim_{r\to0^+}\partial_r^j f(r).$$

Then, for all $s > j + 1/2$, we can consider γ_j as a (y, η)-independent symbol in $S^{j+1/2}(\mathbb{R}^q \times \mathbb{R}^q; H^s(\mathbb{R}_+), \mathbb{C})$.
In fact, all we have to check is that $\|\widetilde{\kappa}_{[\eta]^{-1}}\gamma_j\kappa_{[\eta]}\| = O([\eta]^{j+1/2})$ for the group actions $\widetilde{\kappa}$ on $\mathbb{C}$ and κ on $H^s(\mathbb{R}_+)$. Since the group action on $\mathbb{C}$ is the identity, that on $H^s(\mathbb{R}_+)$ is given by (2.2), everything follows from the observation that

$$\partial_r^j\{[\eta]^{1/2}f([\eta]r)\}|_{r=0} = [\eta]^{j+1/2}\partial_r^j f(0).$$

The following lemma is obvious.

Lemma 2.3. *For $a \in S^\mu(\Omega, \mathbb{R}^q; E, F)$ and $b \in S^\nu(\Omega, \mathbb{R}^q; F, G)$, the symbol c defined by $c(y, \eta) = b(y, \eta)a(y, \eta)$ (point-wise composition of operators) belongs to $S^{\mu+\nu}(\Omega, \mathbb{R}^q; E, G)$, and $D_\eta^\alpha D_y^\beta a$ belongs to $S^{\mu-|\alpha|}(\Omega, \mathbb{R}^q; E, F)$.*

Lemma 2.4. *Let $a = a(y, \eta) \in C^\infty(\Omega \times \mathbb{R}^q, \mathcal{L}(E, F))$, and suppose that $a(y, \lambda\eta) = \lambda^\mu \widetilde{\kappa}_\lambda\, a(y, \eta)\, \kappa_{\lambda^{-1}}$ for all $\lambda \geq 1$, $|\eta| \geq R$. Then $a \in S^\mu_{cl}(\Omega, \mathbb{R}^n; E, F)$, and the symbol semi-norms for a can be estimated in terms of the semi-norms for a in $C^\infty(\Omega \times \mathbb{R}^q, \mathcal{L}(E, F))$.*

Proof. Without loss of generality let $R = 1$. We only have to consider the case of large $|\eta|$. For these, the assumption implies that

$$D_\eta^\alpha D_y^\beta a(y, \eta) = \lambda^{-\mu+|\alpha|}\widetilde{\kappa}_{\lambda^{-1}}\,(D_\eta^\alpha D_y^\beta a)(y, \lambda\eta)\,\kappa_\lambda.$$

Letting $\lambda = [\eta]$, we conclude that

$$\widetilde{\kappa}_{[\eta]^{-1}} D_\eta^\alpha D_y^\beta a(y, \eta)\kappa_{[\eta]} = [\eta]^{\mu-|\alpha|}(D_\eta^\alpha D_y^\beta a)(y, \eta/[\eta]).$$

The norm of the right hand side in $\mathcal{L}(E, F)$ clearly is $O([\eta]^{\mu-|\alpha|})$. Moreover, a is classical, since it is homogeneous of degree μ in the sense of (2.4). □

Definition 2.5. Let $\Omega = \Omega_1 \times \Omega_2 \subseteq \mathbb{R}^q \times \mathbb{R}^q$ be open and $a \in S^\mu(\Omega, \mathbb{R}^q \times \mathbb{R}^l; E, F)$. The parameter-dependent pseudodifferential operator $\operatorname{op} a$ is the operator family $\{\operatorname{op} a(\lambda) : \lambda \in \mathbb{R}^l\}$ defined by

$$[\operatorname{op} a(\lambda) f](y) = \int e^{i(y-\widetilde{y})\eta} a(y, \widetilde{y}, \eta, \lambda) f(\widetilde{y})\, d\widetilde{y}\, đ\eta, \tag{2.5}$$

$f \in C_0^\infty(\Omega_2, E), y \in \Omega_1$. This reduces to

$$[\operatorname{op} a(\lambda) f](y) = \int e^{iy\eta} a(y, \eta)\hat{f}(\eta)\, đ\eta \tag{2.6}$$

for symbols that are independent of y'. Here, $\hat{f}(\eta) = \mathcal{F}_{y\to\eta} f(\eta) = \int e^{-iy\eta} f(y)\, dy$ is the vector-valued Fourier transform of f, and $đ\eta = (2\pi)^{-n} d\eta$.

Definition 2.6. Let E, κ be as in 2.1, $q \in \mathbb{N}$, $s \in \mathbb{R}$. The *wedge Sobolev space* $\mathcal{W}^s(\mathbb{R}^q, E)$ is the completion of $\mathcal{S}(\mathbb{R}^q, E) = \mathcal{S}(\mathbb{R}^q)\hat{\otimes}_\pi E$ in the norm

$$\|u\|_{\mathcal{W}^s(\mathbb{R}^q,E)} = \left(\int \langle\eta\rangle^{2s} \|\kappa_{\langle\eta\rangle^{-1}}\mathcal{F}_{y\to\eta}u(\eta)\|_E^2 \, d\eta\right)^{\frac{1}{2}}.$$

It is a subset of $\mathcal{S}'(\mathbb{R}^q, E)$. There are a few straightforward generalizations: If $\{E_k\}$ is a sequence of Banach spaces, $E_{k+1} \hookrightarrow E_k$, $E = \text{proj} - \lim E_k$, and the group action coincides on all spaces, we let $\mathcal{W}^s(\mathbb{R}^q, E) = \text{proj} - \lim \mathcal{W}^s(\mathbb{R}^q, E_k)$. Similarly we treat inductive limits. For $\Omega \subseteq \mathbb{R}^q$ open we shall write $u \in \mathcal{W}^s_{comp}(\Omega, E)$, if there is a function $\varphi \in C_0^\infty(\Omega)$ such that $u = \varphi u$, and say $u \in \mathcal{W}^s_{loc}(\Omega, E)$, if $u \in \mathcal{D}'(\Omega, E)$ and $\varphi u \in \mathcal{W}^s(\mathbb{R}^q, E)$ for all $\varphi \in C_0^\infty(\mathbb{R}^q)$.

2.7. Elementary properties of wedge Sobolev spaces.
(a) $\mathcal{W}^s(\mathbb{R}^q, H^s(\mathbb{R}_+)) = H^s(\mathbb{R}^{q+1}_+)$.
(b) $\mathcal{W}^s(\mathbb{R}^q, H^s_0(\mathbb{R}_+)) = H^s_0(\mathbb{R}^{q+1}_+)$.
(c) $\mathcal{W}^s(\mathbb{R}^q, \mathbb{C}) = H^s(\mathbb{R}^q)$, using the trivial group action $\kappa_\lambda = id$.

Theorem 2.8. *Let E, F be Banach spaces as in 2.1, $s, \mu \in \mathbb{R}$, and $a \in S^\mu(\mathbb{R}^q_y, \mathbb{R}^q_\eta \times \mathbb{R}^l_\lambda; E, F)$ or $a \in S^\mu(\mathbb{R}^q_y \times \mathbb{R}^q_{\tilde{y}}, \mathbb{R}^q_\eta \times \mathbb{R}^l_\lambda; E, F)$. Then for every $\lambda \in \mathbb{R}^l$*

$$\operatorname{op} a(\lambda) : \mathcal{W}^s_{comp}(\mathbb{R}^q, E) \longrightarrow \mathcal{W}^{s-\mu}_{loc}(\mathbb{R}^q, F)$$

is bounded. If a is independent of y and $\tilde{y}$, then we may omit the subscripts 'comp' and 'loc'.
The mapping op : *symbol* $\mapsto$ *operator is continuous in the corresponding topologies.*

A proof may be found in [19, Section 3.2.1].

Boutet de Monvel's algebra

We start with a review of the relevant spaces and terminology. A central notion in Boutet de Monvel's calculus is the so-called transmission property. It is a condition on the symbols of the pseudodifferential operators that ensures that the operators map functions which are smooth up to the boundary to functions which are smooth up to the boundary.

Definition 2.9.
(a) Let $H^+ = \{(e^+f)\hat{}: f \in \mathcal{S}(\mathbb{R}_+)\}$, $H_0^- = \{(e^-f)\hat{}: f \in \mathcal{S}(\mathbb{R}_-)\}$, where the hat $\hat{}$ indicates the Fourier transform on $\mathbb{R}$, and $e^\pm$ stands for extension by zero to the opposite half axis. H' denotes the space of all polynomials. Then let

$$H = H^+ \oplus H_0^- \oplus H'.$$

Write H_d, $d \in \mathbb{N}$, for the subspace of all functions $f \in H$ with $f(\rho) = O(\langle\rho\rangle^{d-1})$.
(b) Let $U = U' \times \mathbb{R}$, $U' \subseteq \mathbb{R}^{n-1}$ open. A symbol $p \in S^\mu(U, \mathbb{R}^q)$ has the *transmission property at* $r = 0$ if for every $k \in \mathbb{N}$

$$D_r^k p(x', r, \xi', \langle\xi'\rangle \rho)|_{r=0} \in S^\mu(U'_{x'}, \mathbb{R}^{n-1}_{\xi'}) \hat{\otimes}_\pi H_{d,\rho}, \tag{2.7}$$

where $d = \text{entier}(\mu) + 1$. Write $p \in S^\mu_{tr}(U, \mathbb{R}^q)$, $p \in S^\mu_{cl,tr}(U, \mathbb{R}^q)$, etc.

Remark 2.10. Recall that

$$\begin{aligned} \mathcal{S}(\mathbb{R}_+) &= \text{proj} - \lim\nolimits_{\sigma,\tau\in\mathbb{N}} H^{\sigma,\tau}(\mathbb{R}_+), \\ \mathcal{S}'(\mathbb{R}_+) &= \text{ind} - \lim\nolimits_{\sigma,\tau\in\mathbb{N}} H_0^{-\sigma,-\tau}(\mathbb{R}_+). \end{aligned}$$

Making use of the notation of 2.1, we will, in particular, deal with the spaces $S^\mu(U, \mathbb{R}^n; \mathcal{S}'(\mathbb{R}_+), \mathcal{S}(\mathbb{R}_+))$, $S^\mu(U, \mathbb{R}^n; \mathcal{S}'(\mathbb{R}_+), \mathbb{C})$, and $S^\mu(U, \mathbb{R}^n; \mathbb{C}, \mathcal{S}(\mathbb{R}_+))$.

Definition 2.11. Let E, F be Fréchet spaces and suppose both are continuously embedded in the same Hausdorff vector space. The exterior direct sum $E \oplus F$ is Fréchet and has the closed subspace $\Delta = \{(a, -a) : a \in E \cap F\}$. The non-direct sum of E and F then is the Fréchet space $E + F := E \oplus F/\Delta$.

2.12. Parameter-dependent operators and symbols in Boutet de Monvel's calculus. Let $U \subseteq \mathbb{R}^{n-1}$ be open. A *parameter-dependent operator of order* $\mu \in \mathbb{R}$ *and type* $d \in \mathbb{N}$ *in Boutet de Monvel's calculus* on $U \times \mathbb{R}_+$ is a family of operators

$$A(\lambda) : \begin{array}{c} C_0^\infty(U \times \overline{\mathbf{R}}_+)^{n_1} \\ \oplus \\ C_0^\infty(U)^{m_1} \end{array} \longrightarrow \begin{array}{c} C^\infty(U \times \overline{\mathbf{R}}_+)^{n_2} \\ \oplus \\ C^\infty(U)^{m_2} \end{array} \tag{2.8}$$

of the following form:

$$\begin{aligned} A(\lambda) &= \begin{bmatrix} P_+(\lambda) & 0 \\ 0 & 0 \end{bmatrix} \\ &+ \begin{bmatrix} \sum_{j=0}^d (\operatorname{op} g_j(\lambda) + G_j(\lambda))\partial_r^j & \operatorname{op} k(\lambda) + K_0(\lambda) \\ \sum_{j=0}^d (\operatorname{op} t_j(\lambda) + T_j(\lambda))\partial_r^j & \operatorname{op} s(\lambda) + S_0(\lambda) \end{bmatrix}, \end{aligned} \tag{2.9}$$

where

(i) $P(\cdot) = \operatorname{op} p(\cdot)$ with $p \in S^\mu_{tr}(U \times \mathbb{R} \times U \times \mathbb{R}, \mathbb{R}^q; \mathbb{R}^l)$, $P_+ = \mathrm{r}^+ P \mathrm{e}^+$. Here r^+ denotes restriction of functions from $U \times \mathbb{R}$ to $U \times \mathbb{R}_+$, e^+ denotes extension by zero from $U \times \mathbb{R}_+$ to $U \times \mathbb{R}$.

(ii) The symbols g_j, t_j, k, and s belong to the following spaces:

$$\begin{aligned} g_j &\in S^{\mu-j}(U, \mathbb{R}^{n-1} \times \mathbb{R}^l; \mathcal{S}'(\mathbb{R}_+)^{n_1}, \mathcal{S}(\mathbb{R}_+)^{n_2}), \\ t_j &\in S^{\mu-j}(U, \mathbb{R}^{n-1} \times \mathbb{R}^l; \mathcal{S}'(\mathbb{R}_+)^{n_1}, \mathbb{C}^{m_2}), \\ k &\in S^\mu(U, \mathbb{R}^{n-1} \times \mathbb{R}^l; \mathbb{C}^{m_1}, \mathcal{S}(\mathbb{R}_+)^{n_2}), \text{ and} \\ s &\in S^\mu(U, \mathbb{R}^{n-1} \times \mathbb{R}^l; \mathbb{C}^{m_1}, \mathbb{C}^{m_2}). \end{aligned}$$

(iii) for $j = 0, \dots, d$, the operators G_j, T_j, K_0, and S_0 are rapidly decreasing families of integral operators with smooth kernels:

G_j has an integral kernel in $\mathcal{S}(\mathbb{R}^l, C^\infty((U \times \overline{\mathbb{R}}_+) \times (U \times \overline{\mathbb{R}}_+)))$,
T_j has an integral kernel in $\mathcal{S}(\mathbb{R}^l, C^\infty((U \times \overline{\mathbb{R}}_+) \times U))$,
K_0 has an integral kernel in $\mathcal{S}(\mathbb{R}^l, C^\infty(U \times (U \times \overline{\mathbb{R}}_+)))$, and
S_0 has an integral kernel in $\mathcal{S}(\mathbb{R}^l, C^\infty(U \times U))$.

Of course, all these integral kernels take values in matrices of the corresponding sizes.

(iv) ∂_r is the normal derivative, i.e., the derivative with respect to the variable in $\mathbb{R}_+$ on $U \times \mathbb{R}_+$.

We call an operator

$$A_0(\lambda) = \begin{bmatrix} \sum_{j=0}^d G_j(\lambda)\partial_r^j & K_0(\lambda) \\ \sum_{j=0}^d T_j(\lambda)\partial_r^j & S_0(\lambda) \end{bmatrix}, \qquad \lambda \in \mathbb{R}^l,$$

with the above choice of G_j, T_j, K_0, and S_0 a *regularizing parameter-dependent operator of type d in Boutet de Monvel's calulus.* It is a consequence of Theorem 2.8 that the operators in (2.9) indeed have the desired mapping property.

We shall write $A \in \mathcal{B}^{\mu,d}(U \times \mathbb{R}_+; \mathbb{R}^l)$ for a parameter-dependent operator of order μ and type d, and $A \in \mathcal{B}^{-\infty,d}(U \times \mathbb{R}_+; \mathbb{R}^l)$ for a regularizing parameter-dependent operator of type d.

The decomposition $P_+ + G$ is not unique; certain regularizing pseudodifferential operators provide examples for operators that belong to both classes. The topology on $\mathcal{B}^{\mu,d}(U \times \mathbb{R}_+; \mathbb{R}^l)$ and $\mathcal{B}^{-\infty,d}(U \times \mathbb{R}_+; \mathbb{R}^l)$ is that of a non-direct sum of Fréchet spaces.

Given an operator $A \in \mathcal{B}^{\mu,d}(U \times \mathbb{R}_+; \mathbb{R}^l)$ in the notation of (2.9) we let $g = \sum_{j=0}^d g_j \partial_r^j$, and $t = \sum_{j=0}^d t_j \partial_r^j$. We then have a quintuple $a = \{p, g, k, t, s\}$ of symbols for A. It is not unique, but any other choice differs only by a quintuple inducing a regularizing element.

2.13. Boutet de Monvel's algebra on a manifold. Symbol levels. Let X be an n-dimensional C^∞ manifold with boundary Y, embedded in an n-dimensional manifold G without boundary, all not necessarily compact. In the following we shall denote by X the open interior of X, while $\overline{X}$ denotes the closure. Let V_1, V_2 be vector bundles over G, and let W_1, W_2 be vector bundles over Y.

By $\{G_j\}$ denote a locally finite open covering of G, and suppose that the coordinate charts map $X \cap G_j$ to $U_j \times \mathbb{R}_+ \subset \mathbb{R}^n_+$ and $Y \cap G_j$ to $U_j \times \{0\}$ for a suitable open set $U_j \subseteq \mathbb{R}^{n-1}$, unless $G_j \cap Y = \emptyset$.

For a smooth function φ on G write M_φ for the operator of multiplication with the diagonal matrix $\operatorname{diag}\{\varphi, \varphi|_Y\}$. We will say that $A \in \mathcal{B}^{\mu,d}(X; \mathbb{R}^l)$ if

$$(2.10) \qquad A(\lambda) : \begin{matrix} C_0^\infty(\overline{X}, V_1) \\ \oplus \\ C_0^\infty(Y, W_1) \end{matrix} \rightarrow \begin{matrix} C^\infty(\overline{X}, V_2) \\ \oplus \\ C^\infty(Y, W_2) \end{matrix},$$

is an operator with the following properties:

(i) For all $C_0^\infty(G_j)$ functions φ, ψ, supported in the same coordinate neighborhood G_j intersecting the boundary, the push-forward

$$(M_\varphi A(\lambda) M_\psi)_* : \begin{array}{c} C_0^\infty(U_j \times \overline{\mathbf{R}}_+, V_1) \\ \oplus \\ C_0^\infty(U_j, W_1) \end{array} \rightarrow \begin{array}{c} C^\infty(U_j \times \overline{\mathbf{R}}_+, V_2) \\ \oplus \\ C^\infty(U_j, W_2) \end{array},$$

induced by $M_\varphi A(\lambda) M_\psi$ and the coordinate maps, is an operator in $\mathcal{B}^{\mu,d}(U_j \times \mathbb{R}_+; \mathbb{R}^l)$.

(ii) If φ, ψ are as before, but the coordinate chart does not intersect the boundary, then all entries in the matrix $(M_\varphi A(\lambda) M_\psi)_*$ – except for the pseudodifferential part – are regularizing.

(iii) If the supports of the functions $\varphi, \psi \in C_0^\infty(G)$ are disjoint, then $M_\varphi A(\lambda) M_\psi$ is a rapidly decreasing function of λ with values in the regularizing operators of type d.

It remains to define the regularizing elements. A *regularizing operator of type* 0 *in Boutet de Monvel's calculus* is an operator R acting on the above spaces with the property that there are continuous extensions

$$R : \begin{array}{c} L^2(X, V_1) \\ \oplus \\ L^2(Y, W_1) \end{array} \rightarrow \begin{array}{c} C^\infty(\overline{X}, V_2) \\ \oplus \\ C^\infty(Y, W_2) \end{array},$$

$$R^* : \begin{array}{c} L^2(X, V_2) \\ \oplus \\ L^2(Y, W_2) \end{array} \rightarrow \begin{array}{c} C^\infty(\overline{X}, V_1) \\ \oplus \\ C^\infty(Y, W_1) \end{array}.$$

Here R^* is the formal adjoint with respect to the inner product on the respective spaces. A *regularizing operator of type* d is a sum $R = \sum_{j=0}^{d} R_j \begin{bmatrix} \partial_r^j & 0 \\ 0 & I \end{bmatrix}$ with all R_j regularizing of type zero. We write $\mathcal{B}^{-\infty,d}(X)$ for the regularizing elements of type d and $\mathcal{B}^{-\infty,d}(X; \mathbb{R}^q)$ for the parameter-dependent regularizing elements, i.e., the Schwartz functions on $\mathbb{R}^q$ with values in $\mathcal{B}^{-\infty,d}(X)$.

We topologize $\mathcal{B}^{\mu,d}(X; \mathbb{R}^l)$ as the corresponding non-direct sum of Fréchet spaces.

For each coordinate patch G_j intersecting the boundary, $A(\lambda)$ induces an operator

$$A_j(\lambda) = \begin{bmatrix} P_{j+}(\lambda) + G_j(\lambda) & K_j(\lambda) \\ T_j(\lambda) & S_j(\lambda) \end{bmatrix}$$

on $U_j \times \mathbb{R}_+$. We find a quintuple $a_j(\lambda) = \{p_j(\lambda), g_j(\lambda), k_j(\lambda), t_j(\lambda), s_j(\lambda)\}$ of symbols for $P_j(\lambda), G_j(\lambda), K_j(\lambda), T_j(\lambda), S_j(\lambda)$ in the sense of 2.12.

We shall call A *classical*, if all entries in the quintuples $a_j = \{p_j, g_j, k_j, t_j, s_j\}$ are classical elements in the respective symbol classes, i.e., p_j and s_j are classical pseudodifferential symbols, while g_j, k_j, t_j are classical operator-valued symbols. For an interior patch, we have the pseudodifferential symbol for P_j; all other symbols can be taken to be zero. Write $A \in \mathcal{B}^{\mu,d}_{cl}(X; \mathbb{R}^l)$.

Example 2.14. The Dirichlet problem $\binom{\Delta}{\gamma_0}$ is an operator in Boutet de Monvel's calculus of order 2 and type 1: Clearly, the Laplacian Δ is a differential operator of order 2. As we saw in Example 2.2, the operator of evaluation at the boundary, γ_0, is an operator-valued symbol in $S^{1/2}(\mathbb{R}^q, \mathbb{R}^q; H^s(\mathbb{R}_+), \mathbb{C})$, provided $s > 1/2$. It is not so obvious that this is an operator of type 1: Using the integration by parts formula

$$u(0) = \int_0^\infty [\eta] e^{-r[\eta]} u(r) dr + \int_0^\infty e^{-r[\eta]} \partial_r u(r) dr$$

valid for $u \in \mathcal{S}(\mathbb{R}_+)$, we may we may rewrite γ_0 in the form

$$\gamma_0 = t_0 + t_1 \partial_r.$$

Here, $t_0 \in S^{1/2}(\mathbb{R}^q, \mathbb{R}^q; \mathcal{S}'(\mathbb{R}_+), \mathbb{C})$ is given by $t_0 u = \int_0^\infty [\eta] e^{-r[\eta]} u(r) dr$, while the operator-valued symbol $t_1 \in S^{-1/2}(\mathbb{R}^q, \mathbb{R}^q; \mathcal{S}'(\mathbb{R}_+), \mathbb{C})$ is defined by integrating $\partial_r u$ against $e^{-r[\eta]}$. Hence γ_0 is of type 1.

The Dirichlet problem is independent of any parameter, but since it is a *differential* boundary value problem, we may also consider it as a parameter-dependent element. Since the order of γ_0 only is $1/2$, we may even replace γ_0 by $\Lambda\gamma_0$, where Λ is a (parameter-dependent) order reduction of order $3/2$, and still have order 2. Here, the vector bundle W_1 is zero, while V_1, V_2, W_2 can be taken trivial one-dimensional.

Proposition 2.15. *Let $A \in \mathcal{B}^{\mu,d}(X; \mathbb{R}^l)$, $B \in \mathcal{B}^{\mu',d'}(X; \mathbb{R}^l)$, and $\alpha, \beta \in \mathbb{C}$. Then*
(a) *$\alpha A + \beta B \in \mathcal{B}^{\mu'',d''}(X; \mathbb{R}^l)$ for $\mu'' = \max\{\mu, \mu'\}$, $d'' = \max\{d, d'\}$.*
(b) *$A \circ B \in \mathcal{B}^{\mu'',d''}(X; \mathbb{R}^l)$ for $\mu'' = \max\{\mu + \mu'\}$, $d'' = \max\{\mu' + d, d'\}$.*
We assume here that the vector bundles A and B act on are such that the addition and composition make sense.

For a proof see Rempel, Schulze [13, Section 2.3.3.2].

3. Wedge Sobolev spaces

In the following, we let G be a closed compact manifold of dimension n, and let X be an embedded n-dimensional submanifold with boundary, Y. We write $G^\wedge = G \times \mathbb{R}_+$, $X^\wedge = X \times \mathbb{R}_+$, $Y^\wedge = Y \times \mathbb{R}_+$.

3.1. Parameter-dependent order reductions on G. For each $\mu \in \mathbb{R}$ there is a pseudodifferential operator Λ^μ with local parameter-dependent elliptic symbols of order μ, depending on the parameter $\tau \in \mathbb{R}$, such that

$$\Lambda^\mu(\tau) : H^s(G,V) \to H^{s-\mu}(G,V)$$

is an isomorphism for all τ.

In order to construct such an operator one can e.g. start with symbols of the form $\langle \xi, (\tau, C)\rangle^\mu \in S^\mu(\mathbb{R}^n, \mathbb{R}^n_\xi; \mathbb{R}_\tau)$ with a large constant $C > 0$ and patch them together to an operator on the manifold G with the help of a partition of unity and cut-off functions.

Alternatively, one can choose a Hermitean connection on V and consider the operator $(C + \tau^2 - \Delta)^{\frac{\mu}{2}}$, where Δ denotes the Laplacian and C is a large positive constant.

Definition 3.2. For $\beta \in \mathbb{R}$, Γ_β denotes the vertical line $\{z \in \mathbb{C} : \operatorname{Re} z = \beta\}$. We recall that the classical Mellin transform Mu of a complex-valued $C_0^\infty(\mathbb{R}_+)$-function u is given by

$$(Mu)(z) = \int_0^\infty t^{z-1} u(t)\, dt. \tag{3.1}$$

M extends to an isomorphism $M : L^2(\mathbb{R}_+) \to L^2(\Gamma_{1/2})$. Of course, (3.1) also makes sense for functions with values in a Fréchet space E. The fact that $Mu|_{\Gamma_{1/2-\gamma}}(z) = M_{t\to z}(t^{-\gamma}u)(z+\gamma)$ for $u \in C_0^\infty(\mathbb{R}_+)$ motivates the following definition of the *weighted Mellin transform* M_γ:

$$M_\gamma u(z) = M_{t\to z}(t^{-\gamma}u)(z+\gamma), \qquad u \in C_0^\infty(\mathbb{R}_+, E).$$

For a Hilbert space E, the inverse of M_γ is given by

$$[M_\gamma^{-1}h](t) = \frac{1}{2\pi i}\int_{\Gamma_{1/2-\gamma}} t^{-z} h(z)\, dz.$$

3.3. Totally characteristic Sobolev spaces.

(a) Let $\{\Lambda^\mu : \mu \in \mathbb{R}\}$ be a family of parameter-dependent order reductions as in 3.1. For $s, \gamma \in \mathbb{R}$, the space $\mathcal{H}^{s,\gamma}(G^\wedge)$ is the closure of $C_0^\infty(G^\wedge)$ in the norm

$$\|u\|_{\mathcal{H}^{s,\gamma}(G^\wedge)} = \left\{ \int_{\Gamma_{\frac{n+1}{2}-\gamma}} \|\Lambda^s(\operatorname{Im} z) Mu(z)\|^2_{L^2(G)}\, |dz| \right\}^{1/2}. \tag{3.2}$$

Recall that n is the dimension of X and G. The space $\mathcal{H}^{s,\gamma}(G^\wedge)$ is independent of the particular choice of the order reducing family.

(b) For $s = l \in \mathbb{N}$ we obtain the alternative description

$$u \in \mathcal{H}^{l,\gamma}(G^\wedge) \quad \text{iff} \quad t^{n/2-\gamma}(t\partial_t)^k Du(x,t) \in L^2(G^\wedge)$$

for all $k \leq l$ and all differential operators D of order $\leq l-k$ on G, cf. [19, Section 2.1.1, Proposition 2].
(c) We let $\mathcal{H}^{s,\gamma}(X^\wedge) = \{f|_{X^\wedge} : f \in \mathcal{H}^{s,\gamma}(G^\wedge)\}$, endowed with the quotient norm:

$$\|u\|_{\mathcal{H}^{s,\gamma}(X^\wedge)} = \inf\{\|f\|_{\mathcal{H}^{s,\gamma}(G^\wedge)} : f \in \mathcal{H}^{s,\gamma}(G^\wedge),\ f|_{X^\wedge} = u\}.$$

(d) $\mathcal{H}^{s,\gamma}(X^\wedge) \subseteq H^s_{loc}(X^\wedge)$, where the subscript '*loc*' refers to the t-variable only. Moreover, $\mathcal{H}^{s,\gamma}(X^\wedge) = t^\gamma \mathcal{H}^{s,0}(X^\wedge)$; $\mathcal{H}^{0,0}(X^\wedge) = t^{-n/2}L^2(X^\wedge)$.
(e) $\mathcal{H}^{0,0}(X^\wedge)$ has a natural inner product

$$(u,v)_{\mathcal{H}^{0,0}(X^\wedge)} = \frac{1}{2\pi i}\int_{\Gamma_{\frac{n+1}{2}}} (Mu(z), Mv(z))_{L^2(X)}\, dz.$$

(f) If φ is the restriction to $X^\wedge$ of a function in the Schwartz space $\mathcal{S}(G\times\mathbb{R})$, then the operator M_φ of multiplication by φ,

$$M_\varphi : \mathcal{H}^{s,\gamma}(X^\wedge) \to \mathcal{H}^{s,\gamma}(X^\wedge),$$

is bounded for all $s, \gamma \in \mathbb{R}$, and the mapping $\varphi \mapsto M_\varphi$ is continuous in the corresponding topology.

Definition 3.4. Let $\mathcal{F}$ be a subspace of $\mathcal{D}'(X^\wedge)$ or $\mathcal{D}'(G^\wedge)$ with a stronger topology. Suppose that φ is a smooth function on $G \times \overline{\mathbb{R}}_+$ and that multiplication by φ is continuous on $\mathcal{F}$. Then $[\varphi]\mathcal{F}$ denotes the closure of the space $\{\varphi u : u \in \mathcal{F}\}$ in $\mathcal{F}$.

3.5. The spaces H^s_{cone}. Let $\{G_j\}_{j=1}^J$ be a finite covering of G by open sets, $\kappa_j : G_j \to U_j$ the coordinate maps onto bounded open sets in $\mathbb{R}^n$, and $\{\varphi_j\}_{j=1}^J$ a subordinate partition of unity. The maps κ_j induce a push-forward of functions and distributions: For a function u on G_j

$$(\kappa_{j*}u)(x) = u(\kappa_j^{-1}(x)), \qquad x \in U_j, \tag{3.3}$$

for a distribution u ask that $(\kappa_{j*}u)(\varphi) = u(\varphi \circ \kappa_j)$, $\varphi \in C_0^\infty(U_j)$. For $j = 1, \ldots, J$, consider the diffeomorphism

$$\chi_j : U_j \times \mathbb{R} \to \{(x[t], t) : x \in U_j, t \in \mathbb{R}\} =: C_j \subset \mathbb{R}^{n+1},$$

given by $\chi_j(x,t) = (x[t], t)$. Its inverse is $\chi_j^{-1}(y,t) = (y/[t], t)$. For $s \in \mathbb{R}$ we define $H^s_{cone}(G\times\mathbb{R})$ as the set of all $u \in H^s_{loc}(G\times\mathbb{R})$ such that, for $j = 1, \ldots, J$, the push-forward $(\chi_j\kappa_j)_*(\varphi_j u)$, which may be regarded as a distribution on $\mathbb{R}^{n+1}$ after extension by zero, is an element of $H^s(\mathbb{R}^{n+1})$. The space $H^s_{cone}(G\times\mathbb{R})$ is endowed with the corresponding Hilbert space topology. We let

$$H^s_{cone}(X^\wedge) = \{u|_{X\times\mathbb{R}_+} : u \in H^s_{cone}(G\times\mathbb{R})\}. \tag{3.4}$$

For more details see Schrohe, Schulze [16, Section 4.2]. The subscript "cone" is motivated by the fact that, away from zero, these are the Sobolev spaces for an infinite cone with center at the origin and cross-section X. In particular, the space $H^s_{cone}(S^n \times \mathbb{R}_+)$ coincides with $H^s(\mathbb{R}^{n+1} \setminus \{0\})$.

Definition 3.6. For $s, \gamma \in \mathbb{R}$ and $\omega \in C_0^\infty(\overline{\mathbb{R}}_+)$ with $\omega(r) \equiv 1$ near $r = 0$, let

$$(3.5)\quad \mathcal{K}^{s,\gamma}(X^\wedge) = \{u \in \mathcal{D}'(X^\wedge) : \omega u \in \mathcal{H}^{s,\gamma}(X^\wedge),\ (1-\omega)\,u \in H^s_{cone}(X^\wedge)\}.$$

The definition is independent of the choice of ω by 3.3 (f). In the notation of Definition 3.4,

$$(3.6)\qquad \mathcal{K}^{s,\gamma}(X^\wedge) = [\omega]\mathcal{H}^{s,\gamma}(X^\wedge) + [1-\omega]\,H^s_{cone}(X^\wedge).$$

We endow it with the Banach topology

$$\|u\|_{\mathcal{K}^{s,\gamma}(X^\wedge)} = \|\omega u\|_{\mathcal{H}^{s,\gamma}(X^\wedge)} + \|(1-\omega)u\|_{H^s_{cone}(X^\wedge)}.$$

In fact, this is a Hilbert topology with the inner product inherited from $\mathcal{H}^{s,\gamma}$ and H^s_{cone}.

Theorem 3.7. *For $s > 1/2$ and $\gamma \in \mathbb{R}$ the restriction $\gamma_0 u = u|_{Y^\wedge}$ of u to $Y^\wedge$ induces a continuous operator*

$$\gamma_0 : \mathcal{K}^{s,\gamma}(X^\wedge) \to \mathcal{K}^{s-1/2,\gamma-1/2}(Y^\wedge).$$

By r denote the normal coordinate in a neighborhood of Y. Then the operators $\gamma_j : u \mapsto \partial_r^j u|_{Y^\wedge}$ define continuous mappings

$$(3.7)\qquad \gamma_j : \mathcal{K}^{s,\gamma}(X^\wedge) \to \mathcal{K}^{s-j-1/2,\gamma-1/2}(Y^\wedge).$$

Proof. For one thing this can be deduced from the trace theorem for the usual Sobolev spaces. Note that the shift in the weight $\gamma \mapsto \gamma - 1/2$ is due to the fact that $\dim Y = n-1$. We shall give an independent proof in 4.4, below. □

The following lemma is obvious after 3.3 (d):

Lemma 3.8. $\mathcal{K}^{0,0}(X^\wedge) = \mathcal{H}^{0,0}(X^\wedge) = t^{-n/2}L^2(X^\wedge)$.

Lemma 3.9. *A strongly continuous group action κ_λ can be defined on $\mathcal{K}^{s,\gamma}(X^\wedge)$ by*

$$(\kappa_\lambda f)(x,t) = \lambda^{\frac{n+1}{2}} f(x, \lambda t), \qquad f \in \mathcal{K}^{s,\gamma}(X^\wedge),\ s \geq 0.$$

This action is unitary on $\mathcal{K}^{0,0}(X^\wedge)$ and extends to distributions by $(\kappa_\lambda u)(\varphi) = u(\kappa_{\lambda^{-1}}\varphi)$ for $u \in \mathcal{D}'(X^\wedge)$, $\varphi \in C_0^\infty(X^\wedge)$.

Proof. It is lengthy but straightforward to see that κ is strongly continuous; it is unitary on $\mathcal{K}^{0,0}(X^\wedge)$ in view of Lemma 3.8. □

Remark 3.10. The definitions of the spaces $\mathcal{H}^{s,\gamma}$ and $\mathcal{K}^{s,\gamma}$ also make sense for functions and distributions taking values in a vector bundle V. We shall then write $\mathcal{H}^{s,\gamma}(X^\wedge, V)$ and $\mathcal{K}^{s,\gamma}(X^\wedge, V)$, respectively. In later constructions we will often have to deal with direct sums

$$\mathcal{K}^{s,\gamma}(X^\wedge, V) \oplus \mathcal{K}^{s-1/2,\gamma-1/2}(Y^\wedge, W)$$

for vector bundles V and W over X and Y, respectively. On these spaces we use the natural group action

$$\kappa_\lambda(u,v) = (\lambda^{\frac{n+1}{2}} u(\cdot\,,\lambda\cdot), \lambda^{\frac{n}{2}} v(\cdot\,,\lambda\cdot)).$$

Definition 3.11. For real s and γ we let $\mathcal{W}^{s,\gamma}(X^\wedge \times \mathbb{R}^q) = \mathcal{W}^s(\mathbb{R}^q, \mathcal{K}^{s,\gamma}(X^\wedge))$.

Theorem 3.12. *The restriction operator γ_0 induces a continuous map*

$$\gamma_0 : \mathcal{W}^{s,\gamma}(X^\wedge \times \mathbb{R}^q) \to \mathcal{W}^{s-1/2,\gamma-1/2}(Y^\wedge \times \mathbb{R}^q).$$

Proof. We know from Theorem 3.7 that $\gamma_0 : \mathcal{K}^{s,\gamma}(X^\wedge) \to \mathcal{K}^{s-1/2,\gamma-1/2}(Y^\wedge)$ is a bounded operator. So we may consider it once more an operator-valued symbol, independent of y, η. Just as in Example 2.2 one checks that $\gamma_0 \in S^{1/2}(\mathbb{R}^q \times \mathbb{R}^q; \mathcal{K}^{s,\gamma}(X^\wedge), \mathcal{K}^{s-1/2,\gamma-1/2}(Y^\wedge))$. Now Theorem 2.8 gives the assertion. □

Proposition 3.13. *Let $\varphi \in \mathcal{S}(\overline{X}^\wedge \times \mathbb{R}^q)$. Then the operator of multiplication by φ furnishes a bounded operator on $\mathcal{W}^{s,\gamma}(X^\wedge \times \mathbb{R}^q)$ for all $s, \gamma \in \mathbb{R}$. Its norm depends continuously on the semi-norms for φ in $\mathcal{S}(\overline{X}^\wedge \times \mathbb{R}^q)$.*

Proof. We shall use a tensor product argument based on the identity

$$\mathcal{S}(\overline{X}^\wedge \times \mathbb{R}^q) = \mathcal{S}(\overline{X}^\wedge) \hat{\otimes}_\pi \mathcal{S}(\mathbb{R}^q).$$

Let $\varphi = \sigma \otimes \psi$ with $\sigma \in \mathcal{S}(\overline{X}^\wedge)$ and $\psi \in \mathcal{S}(\mathbb{R}^q)$ be a pure tensor. We shall show the separate continuity of the multiplications. Since both $\mathcal{S}(\overline{X}^\wedge)$ and $\mathcal{S}(\mathbb{R}^q)$ are Fréchet spaces this will imply the joint continuity and establish the proof.
Let us first deal with multiplication by σ, denoted for the moment by M_σ. We may consider this multiplication as the application of a pseudodifferential operator with the (y,η)-independent operator-valued symbol $a(y,\eta) = M_\sigma$. Let us check that a is an element of $S^0(\mathbb{R}^q \times \mathbb{R}^q; \mathcal{K}^{s,\gamma}(X^\wedge), \mathcal{K}^{s,\gamma}(X^\wedge))$ for all s. First of all, an application of 3.3 (b) together with interpolation shows that M_σ is bounded on $\mathcal{K}^{s,\gamma}(X^\wedge)$. In view of the independence of y and η we now only have to estimate the norm in $\mathcal{L}(\mathcal{K}^{s,\gamma}(X^\wedge))$ of the operator $\widetilde{\kappa}_{[\eta]^{-1}} a(y,\eta) \kappa_{[\eta]}$. This in turn simply is multiplication by $\sigma([\eta]^{-1}\cdot)$, which is uniformly bounded by another application of 3.3 (b).

Next let us treat multiplication by $\psi = \psi(y)$ and show that it furnishes a bounded operator on $\mathcal{W}^s(\mathbb{R}^q, E)$ for every Banach space E with group action κ. In fact, since the wedge Sobolev spaces are defined as the completion of $\mathcal{S}(\mathbb{R}^q, E)$ in the corresponding norm, it is sufficient to show that, for a pure tensor $u = u_0 \otimes e$ in $\mathcal{S}(\mathbb{R}^q) \otimes E$, we have

$$\|\psi u\|_{\mathcal{W}^s(\mathbb{R}^q,E)} \leq C \|u\|_{\mathcal{W}^s(\mathbb{R}^q,E)}$$

with a constant independent of u. Choose an integer $l > q/2$. With the help of Peetre's inequality and (2.1), in particular the fact that

$$\begin{aligned} \|\kappa_{[\eta+\xi]^{-1}}\kappa_{[\eta]}\|_E &= \|\kappa_{[\eta+\xi]^{-1}[\eta]}\|_E \\ &\leq C \max\{[\eta+\xi]^{-1}[\eta], [\eta+\xi][\eta]^{-1}\}^M \leq C'[\xi]^M, \end{aligned}$$

we get the following estimate

$$\begin{aligned} &\|\psi u\|^2_{\mathcal{W}^s(\mathbb{R}^q,E)} \\ &= \int [\eta]^{2s} |\mathcal{F}(\psi u_0)(\eta)|^2 \|\kappa_{[\eta]^{-1}} e\|^2_E \, d\eta \\ &= (2\pi)^{q/2} \int [\eta]^{2s} \left| \int \mathcal{F}u_0(\eta - \xi)\mathcal{F}\psi(\xi)\, d\xi \right|^2 \|\kappa_{[\eta]^{-1}} e\|^2_E \, d\eta \\ &\leq (2\pi)^{q/2} \int [\eta]^{2s} \int |\mathcal{F}u_0(\eta-\xi)\mathcal{F}\psi(\xi)[\xi]^l|^2 \, d\xi \; \|\kappa_{[\eta]^{-1}} e\|^2_E \, d\eta \int [\xi]^{-2l} \, d\xi \\ &= C \iint [\eta+\xi]^{2s} |\mathcal{F}u_0(\eta)|^2 |\mathcal{F}\psi(\xi)|^2 [\xi]^{2l} \|\kappa_{[\eta+\xi]^{-1}} e\|^2_E \, d\eta \, d\xi \\ &\leq C' \iint [\eta]^{2s} [\xi]^{2|s|+2l+2M} |\mathcal{F}u_0(\eta)|^2 |\mathcal{F}\psi(\xi)|^2 \|\kappa_{[\eta]^{-1}} e\|^2_E \, d\eta \, d\xi \\ &\leq C'' \|\psi\|_{H^{|s|+l+M}(\mathbb{R}^q)} \, \|u\|^2_{\mathcal{W}^s(\mathbb{R}^q,E)}. \end{aligned}$$

Here the first inequality is Cauchy-Schwarz'. □

4. Operator-valued Mellin symbols

As before, we let G be a closed compact manifold of dimension n, and let X be an embedded n-dimensional submanifold with boundary, Y.

Definition 4.1.
(a) For $\mu \in \mathbb{R}, d \in \mathbb{N}$, we define $M_O^{\mu,d}(X;\mathbb{R}^q)$ as the space of all functions

$$a \in \mathcal{A}\left(\mathbb{C}, \mathcal{B}^{\mu,d}(X;\mathbb{R}^q)\right)$$

with the following property: Given $c_1 < c_2$ in $\mathbb{R}$

$$a(\beta + i\tau) \in \mathcal{B}^{\mu,d}(X;\mathbb{R}^q \times \mathbb{R}_\tau), \tag{4.1}$$

uniformly for all $\beta \in [c_1, c_2]$.

We call the elements of $M_O^{\mu,d}(X;\mathbb{R}^q)$ holomorphic Mellin symbols of order μ and type d. We assume that the vector bundles $a(z)$ is acting on are independent of z. The topology of $M_O^{\mu,d}(X)$ is given by the semi-norm systems for the topology of $\mathcal{A}\left(\mathbb{C},\mathcal{B}^{\mu,d}\left(X;\mathbb{R}^q\right)\right)$ and, for families $\{a_\beta : \beta \in \mathbb{R}\}$, the topology of uniform convergence on compact subsets of $\mathbb{R}_\beta$ in $\mathcal{B}^{\mu,d}(X;\mathbb{R}^q \times \mathbb{R}_\tau)$. Clearly, $M_O^{\mu,d}\left(X;\mathbb{R}^q\right)$ is a Fréchet space with this topology.
(b) $M_{O,cl}^{\mu,d}(X;\mathbb{R}^q)$ is the corresponding space with $\mathcal{B}^{\mu,d}(X;\mathbb{R}^q)$ replaced by $\mathcal{B}_{cl}^{\mu,d}(X;\mathbb{R}^q)$.

Example 4.2. Let $\mu \in \mathbb{N}$ and let $A_k \in \mathcal{B}^{\mu-k,d}(X)$, $k = 0,\dots,\mu$, be *differential* boundary value problems. Then

$$a(z) = \sum_{k=0}^{\mu} A_k z^k \in M_O^{\mu,d}\left(X\right).$$

4.3. Mellin symbols and operators. Let $f \in C^\infty(\mathbb{R}_+ \times \mathbb{R}_+, \mathcal{B}^{\mu,d}(X;\Gamma_{1/2-\gamma}))$. For each fixed $(t,t',z) \in \mathbb{R}_+ \times \mathbb{R}_+ \times \Gamma_{1/2-\gamma}$, we have a boundary value problem

$$f(t,t',z) : \begin{array}{c} C_0^\infty(\overline{X},V_1) \\ \oplus \\ C_0^\infty(Y,W_1) \end{array} \rightarrow \begin{array}{c} C^\infty(\overline{X},V_2) \\ \oplus \\ C^\infty(Y,W_2) \end{array}$$

in Boutet de Monvel's calculus.
For $u \in C_0^\infty(\overline{X}^\wedge,V_1) \oplus C_0^\infty(\overline{Y}^\wedge,W_1) = C_0^\infty(\mathbb{R}_+, C^\infty(\overline{X},V_1) \oplus C^\infty(Y,W_1))$ we define the Mellin operator $\mathrm{op}_M^\gamma f$ by

$$\{\mathrm{op}_M^\gamma f\}u(t) = \frac{1}{2\pi i}\int_{\Gamma_{1/2-\gamma}} \int_0^\infty (t/t')^{-z} f(t,t',z)u(t')dt'/t'dz.$$

If f is independent of t', this reduces to

$$\{\mathrm{op}_M^\gamma f\}u(t) = \frac{1}{2\pi i}\int_{\Gamma_{1/2-\gamma}} t^{-z} f(t,z)Mu(z)dz.$$

It is easy to see the continuity of

$$\mathrm{op}_M^\gamma f : \begin{array}{c} C_0^\infty(\overline{X}^\wedge,V_1) \\ \oplus \\ C_0^\infty(Y^\wedge,W_1) \end{array} \rightarrow \begin{array}{c} C^\infty(\overline{X}^\wedge,V_2) \\ \oplus \\ C^\infty(Y^\wedge,W_2) \end{array}.$$

For $f \in C^\infty(\overline{\mathbb{R}}_+ \times \overline{\mathbb{R}}_+, \mathcal{B}^{\mu,d}(X;\Gamma_{1/2-\gamma}))$ and $\omega_1,\omega_2 \in C_0^\infty(\mathbb{R}_+)$ we obtain a bounded extension

$$(4.2) \qquad \omega_1\{\mathrm{op}_M^\gamma f\}\omega_2 : \begin{array}{c} \mathcal{K}^{s,\gamma+\frac{n}{2}}(X^\wedge,V_1) \\ \oplus \\ \mathcal{K}^{s,\gamma+\frac{n-1}{2}}(Y^\wedge,W_1) \end{array} \rightarrow \begin{array}{c} \mathcal{K}^{s-\mu,\gamma+\frac{n}{2}}(X^\wedge,V_2) \\ \oplus \\ \mathcal{K}^{s-\mu,\gamma+\frac{n-1}{2}}(Y^\wedge,W_2) \end{array}$$

provided $s > d - 1/2$. A proof is given in [16, Proposition 2.1.5].

In the following we shall use the abbreviations

$$\begin{aligned}\mathcal{K}_1^{s,\gamma} &= \mathcal{K}^{s,\gamma+\frac{n}{2}}(X^\wedge, V_1) \oplus \mathcal{K}^{s,\gamma+\frac{n-1}{2}}(Y^\wedge, W_1) \text{ and} \\ \mathcal{K}_2^{s,\gamma} &= \mathcal{K}^{s-\mu,\gamma+\frac{n}{2}-\mu}(X^\wedge, V_2) \oplus \mathcal{K}^{s-\mu,\gamma+\frac{n-1}{2}-\mu}(Y, W_2).\end{aligned}$$

4.4. Alternative proof of Theorem 3.7. We consider the operator of evaluation at the boundary γ_0. As we saw in Example 2.14, it is a parameter-dependent operator in Boutet de Monvel's calculus of order 1/2 and type 1. We may therefore regard it as a Mellin operator with a Mellin symbol independent of t, t', and z. The mapping properties (4.2), applied with the choice V_1, W_2 = trivial one-dimensional, W_1, V_2 = zero, show that for every choice of cut-off functions ω_1, ω_2 near zero and $s > 1/2$,

$$\omega_1\gamma_0\omega_2 : \mathcal{K}^{s,\gamma}(X^\wedge) \to \mathcal{K}^{s-1/2,\gamma-1/2}(Y^\wedge)$$

is bounded. Away from zero, the spaces $\mathcal{K}^{\cdot,\cdot}$ coincide with usual Sobolev spaces on the cone, hence the result there follows from the usual trace theorem.

Proposition 4.5. *Given $\mu, \mu' \in \mathbf{Z}$ and $d, d' \in \mathbb{N}$, let $\mu'' = \mu + \mu'$ and $d'' = \max\{\mu' + d, d'\}$. Then there is a continuous multiplication*

$$M_O^{\mu,d}(X;\mathbb{R}^q) \times M_O^{\mu',d'}(X;\mathbb{R}^q) \to M_O^{\mu'',d''}(X;\mathbb{R}^q)$$

given by the point-wise composition in Boutet de Monvel's calculus: $(a, b) \mapsto c$ with $c(z,\eta) = a(z,\eta) \circ b(z,\eta)$.

The proof is straightforward from the definition and Proposition 2.15.

4.6. Operator-valued Mellin symbols. Let $\gamma, \mu \in \mathbb{R}$, $\Omega \subseteq \mathbb{R}^q$, and $f \in C^\infty(\overline{\mathbb{R}}_+ \times \Omega, \mathcal{B}^{\mu,d}(X; \Gamma_{1/2-\gamma} \times \mathbb{R}^q))$. Recall that $[\cdot]$ is a smooth positive function on $\mathbb{R}^q$ coinciding with $|\cdot|$ outside a neighborhood of zero.
Given $\omega_1, \omega_2 \in C_0^\infty(\overline{\mathbb{R}}_+)$ define

$$a(y,\eta) = \omega_1(t[\eta])t^{-\mu}\{\mathrm{op}_M^\gamma f(t,y,z,t\eta)\}\omega_2(t[\eta]).$$

According to (4.2) this furnishes a function a on $\Omega \times \mathbb{R}^q$ with values in $\mathcal{L}(\mathcal{K}_1^{s,\gamma}(X^\wedge), \mathcal{K}_2^{s,\gamma}(X^\wedge))$ for all $s > d - 1/2$. We will show that a in fact is an element of $S^\mu(\Omega, \mathbb{R}^q; \mathcal{K}_1^{s,\gamma}, \mathcal{K}_2^{s,\gamma})$. The proof is based on Proposition 4.8, below, and a tensor product argument given in Corollary 4.9. We shall keep the notation $\mathcal{K}_1^{s,\gamma}, \mathcal{K}_2^{s,\gamma}, \gamma, \mu, f, a, \omega_1, \omega_2$ fixed.

First let us note the following:

Lemma 4.7. *If f is independent of t, then there is a $C > 0$ such that*

$$a(y,\lambda\eta) = \lambda^\mu \kappa_\lambda a(y,\eta)\kappa_{\lambda^{-1}}$$

for all $\lambda \geq 1$, and $|\eta| \geq C$.

Proof. We have

$$\kappa_\lambda\{\mathrm{op}_M^\gamma f(y,z,t\eta)\} = \{\mathrm{op}_M^\gamma f(y,z,\lambda t\eta)\}\kappa_\lambda.$$

Next choose C so large that $[\eta] = |\eta|$ for $|\eta| \geq C$. For $u \in C_0^\infty(\mathbb{R}_+)$ and $\lambda \geq 1$ this implies that

$$\begin{aligned}&\kappa_\lambda\{\omega_1(t[\eta])t^{-\mu}\{\mathrm{op}_M^\gamma f(y,z,t\eta)\}\omega_2(t[\eta])\kappa_{\lambda^{-1}}u\}\\ &= \omega_1(\lambda t[\eta])(\lambda t)^{-\mu}\{\mathrm{op}_M^\gamma f(y,z,t\lambda\eta)\}\omega_2(\lambda t[\eta])u.\end{aligned}$$

Since $\omega_j(\lambda t[\eta]) = \omega_j(t[\lambda\eta])$, $j=1,2$, this gives the desired result. □

The proposition, below, shows the assertion for the case where the symbol f is independent of t.

Proposition 4.8. *Let $g = g(y,z,\eta) \in C^\infty(\Omega, \mathcal{B}^{\mu,d}(X;\mathbb{R}^q))$ be independent of t. Then the function b defined by*

$$b(y,\eta) = \omega_1(t[\eta])t^{-\mu}\{\mathrm{op}_M^\gamma g(y,z,t\eta)\}\omega_2(t[\eta])$$

is an element of $S_{cl}^\mu(\Omega,\mathbb{R}^q;\mathcal{K}_1^{s,\gamma},\mathcal{K}_2^{s,\gamma})$, and the symbol semi-norms for b can be estimated in terms of those for g.

Proof. For fixed y and η, the operator $b(y,\eta)$ is an element of $\mathcal{L}(\mathcal{K}_1^{s,\gamma},\mathcal{K}_2^{s,\gamma})$ by 4.3. Moreover, it is a smooth function of y and η, and its semi-norms in $C^\infty(\Omega\times\mathbb{R}^q,\mathcal{L}(\mathcal{K}_1^{s,\gamma},\mathcal{K}_2^{s,\gamma}))$ depend continuously on those for g. According to the lemma above it is homogeneous of degree μ for large $|\eta|$. The assertion therefore follows from Lemma 2.4. □

Corollary 4.9. It is now easy to see that a is an element of $S^\mu(\Omega,\mathbb{R}^q;\mathcal{K}_1^{s,\gamma},\mathcal{K}_2^{s,\gamma})$ for all $s > d-1/2$. Indeed, we use the fact that

$$C^\infty(\overline{\mathbb{R}}_+\times\Omega,\mathcal{B}^{\mu,d}(X;\Gamma_{1/2-\gamma}\times\mathbb{R}^q)) = C^\infty(\overline{\mathbb{R}}_+)\hat{\otimes}_\pi C^\infty(\Omega,\mathcal{B}^{\mu,d}(X;\Gamma_{1/2-\gamma}\times\mathbb{R}^q)).$$

Employing the continuity of the mapping $g\mapsto b$ in Proposition 4.9 it is therefore sufficient to consider the case where

$$f(t,y,z,\eta) = \varphi(t)g(y,z,\eta)$$

with $\varphi\in C^\infty(\overline{\mathbb{R}}_+)$ and $g\in C^\infty(\Omega,\mathcal{B}^{\mu,d}(X;\Gamma_{1/2-\gamma}\times\mathbb{R}^q))$ independent of t. Choose a function $\omega\in C_0^\infty(\overline{\mathbb{R}}_+)$ with $\omega(t)\omega_1(t[\eta]) = \omega_1(t[\eta])$. This is possible since $[\eta]$ is bounded away from zero. We have

$$a(y,\eta) = M_{\omega\varphi}\,\omega_1(t[\eta])\{\mathrm{op}_M^\gamma g(y,z,t\eta)\}\omega_2(t[\eta]).$$

Here, $M_{\omega\varphi}$ denotes the operator of multiplication by $\omega\varphi$. We note that

$$\kappa_{[\eta]^{-1}}M_{\omega\varphi}\kappa_{[\eta]} = M_\psi,$$

where $\psi(t) = \omega([\eta]^{-1}t)\varphi([\eta]^{-1}t)$. The norm of this operator on $\mathcal{K}_2^{s,\gamma}$ is uniformly bounded in η; it can be estimated in terms of the semi-norms for φ. Therefore $M_{\omega\varphi}$ furnishes an element in $S^0(\Omega, \mathbb{R}^q; \mathcal{K}_2^{s,\gamma}, \mathcal{K}_2^{s,\gamma})$, and we get the statement from Lemma 2.3.

Theorem 4.10. *Let $\gamma, \mu \in \mathbb{R}$, $\Omega \subseteq \mathbb{R}^q$, $f \in C^\infty(\overline{\mathbb{R}}_+ \times \Omega, \mathcal{B}^{\mu,d}(X; \Gamma_{1/2-\gamma} \times \mathbb{R}^q))$. Then the operator*

$$\operatorname{op}\{\omega_1(t[\eta])t^{-\mu}\{\operatorname{op}_M^\gamma f(t,y,z,t\eta)\}\omega_2(t[\eta])\} : \mathcal{W}^s_{comp}(\mathbb{R}^q, \mathcal{K}_1^{s,\gamma}) \to \mathcal{W}^{s-\mu}_{loc}(\mathbb{R}^q, \mathcal{K}_2^{s,\gamma})$$

is continuous.

Proof. This now is immediate from Theorem 2.8. □

Lemma 4.11. *We use the notation of Theorem 4.10 and let $\beta \in \mathbb{R}$. Then*

$$\omega_1(t[\eta])\{\operatorname{op}_M^\gamma f(t,y,z,t\eta)\}\omega_2(t[\eta])t^\beta = \omega_1(t[\eta])t^\beta\{\operatorname{op}_M^{\gamma-\beta} T^{-\beta} f(t,y,z,t\eta)\}\omega_2(t[\eta]).$$

In case f even is an element in $C^\infty(\overline{\mathbb{R}}_+ \times \Omega, M_O^{\mu,d}(X; \mathbb{R}^q))$ we additionally have

$$\omega_1(t[\eta])\{\operatorname{op}_M^\gamma f(t,y,z,t\eta)\}\omega_2(t[\eta])t^\beta = \omega_1(t[\eta])t^\beta\{\operatorname{op}_M^\gamma T^{-\beta} f(t,y,z,t\eta)\}\omega_2(t[\eta]).$$

Here we consider both sides as operators on $C_0^\infty(\mathbb{R}_+, C^\infty(X))$; $T^{-\beta}$ is the translation operator defined by $T^{-\beta}f(t,y,z,t\eta) = f(t,y,z-\beta,t\eta)$.

Proof. Using a tensor product argument, it is sufficient to treat the case where f is independent of t and y, i.e., $f \in \mathcal{B}^{\mu,d}(X; \Gamma_{1/2-\gamma} \times \mathbb{R}^q)$. But then

$$\begin{aligned}
\{\operatorname{op}_M^\gamma f(z,t\eta)\}t^\beta u(t) &= \frac{1}{2\pi i}\int_{\Gamma_{1/2-\gamma}}\int_0^\infty (t/t')^{-z} f(z,t\eta) t'^\beta u(t')\,dt'/t'dz \\
&= t^\beta \frac{1}{2\pi i}\int_{\Gamma_{1/2-\gamma}}\int_0^\infty (t/t')^{-(z+\beta)} f(z,t\eta) u(t')\,dt'/t'dz \\
&= t^\beta \frac{1}{2\pi i}\int_{\Gamma_{1/2-\gamma+\beta}}\int_0^\infty (t/t')^{-z} T^{-\beta} f(z,t\eta) u(t')\,dt'/t'dz,
\end{aligned}$$

so the first assertion is obvious. In case f is holomorphic, Cauchy's theorem allows us to shift the contour of integration, and we obtain the second statement. □

Mellin quantization

Definition 4.12. A symbol $p = p(t,y,\tau,\eta)$ in $C^\infty(\mathbb{R}_+ \times \Omega, \mathcal{B}^{\mu,d}(X; \mathbb{R}_\tau \times \mathbb{R}^q_\eta))$ is called *edge-degenerate*, if there is a symbol $\tilde{p}$ in $C^\infty(\overline{\mathbb{R}}_+ \times \Omega, \mathcal{B}^{\mu,d}(X; \mathbb{R}_\tau \times \mathbb{R}^q_\eta))$ with $p(t,y,\tau,\eta) = \tilde{p}(t,y,t\tau,t\eta)$.

We shall now show that given an edge-degenerate symbol we can find a Mellin symbol which induces the same operator up to a smoothing perturbation and vice versa. We start with an analysis of the following simple function.

Lemma 4.13. *For $t,t'>0$ let $T(t,t')=\dfrac{t-t'}{\ln t-\ln t'}$. Then T is a smooth positive function on $\mathbb{R}_+\times\mathbb{R}_+$, $T(t,t)=t$. Moreover:*

(a) *Write $x=t/t'$. We have $t'\partial_{t'}=-x\partial_x$ and $t'^{-1}T(t,t')=\dfrac{x-1}{\ln x}$.*

(b) *For each $k\in\mathbb{N}$ the functions*

$$(t'\partial_{t'})^k[t'^{-1}T(t,t')]|_{t'=t},\qquad (t'\partial_{t'})^k[t^{-1}T(t,t')]|_{t'=t},$$
$$(t'\partial_{t'})^k[t'T(t,t')^{-1}]|_{t'=t},\ \text{and}\ \ (t'\partial_{t'})^k[t'T(t,t')^{-1}]|_{t'=t}$$

are constant in t.

Note that T cannot be continued to a function in $C^\infty(\overline{\mathbb{R}}_+\times\overline{\mathbb{R}}_+)$.

Proof. (a) is trivial. For (b) write $\varphi(x)=(x-1)/\ln x$; this is a smooth function on $\mathbb{R}_+$, $\varphi(1)=1$. The observation that $(t'\partial_{t'})^k[t'^{-1}T(t,t')]|_{t=t'}=(-x\partial_x)^k\varphi(x)|_{x=1}$ shows the first claim. For the second note that $t^{-1}T(t,t')=\varphi(x)/x$, while the third and fourth follow by replacing φ by $1/\varphi$. □

Proposition 4.14. *For $p\in C^\infty(\mathbb{R}_+\times\Omega,\mathcal{B}^{\mu,d}(X;\mathbb{R}\times\mathbb{R}^q_\eta))$ define $g\in C^\infty(\mathbb{R}_+\times\mathbb{R}_+,\mathcal{B}^{\mu,d}(X;\Gamma_0\times\mathbb{R}^q_\eta))$ by*

$$g(t,t',y,i\tau,\eta)=p(t,y,-T(t,t')^{-1}\tau,\eta)t'T(t,t')^{-1}. \tag{4.3}$$

Then $\operatorname{op}_t p(t,y,\tau,\eta)=\operatorname{op}_M^{1/2}g(t,t',y,i\tau,\eta)$.
Conversely let $f\in C^\infty(\mathbb{R}_+\times\Omega,\mathcal{B}^{\mu,d}(X;\Gamma_0\times\mathbb{R}^q_\eta))$ and define $q\in C^\infty(\mathbb{R}_+\times\mathbb{R}_+,\mathcal{B}^{\mu,d}(X;\mathbb{R}_\tau\times\mathbb{R}^q_\eta))$ by

$$q(t,t',y,\tau,\eta)=f(t,y,-iT(t,t')\tau,\eta)T(t,t')/t'. \tag{4.4}$$

Then $\operatorname{op}_t q(t,t',y,\tau,\eta)=\operatorname{op}_M^{1/2}f(t,y,i\tau,\eta)$.

The subscript t with op indicates that the pseudodifferential action is with respect to t and the covariable τ only.

Proof. The proof is a straightforward computation. For completeness let us sketch (4.3), omitting for better legibility the variables x and y.

$$\begin{aligned}
&\{\operatorname{op}_t p(t,\tau,\eta)\}u(t,\eta)\\
&=(2\pi)^{-1}\iint e^{i(t-t')\tau}p(t,\tau,\eta)u(t')\,dt'd\tau\\
&=(2\pi)^{-1}\int\int_0^\infty (t/t')^{iT(t,t')\tau}p(t,\tau,\eta)u(t')\,dt'd\tau\\
&=(2\pi)^{-1}\int\int_0^\infty (t/t')^{i\tau}p(t,T(t,t')^{-1}\tau,\eta)t'T(t,t')^{-1}u(t')\,dt'/t'd\tau.
\end{aligned}$$

The proof of the second identity is analogous. □

As a preparation for the proof of Theorem 4.17, below, we need the following well-known facts. For a proof see e.g. Schrohe, Schulze [16, 2.1.12, 2.3.3].

Lemma 4.15. *Given a sequence* $f_j \in C^\infty(\mathbb{R}_+ \times \mathbb{R}_+ \times \Omega, \mathcal{B}^{\mu_j,d}(X;\Gamma_0 \times \mathbb{R}^q))$ *with* $\mu_j \to -\infty$, *there is a symbol* $f \in C^\infty(\mathbb{R}_+ \times \Omega, \mathcal{B}^{\mu,d}(X;\Gamma_0 \times \mathbb{R}^q))$, $\mu = \max\{\mu_j\}$ *such that* $f \sim \sum f_j$; *the symbol* f *is unique modulo* $C^\infty(\mathbb{R}_+ \times \Omega, \mathcal{B}^{-\infty,d}(X;\Gamma_0 \times \mathbb{R}^q))$. *If the symbols* f_j *even belong to* $C^\infty(\overline{\mathbb{R}}_+ \times \Omega, \mathcal{B}^{\mu_j,d}(X;\Gamma_0 \times \mathbb{R}^q))$, *then we find* $f \in C^\infty(\overline{\mathbb{R}}_+ \times \Omega, \mathcal{B}^{\mu,d}(X;\Gamma_0 \times \mathbb{R}^q))$; *it is unique modulo* $C^\infty(\overline{\mathbb{R}}_+ \times \Omega, \mathcal{B}^{-\infty,d}(X;\Gamma_0 \times \mathbb{R}^q))$.

Lemma 4.16. *Given* $f \in C^\infty(\mathbb{R}_+ \times \mathbb{R}_+ \times \Omega, \mathcal{B}^{\mu,d}(X;\Gamma_0 \times \mathbb{R}^q))$ *there is a symbol* $g \in C^\infty(\mathbb{R}_+ \times \Omega, \mathcal{B}^{\mu,d}(X;\Gamma_0 \times \mathbb{R}^q))$ *with*

$$\operatorname{op}_M^{1/2} f \equiv \operatorname{op}_M^{1/2} g \mod C^\infty(\Omega, \mathcal{B}^{-\infty,d}(X^\wedge;\mathbb{R}^q)); \tag{4.5}$$

it has the asymptotic expansion

$$g(t,y,z,\eta) \sim \sum_{k=0}^{\infty} \frac{1}{k!}(-t'\partial_{t'})^k \partial_z^k f(t,t',y,z,\eta)|_{t'=t}. \tag{4.6}$$

Conversely, every symbol with this asymptotic expansion satisfies relation (4.5).

Theorem 4.17. *Let* $p \in C^\infty(\mathbb{R}_+ \times \Omega, \mathcal{B}^{\mu,d}(X;\mathbb{R} \times \mathbb{R}^q_\eta))$ *be edge-degenerate. Then there is a symbol* $f \in C^\infty(\overline{\mathbb{R}}_+ \times \Omega, \mathcal{B}^{\mu,d}(X;\Gamma_0 \times \mathbb{R}^q_\eta))$ *with*

$$\operatorname{op}_t p(t,y,\tau,\eta) = \operatorname{op}_M^{1/2} f(t,y,i\tau,t\eta) \mod C^\infty(\Omega, \mathcal{B}^{-\infty,d}(X^\wedge;\mathbb{R}^q)). \tag{4.7}$$

Conversely, given $f \in C^\infty(\overline{\mathbb{R}}_+ \times \Omega, \mathcal{B}^{\mu,d}(X;\Gamma_0 \times \mathbb{R}^q_\eta))$ *there is an edge-degenerate boundary value problem* p *such that relation* (4.7) *holds. The corresponding statement holds for classical symbols, i.e., for* $\mathcal{B}^{\mu,d}$ *replaced by* $\mathcal{B}^{\mu,d}_{cl}$.

Proof. Let $p(t,y,\tau,\eta) = \tilde{p}(t,y,t\tau,t\eta)$ with $\tilde{p} \in C^\infty(\overline{\mathbb{R}}_+ \times \Omega, \mathcal{B}^{\mu,d}(X;\mathbb{R} \times \mathbb{R}^q))$. We know from Proposition 4.14 and Lemma 4.16 that

$$\operatorname{op}_t p(t,y,\tau,\eta) \equiv \operatorname{op}_M^{1/2} g(t,t',y,i\tau,\eta) \equiv \operatorname{op}_M^{1/2} \tilde{g}(t,y,i\tau,\eta),$$

where

$$\begin{aligned}
\tilde{g}(t,y,i\tau,\eta) &\sim \sum_{k=0}^{\infty} \frac{1}{k!}(-t'\partial_{t'})^k D_\tau^k g(t,t',i\tau)|_{t'=t} \\
&\sim \sum_{k=0}^{\infty} \frac{1}{k!}(-t'\partial_{t'})^k D_\tau^k \{p(t,y,-T(t,t')^{-1}\tau,\eta)t'T(t,t')^{-1}\}|_{t'=t} \\
&\sim \sum_{k=0}^{\infty} \frac{1}{k!}(-t'\partial_{t'})^k D_\tau^k \{\tilde{p}(t,y,-T(t,t')^{-1}t\tau,t\eta)t'T(t,t')^{-1}\}|_{t'=t}.
\end{aligned}$$

Next we prove that, for each k, the function f_k defined by

$$f_k(t,y,\tau,\eta) = (-t'\partial_{t'})^k D_\tau^k\{\widetilde{p}(t,y,-T(t,t')t\tau,\eta)t'T(t,t')^{-1}\}|_{t'=t}$$

is an element of $C^\infty(\overline{\mathbb{R}}_+\times\Omega,\mathcal{B}^{\mu,d}(X;\mathbb{R}\times\mathbb{R}^q))$. In fact, Leibniz' formula implies that

$$\begin{aligned}
&(-t'\partial_{t'})^k D_\tau^k\left\{\widetilde{p}(t,y,-T(t,t')^{-1}t\tau,\eta)t'T(t,t')^{-1}\right\}\\
&= \sum_{k_1+k_2+k_3=k} c_{k_1,k_2,k_3}(-t'\partial_{t'})^{k_1}\left\{D_\tau^k\widetilde{p}(t,y,-T(t,t')^{-1}t\tau,\eta)\right\}\\
&\cdot(-t'\partial_{t'})^{k_2}\left\{t'T(t,t')^{-1}\right\}(-t'\partial_{t'})^{k_3}\left\{-T(t,t')^{-1}t\right\}^k,
\end{aligned}$$

hence Lemma 4.13 shows that we only have to check the derivatives

$$(-t'\partial_{t'})^{k_1}\left\{D_\tau^k\widetilde{p}(t,y,-T(t,t')^{-1}t\tau,\eta)\right\}.$$

For $k_1=1$, this is just $D_\tau^{k+1}\widetilde{p}(t,y,-T(t,t')^{-1}t\tau,\eta)\tau t'\partial_{t'}T(t,t')^{-1}t$. Together with iteration, Lemma 4.13 again yields the smoothness.

According to Lemma 4.15 we can find an $f\in C^\infty(\overline{\mathbb{R}}_+\times\Omega,\mathcal{B}^{\mu,d}(X;\Gamma_0\times\mathbb{R}^q))$ with $f\sim\sum_{k=0}^\infty f_k$. Then $f(t,y,\tau,t\eta)\sim\sum f_k(t,y,\tau,t\eta)$ in $C^\infty(\mathbb{R}_+\times\Omega,\mathcal{B}^{\mu,d}(X;\Gamma_0\times\mathbb{R}^q))$, and hence

$$\mathrm{op}_M^{1/2}f(t,y,i\tau,t\eta)\equiv\mathrm{op}_t p(t,y,\tau,\eta)\mod C^\infty(\Omega,\mathcal{B}^{-\infty,d}(X^\wedge;\mathbb{R}^q)).$$

Clearly, the same argument applies with $\mathcal{B}^{\mu,d}$ replaced by $\mathcal{B}_{cl}^{\mu,d}$.

The converse statement follows in the same way, using the second part of Proposition 4.14 and the asymptotic expansion formula for pseudodifferential double symbols. □

4.18. Mellin quantization for arbitrary weights. We have solved the question how to associate to an edge-degenerate boundary value problem $p\in C^\infty(\mathbb{R}_+\times\Omega,\mathcal{B}^{\mu,d}(X;\mathbb{R}\times\mathbb{R}_\eta^q))$ a Mellin symbol $f_{1/2}\in C^\infty(\overline{\mathbf{R}}_+\times\Omega,\mathcal{B}^{\mu,d}(X;\Gamma_0\times\mathbb{R}_\eta^q))$ with $\mathrm{op}_t p(t,y,\tau,\eta)\equiv\mathrm{op}_M^{1/2}f_{1/2}(t,y,i\tau,t\eta)\mod C^\infty(\Omega,\mathcal{B}^{-\infty,d}(X^\wedge;\mathbb{R}_\eta^q))$. This allows us to treat the case of arbitrary weights.

Theorem 4.19. *For every edge-degenerate $p\in C^\infty(\mathbb{R}_+\times\Omega,\mathcal{B}^{\mu,d}(X;\mathbb{R}\times\mathbb{R}_\eta^q))$ and every $\gamma\in\mathbb{R}$ there is an $f_\gamma\in C^\infty(\overline{\mathbb{R}}_+\times\Omega,\mathcal{B}^{\mu,d}(X;\Gamma_{1/2-\gamma}\times\mathbb{R}_\eta^q))$ such that*

$$\mathrm{op}_M^\gamma f_\gamma(t,y,1/2-\gamma+i\tau,t\eta)\equiv\mathrm{op}_t p(t,y,\tau,\eta)\tag{4.8}$$

modulo $C^\infty(\Omega,\mathcal{B}^{-\infty,d}(X^\wedge;\mathbb{R}^q))$. The corresponding statement holds for classical symbols, i.e., for $\mathcal{B}^{\mu,d}$ replaced by $\mathcal{B}_{cl}^{\mu,d}$.

Proof. The Mellin symbol f_γ can be computed in terms of the function $f=f_{1/2}$ in Theorem 4.17. The definition of op_M^γ shows that

$$\mathrm{op}_t p(t,y,\tau,\eta)\equiv\mathrm{op}_M^{1/2}f_{1/2}(t,y,i\tau,t\eta)=\mathrm{op}_M^\gamma g_\gamma(t,t',y,1/2-\gamma+i\tau,t\eta),$$

where $g_\gamma(t,t',y,1/2-\gamma+i\tau,\eta) = (t/t')^{1/2-\gamma} f_{1/2}(t,y,i\tau,\eta)$. We can convert g_γ to a t'-independent symbol f_γ with

$$\begin{aligned} f_\gamma(t,y,1/2-\gamma+i\tau,\eta) &\sim \sum_{k=0}^{\infty} \frac{1}{k!}(-t'\partial_{t'})^k D_\tau^k g_\gamma(t,t',y,1/2-\gamma+i\tau,\eta)|_{t'=t} \\ &\sim \sum_{k=0}^{\infty} \frac{1}{k!}(-t'\partial_{t'})^k (t/t')^{1/2-\gamma}|_{t'=t}\, D_\tau^k f_{1/2}(t,y,i\tau,\eta) \qquad (4.9)\\ &\sim \sum_{k=0}^{\infty} \frac{1}{k!}(1/2-\gamma)^k D_\tau^k f_{1/2}(t,y,i\tau,\eta). \end{aligned}$$

Here we used that $(-t'\partial_{t'})^k (t/t')^{1/2-\gamma}|_{t'=t} = (x\partial_x)^k x^{1/2-\gamma}|_{x=1} = (1/2-\gamma)^k$. Since $f_{1/2}$ is smooth up to $t=0$, the asymptotic summation can be carried out in $C^\infty(\overline{\mathbf{R}}_+ \times \Omega, \mathcal{B}^{\mu,d}(X;\Gamma_{1/2-\gamma} \times \mathbb{R}^q_\eta))$, and this is all we need.
If p is classical, then so is $f_{1/2}$ by Theorem 4.17, hence f_γ will be classical. □

Kernel cut-off

We shall now analyse the behavior of symbols $f \in C^\infty(\overline{\mathbb{R}}_+ \times \Omega, \mathcal{B}^{\mu,d}(X;\Gamma_{1/2-\gamma} \times \mathbb{R}^q))$ under operations of the type

$$f \mapsto M_{\rho\to z}\varphi(\rho) M^{-1}_{\gamma,\zeta\to\rho} f(t,y,\zeta,\eta).$$

Here, φ is either a function in $C_0^\infty(\mathbb{R}_+)$ or of the form $1-\psi$ with $\psi \in C_0^\infty(\mathbb{R}_+)$. For the proof, the specific choice of γ is of little importance. We therefore let $\gamma = 1/2$, so we can work conveniently on the imaginary axis $i\mathbb{R} = \Gamma_0$.

Theorem 4.20. *Let $\psi \in C_0^\infty(\mathbb{R}_+)$ with $\psi(\rho) \equiv 1$ near $\rho = 1$. Let $f \in C^\infty(\overline{\mathbb{R}}_+ \times \Omega, \mathcal{B}^{\mu,d}(X;\Gamma_0 \times \mathbb{R}^q))$. Then the operator-valued function $f_{1-\psi}$ defined by*

$$f_{1-\psi}(t,y,z,\eta) = M_{\rho\to z}(1-\psi(\rho)) M^{-1}_{1/2,\zeta\to\rho} f(t,y,\zeta,\eta)$$

is an element of $C^\infty(\overline{\mathbb{R}}_+ \times \mathbb{R}^q; \mathcal{B}^{-\infty,d}(X;\Gamma_0))$. Moreover, the mapping $(\psi, f) \mapsto f_{1-\psi}$ is separately continuous from $C_0^\infty(\mathbb{R}_+) \times C^\infty(\overline{\mathbb{R}}_+ \times \Omega, \mathcal{B}^{\mu,d}(X;\Gamma_0 \times \mathbb{R}^q))$ to $C^\infty(\overline{\mathbb{R}}_+ \times \Omega; \mathcal{B}^{-\infty,d}(X;\Gamma_0 \times \mathbb{R}^q))$.

Proof. Using a tensor product argument as above it is sufficient to treat the case where f is independent of t and y, i.e., $f = f(z,\eta) \in \mathcal{B}^{\mu,d}(X;\Gamma_0 \times \mathbb{R}^q_\eta)$. First note that $\mathcal{B}^{-\infty,d}(X;\Gamma_0 \times \mathbb{R}^q)) = \mathcal{S}(\Gamma_0, \mathcal{B}^{-\infty,d}(X;\mathbb{R}^q))$. In view of the identity

$$M_{\rho\to z}\left(\ln^M \rho\,(-\rho\partial_\rho)^N h\right) = \left(\frac{d}{dz}\right)^M z^N (Mh)(z)$$

valid for, say, $h \in C_0^\infty(\mathbb{R}_+)$, we only have to check that, for all $M, N \in \mathbb{N}$, and each semi-norm p_j on $\mathcal{B}^{\mu-j,d}(X;\mathbb{R}^q)$, the semi-norms

$$\left\| p_j\left(\ln^M \rho(\rho\partial_\rho)^N \left\{(1-\psi(\rho))(M^{-1}_{1/2} f)(\rho,\eta)\right\}\right)\right\|_{L^2(\mathbb{R}_\rho)} \qquad (4.10)$$

are finite and depend continuously on the semi-norms for f and ψ, respectively. For fixed ρ,

$$\begin{aligned}2\pi(1-\psi(\rho))(M_{1/2}^{-1}f)(\rho,\eta) &= \int (1-\psi(\rho))\rho^{-i\tau}f(i\tau,\eta)\,d\tau \\ &= (1-\psi(\rho))\ \ln^{-L}\rho \int (i\partial_\tau)^L\rho^{-i\tau}f(i\tau,\eta)\,d\tau \\ &= (1-\psi(\rho))\ \ln^{-L}\rho \int \rho^{-i\tau}(-i\partial_\tau)^L f(i\tau,\eta)\,d\tau\end{aligned}$$

after integration by parts. Since $f\in\mathcal{B}^{\mu,d}(X;\Gamma_0\times\mathbb{R}^q)$, we conclude that $(1-\psi(\rho))(M_{1/2}^{-1}f)(\rho,\eta)\in\mathcal{B}^{\mu-L,d}(X;\mathbb{R}^q)$ for arbitrary L, so it belongs to $\mathcal{B}^{-\infty,d}(X;\mathbb{R}^q)$. Next write for large L

$$\begin{aligned}&\ln^M\rho\,(\rho\partial_\rho)^N[(1-\psi(\rho))(M_{1/2}^{-1}f)(\rho,\eta)] \\ &= \frac{1}{2\pi}\int \ln^M\rho\,(\rho\partial_\rho)^N[\rho^{-i\tau}(1-\psi(\rho))\ln^{-L}\rho]\,(\partial_\tau^L f)(i\tau,\eta)\,d\tau.\end{aligned}$$

Denoting $\psi_j(\rho):=(\rho\partial_\rho)^j(1-\psi(\rho))$, we conclude from Leibniz' rule that the integral is a linear combination of terms of the form

$$\ln^{M-L-j_3}\rho\ \psi_{j_2}(\rho)\int_{-\infty}^{\infty}\rho^{-i\tau}\tau^{j_1}(\partial_\tau^L f)(i\tau,\eta)\,d\tau, \tag{4.11}$$

where $j_1+j_2+j_3=N$. For a semi-norm p_j on $\mathcal{B}^{\mu-j,d}(X;\mathbb{R}^q)$ and fixed M,N, choose $L>M+N+j+2$. Then $M-L-j_3<0$; moreover, $(1+\tau^2)\tau^{j_1}(\partial_\tau^L f)(i\tau,\eta)\in\mathcal{B}^{\mu-j,d}(X;\mathbb{R}_\tau\times\mathbb{R}^q_\eta)$, so that

$$p_j\left(\int_{-\infty}^{\infty}\rho^{-i\tau}(\tau^{j_1}\partial_\tau^L f(i\tau,\eta))\,d\tau\right)\le C$$

with a constant $C=C(L,j_1,j)$ independent of ρ. We conclude that the semi-norm in (4.10) can be estimated by finitely many expressions

$$\text{const.}\cdot\left\{\int_0^\infty\left|\ln^{M-L-j_3}\rho\ \psi_{j_2}(\rho)\right|^2\frac{d\rho}{\rho}\right\}^{1/2}<\infty.$$

Thus all the semi-norms in (4.10) are finite; they continuously depend on the semi-norms for f in $\mathcal{B}^{\mu,d}(X;\Gamma_0\times\mathbb{R}^q)$ and on those for ψ in $\mathcal{D}_K$, $K\subset\mathbb{R}_+$ compact. Here, $\mathcal{D}_K$ denotes those elements in $C_0^\infty(\mathbb{R}_+)$ that have support in K. □

Theorem 4.21. *Let $\varphi\in C_0^\infty(\mathbb{R}_+)$ and $f\in C^\infty(\overline{\mathbb{R}}_+\times\Omega,\mathcal{B}^{\mu,d}(X;\Gamma_0\times\mathbb{R}^q))$. Then the operator-valued function f_φ defined by*

$$f_\varphi(t,y,z,\eta)=M_{\rho\to z}\varphi(\rho)M_{1/2,\zeta\to\rho}^{-1}f(t,y,\zeta,\eta)$$

is an element of $C^\infty(\overline{\mathbb{R}}_+ \times \mathbb{R}^q; M_O^{\mu,d}(X;\mathbb{R}^q))$. *Moreover, the mapping* $(\varphi, f) \mapsto f_\varphi$ *is separately continuous from* $C_0^\infty(\mathbb{R}_+) \times C^\infty(\overline{\mathbb{R}}_+ \times \Omega, \mathcal{B}^{\mu,d}(X;\Gamma_0 \times \mathbb{R}^q))$ *to* $C^\infty(\overline{\mathbb{R}}_+ \times \Omega; M_O^{\mu,d}(X;\mathbb{R}^q))$.

4.22. Outline of the proof. Using a tensor product argument as in Corollary 4.9, it is no restriction to assume that f is independent of t and y, i.e., $f \in \mathcal{B}^{\mu,d}(X;\Gamma_0 \times \mathbb{R}^q)$. We shall first see quite easily in Lemma 4.23 that f_φ is an operator-valued function in $\mathcal{A}(\mathbb{C}, \mathcal{B}^{\mu,d}(X;\mathbb{R}^q))$. It is more difficult to show that it also defines a family of parameter-dependent operators along each line Γ_β, uniformly for β in compact intervals, in the sense of Definition 4.1. To this end we proceed in the following steps, keeping the notation f, φ, f_φ fixed:

(i) For non-direct sums, it is sufficient to consider each summand separately.

(ii) Show the assertion for regularizing elements.

(iii) Reduce to the case $X = \mathbb{R}^n_+$.

(iv) We then only have to deal with operator-valued symbols of 5 types, with

(v) an additional consideration concerning the transmission property.

Lemma 4.23. *The function* f_φ *is an element of* $\mathcal{A}(\mathbb{C}, \mathcal{B}^{\mu,d}(X;\mathbb{R}^q))$, *and the mapping*

$$C_0^\infty(\mathbb{R}_+) \times \mathcal{B}^{\mu,d}(X;\Gamma_0 \times \mathbb{R}^q) \to \mathcal{A}(\mathbb{C}, \mathcal{B}^{\mu,d}(X;\mathbb{R}^q))$$

given by $(\varphi, f) \mapsto f_\varphi$ *is separately continuous.*

Proof. By definition, $\mathcal{E}'(\mathbb{R}_+, \mathcal{B}^{\mu,d}(X;\mathbb{R}^q)) = \mathcal{L}(C^\infty(\mathbb{R}_+), \mathcal{B}^{\mu,d}(X;\mathbb{R}^q))$ with the topology of bounded convergence. Let us start by showing that $\varphi M_{1/2}^{-1} f \in \mathcal{E}'(\mathbb{R}_+, \mathcal{B}^{\mu,d}(X;\mathbb{R}^q))$ and that the mapping is separately continuous in φ and f.
Let $\psi \in C^\infty(\mathbb{R}_+)$ and denote by $\langle\,\cdot\,,\,\cdot\,\rangle$ the evident $\mathcal{B}^{\mu,d}(X;\mathbb{R}^q)$-valued pairing which extends the $L^2(\mathbb{R}_+, \frac{d\rho}{\rho})$ bilinear form. Then

$$\begin{aligned}
&\langle \varphi M^{-1}_{1/2,\tau\to\rho} f, \psi\rangle \\
&= \langle \varphi \int_{-\infty}^{\infty} \rho^{-i\tau} f(i\tau,\eta)\,d\tau, \psi\rangle \\
&= \langle \int_{-\infty}^{\infty} \rho^{-1-i\tau} f(i\tau,\eta)\,d\tau, \rho\varphi\psi\rangle \qquad (4.12)\\
&= \langle (-\rho\partial_\rho)^N \int_{-\infty}^{\infty} \rho^{-1-i\tau}(1+i\tau)^{-N} f(i\tau,\eta)\,d\tau, \rho\varphi\psi\rangle \\
&= \int_0^\infty \int_{-\infty}^{\infty} \rho^{-i\tau}(1+i\tau)^{-N} f(i\tau,\eta)\,d\tau\, \rho^{-1}(\rho\partial_\rho)^N(\rho\varphi(\rho)\psi(\rho))\frac{d\rho}{\rho}.
\end{aligned}$$

The last integral is an L^1-integral with values in $\mathcal{B}^{\mu,d}(X;\mathbb{R}^q)$, provided N is sufficiently large. This follows from the fact that, for every semi-norm q on $\mathcal{B}^{\mu,d}(X;\mathbb{R}^q)$, we have $q(f(i\tau,\eta)) = O(\langle\tau,\eta\rangle^\mu)$.
Moreover, if the semi-norms for ψ in $C^\infty(\mathbb{R}_+)$ tend to zero, then the last integral tends to zero in all semi-norms of $\mathcal{B}^{\mu,d}(X;\mathbb{R}^q)$. So it indeed defines an element of $\mathcal{E}'(\mathbb{R}_+,\mathcal{B}^{\mu,d}(X;\mathbb{R}^q))$.
Now let us show separate continuity. As ψ varies over a bounded set in $C^\infty(\mathbb{R}_+)$, the integral in (4.13) can be estimated in terms of finitely many semi-norms for $f \in \mathcal{B}^{\mu,d}(X;\Gamma_0\times\mathbb{R}^q)$ and finitely many semi-norms for $\varphi\in\mathcal{D}_K, K\subset\mathbb{R}_+$ compact. Finally note that the Mellin transform yields a continuous map from $\mathcal{E}'(\mathbb{R}_+,\mathcal{B}^{\mu,d}(X;\mathbb{R}^q))$ to $\mathcal{A}(\mathbb{C},\mathcal{B}^{\mu,d}(X;\mathbb{R}^q))$. Indeed, this follows from the relations

$$\begin{aligned}\mathcal{E}'(\mathbb{R}_+,\mathcal{B}^{\mu,d}(X;\mathbb{R}^q)) &= \mathcal{E}'(\mathbb{R}_+)\hat{\otimes}_\pi\mathcal{B}^{\mu,d}(X;\mathbb{R}^q) \quad \text{and}\\ \mathcal{A}(\mathbb{C},\mathcal{B}^{\mu,d}(X;\mathbb{R}^q)) &= \mathcal{A}(\mathbb{C})\hat{\otimes}_\pi\mathcal{B}^{\mu,d}(X;\mathbb{R}^q),\end{aligned}$$

together with the well-known fact that the Mellin transform maps $\mathcal{E}'(\mathbb{R}_+)$ to $\mathcal{A}(\mathbb{C})$ continuously. □

The next proposition settles Step (i) of the Outline 4.22.

Lemma 4.24. *Let $\mathcal{E},\mathcal{F}$ and $\mathcal{Y}$ be Fréchet spaces, and assume that $\mathcal{E}$ and $\mathcal{F}$ are embedded in a common vector space $\mathcal{X}$. Suppose $T:\mathcal{E}+\mathcal{F}\to\mathcal{Y}$ is a linear map, and the restrictions*

$$T:\mathcal{E}\to\mathcal{Y},\qquad T:\mathcal{F}\to\mathcal{Y}$$

are continuous in the topologies of $\mathcal{E}$ and $\mathcal{F}$. Then

$$T:\mathcal{E}+\mathcal{F}\to\mathcal{Y}$$

is continuous in the topology of the non-direct sum.

Proof. Let $\{p_1,p_2,\ldots\}$, $\{q_1,q_2,\ldots\}$ be increasing systems of semi-norms for $\mathcal{E}$ and $\mathcal{F}$, respectively. Denote the translation invariant metric in $\mathcal{Y}$ by d. Then a system of semi-norms for $\mathcal{E}+\mathcal{F}$ is given by $r_j(x) = \inf\{p_j(e)+q_j(f) : e+f=x\}$. So suppose $x_0\in\mathcal{E}+\mathcal{F}$ and $V\subseteq\mathcal{Y}$ is an ε-ball about Tx_0. Then there is a $j\in\mathbb{N}$ and a $\delta>0$ such that $d(Te,0)<\frac{\varepsilon}{2}$ and $d(Tf,0)<\frac{\varepsilon}{2}$, provided that $e\in\mathcal{E}$, $f\in\mathcal{F}$, $p_j(e)<\delta$ and $q_j(f)<\delta$. This implies that $Tx\in V$ for all x with $r_j(x-x_0)<\delta$: In this case we can find $e_1\in\mathcal{E}$, $f_1\in\mathcal{F}$ such that $e_1+f_1 = x-x_0$ and $p_j(e_1)+q_j(f_1)<\delta$. Hence $d(Tx,Tx_0) = d(T(x-x_0),0)\le d(T(e_1),0)+d(T(f_1),0)<\varepsilon$. □

In order to settle Step (ii) of the Outline 4.22, we first note that an operator in $\mathcal{B}^{-\infty,d}(X;\Gamma_0\times\mathbb{R}^q)$ can also be viewed as an element of $\mathcal{S}(\Gamma_0\times\mathbb{R}^q, C^\infty(X\times X))$, where $C^\infty(X\times X)$ is the Fréchet space of smooth kernel sections.

Lemma 4.25. *Let $h\in\mathcal{S}(\Gamma_0)$, $s\in\mathcal{S}(\Gamma_0,C^\infty(X\times X))$. Then*

(a) $H := M(\varphi M_{1/2}^{-1}h) \in \mathcal{A}(\mathbb{C})$, *and* $H|_{\Gamma_\beta} \in \mathcal{S}(\Gamma_\beta)$ *for every* β, *with estimates uniformly in* β *for* β *in compact intervals. The corresponding induced mapping* $(\varphi, h) \mapsto H$ *from* $C_0^\infty(\mathbb{R}_+) \times \mathcal{S}(\Gamma_0)$ *into this subspace of* $\mathcal{A}(\mathbb{C})$ *is separately continuous.*

(b) $F := M(\varphi M_{1/2}^{-1}s) \in \mathcal{A}(\mathbb{C}, C^\infty(X \times X))$, $F|_{\Gamma_\beta} \in \mathcal{S}(\Gamma_\beta, C^\infty(X \times X))$ *for every* β, *with estimates uniformly in* β *for* β *in compact intervals. The mapping* $(\varphi, s) \mapsto F$ *is separately continuous from* $C_0^\infty(\mathbb{R}_+) \times \mathcal{S}(\Gamma_0, C^\infty(X \times X))$ *to this subspace of* $\mathcal{A}(\mathbb{C}, C^\infty(X \times X))$.

Proof. (a) By the Mellin inversion formula in 3.2, $M_{1/2}^{-1}h(t) = \frac{1}{2\pi}\int_{-\infty}^{\infty} t^{-i\tau}h(i\tau)d\tau$. The integral converges; we can differentiate under the integral sign for the derivatives to see that it furnishes a smooth function.
Hence $\varphi M_{1/2}^{-1}h \in C_0^\infty(\mathbb{R}_+)$. It is easy to check that its Mellin transform therefore is rapidly decreasing on each line Γ_β, uniformly for β in compact intervals. Clearly, the mapping $(\varphi, g) \mapsto \varphi g$ is separately continuous from $C_0^\infty(\mathbb{R}_+) \times C^\infty(\mathbb{R}_+)$ to $C_0^\infty(\mathbb{R}_+)$, and the Mellin transform is continuous from $C_0^\infty(\mathbb{R}_+)$ to the subspace of $\mathcal{A}(\mathbb{C})$ consisting of functions that restrict to $\mathcal{S}(\Gamma_\beta)$, uniformly for β in compact intervals, i.e., the space $M_O^{-\infty}$ for $\dim X = 0$. So the separate continuity follows.
(b) follows from (a), noting that $\mathcal{S}(\Gamma_\beta, C^\infty(X \times X)) = \mathcal{S}(\Gamma_\beta)\hat{\otimes}_\pi C^\infty(X \times X)$ and $\mathcal{A}(\mathbb{C}, C^\infty(X \times X)) = \mathcal{A}(\mathbb{C})\hat{\otimes}_\pi C^\infty(X \times X)$. For the continuity assertion we use the continuity of the Mellin transform from $C_0^\infty(\mathbb{R}_+, C^\infty(X \times X))$ to the corresponding subspace of $\mathcal{A}(\mathbb{C}, C^\infty(X \times X))$. □

Since the topology of $\mathcal{B}^{\mu,d}(X; \mathbb{R}^q)$ is precisely that of a non-direct sum of the regularizing terms and the local terms, Step (iii) of the Outline 4.22 is immediate. We next attack Step (iv). Notice that all entries of the symbol quintuple in 2.12 are operator-valued symbols.

Lemma 4.26. *Let* E, F *be Banach spaces with strongly continuous group actions* κ, $\tilde{\kappa}$. *Let* $\mu \in \mathbb{R}$, $m, k, q \in \mathbb{N}$, *and*

$$a = a(x, \xi, z, \eta) \in S^\mu(\mathbb{R}^m_x, \mathbb{R}^k_\xi \times \Gamma_{0,z} \times \mathbb{R}^q_\eta; E, F).$$

Then the function

$$A = A(z) = M_{\rho\to z}(\varphi(\rho)M_{1/2, z\to\rho}^{-1}a)$$

is analytic on $\mathbb{C}$ *with values in* $S^\mu(\mathbb{R}^m, \mathbb{R}^{k+q}; E, F)$. *Moreover, for all* $\beta \in \mathbb{R}$,

$$A|_{\Gamma_\beta} \in S^\mu(\mathbb{R}^m, \mathbb{R}^k \times \Gamma_\beta \times \mathbb{R}^q; E, F), \tag{4.13}$$

uniformly for β *in compact intervals. The mapping* $(\varphi, a) \mapsto A$ *from* $C_0^\infty(\mathbb{R}_+) \times S^\mu(\mathbb{R}^m, \mathbb{R}^k \times \Gamma_0 \times \mathbb{R}^q; E, F)$ *to this Fréchet subspace of* $\mathcal{A}(\mathbb{C}, S^\mu(\mathbb{R}^m, \mathbb{R}^{k+q}; E, F))$ *is separately continuous.*

Proof. In view of the identities

$$S^\mu(\mathbb{R}^m, \mathbb{R}^k \times \Gamma_0 \times \mathbb{R}^q; E, F) \quad = \quad C^\infty(\mathbb{R}^m)\hat{\otimes}_\pi S^\mu(\mathbb{R}^0, \mathbb{R}^k \times \Gamma_0 \times \mathbb{R}^q; E, F),$$

$$\mathcal{A}(\mathbb{C}, S^\mu(\mathbb{R}^m, \mathbb{R}^{k+q}; E, F)) \quad = \quad C^\infty(\mathbb{R}^m)\hat{\otimes}_\pi \mathcal{A}(\mathbb{C}, S^\mu(\mathbb{R}^0, \mathbb{R}^{k+q}; E, F)),$$

we may assume that $m = 0$, i.e., $a \in S^\mu(\mathbb{R}^k \times \Gamma_0 \times \mathbb{R}^q; E, F)$ is independent of x. From Lemma 4.23 we know that $A = M(\varphi M_{1/2}^{-1} a) \in \mathcal{A}(\mathbb{C}, S^\mu(\mathbb{R}^{k+q}; E, F))$. This proves the first part of the statement.
Next consider $A|_{\Gamma_\beta}$. We may assume $\beta = 0$ due to the relation $(Mf)(z+\beta) = M_{t\to z}(t^\beta f)(z)$: Replacing $A|_{\Gamma_\beta}$ by $A|_{\Gamma_0}$ corresponds to replacing $\varphi(t)$ by $t^{-\beta}\varphi(t) \in C_0^\infty(\mathbb{R}_+)$. For the analysis of $A|_{\Gamma_0}$ it is more convenient to switch from the Mellin to the Fourier transform. We write the variable on Γ_0 in the form $z = i\tau$, $\tau \in \mathbb{R}$, and let $p(\tau, \eta) = a(i\tau, \eta)$. A simple computation gives

$$(M_{1/2}\varphi M_{1/2}^{-1} a)(i\tau, \eta) \quad = \quad (\mathcal{F}_{\widetilde{\tau}\to r}\varphi(e^{-r})\mathcal{F}_{r\to\tau}^{-1} p)(\tau, \eta).$$

The symbol p is an element of $S^\mu(\mathbb{R}^{k+1+q}; E, F)$, and $r \mapsto \varphi(e^{-r}) = \psi(r)$ is a function in $C_0^\infty(\mathbb{R})$. So our task is reduced to showing that $q = \mathcal{F}\psi(r)\mathcal{F}^{-1}p \in S^\mu(\mathbb{R}^{k+1+q}; E, F)$. We abbreviate $v = (\xi, \tau, \eta)$ and consider a derivative $D_v^\beta q = D_\xi^{\beta_1} D_\tau^{\beta_2} D_\eta^{\beta_3} q$. We then estimate

$$\begin{aligned}
&\left\|\widetilde{\kappa}_{[v]^{-1}}\left\{D_v^\beta[\mathcal{F}_{r\to\tau}\psi(r)\mathcal{F}^{-1}p](v)\right\}\kappa_{[v]}\right\|_{\mathcal{L}(E,F)} \\
&= \left\|\widetilde{\kappa}_{[v]^{-1}} D_v^\beta(\hat{\psi} * p)(v)\kappa_{[v]}\right\|_{\mathcal{L}(E,F)} \\
&= \left\|\int_{-\infty}^{\infty}\widetilde{\kappa}_{[\xi,\sigma,\eta][v]^{-1}}\hat{\psi}(\tau-\sigma)\widetilde{\kappa}_{[\xi,\sigma,\eta]^{-1}}\right. \\
&\qquad \left.\cdot D_\xi^{\beta_1} D_\sigma^{\beta_2} D_\eta^{\beta_3} p(\xi,\sigma,\eta)\kappa_{[\xi,\sigma,\eta]}\kappa_{[v][\xi,\sigma,\eta]^{-1}}\,d\sigma\right\|_{\mathcal{L}(E,F)} \qquad (4.14)\\
&\leq \int_{-\infty}^{\infty}\left\|\widetilde{\kappa}_{[\xi,\sigma,\eta][v]^{-1}}\right\|_{\mathcal{L}(F)}|\hat{\psi}(\tau-\sigma)| \\
&\qquad \cdot\left\|\widetilde{\kappa}_{[\xi,\sigma,\eta]^{-1}} D_\xi^{\beta_1} D_\sigma^{\beta_2} D_\eta^{\beta_3} p(\xi,\sigma,\eta)\kappa_{[\xi,\sigma,\eta]}\right\|_{\mathcal{L}(E,F)}\left\|\kappa_{[v][\xi,\sigma,\eta]^{-1}}\right\|_{\mathcal{L}(E)}d\sigma.
\end{aligned}$$

Here we have used the fact that (scalar) multiplication by $\hat{\psi}(\tau-\sigma)$ commutes with the action of $\widetilde{\kappa}$. According to (2.1) there are constants c and M such that

$$\|\kappa_{[\xi,\sigma,\eta]^{-1}[v]}\|_{\mathcal{L}(E)},\ \|\widetilde{\kappa}_{[\xi,\sigma,\tau][v]^{-1}}\|_{\mathcal{L}(F)} \quad \leq \quad cL(\xi,\sigma,\eta,v)^M,$$

where $L(\xi,\sigma,\eta,v) = \max\{[\xi,\sigma,\eta]^{-1}[v], [\xi,\sigma,\eta][v]^{-1}\}$. Peetre's inequality implies that

$$[\xi,\sigma,\eta]^{-1}[v] \leq C[(\xi,\sigma,\eta)-(\xi,\tau,\eta)] = C[\sigma-\tau]$$

and, by symmetry, $[\xi,\sigma,\eta][v]^{-1} \leq C[\sigma-\tau]$ for a suitable constant C. Together with the facts that

$$\|\widetilde{\kappa}_{[\xi,\sigma,\eta]^{-1}}(D_\xi^{\beta_1} D_\sigma^{\beta_2} D_\eta^{\beta_3} p)(\xi,\sigma,\eta)\kappa_{[\xi,\sigma,\eta]}\|_{\mathcal{L}(E,F)} = O\left([\xi,\sigma,\eta]^{\mu-|\beta|}\right)$$

and that $\hat{\psi}$ is rapidly decreasing we conclude with Peetre's inequality that the final integral above is $O([v]^{\mu-|\beta|})$.
This shows our claim. Clearly, all estimates depend continuously on φ and p, thus they depend continuously on a, and the corresponding mapping is separately continuous. $\square$

We now complete the proof of Theorem 4.21 with Step (v) of the outline, i.e., the observation that the transmission property is preserved under the construction. This is the contents of the following lemma.

Lemma 4.27. *Let* $p \in S^{\mu}_{tr}(\mathbb{R}^n, \mathbb{R}^n \times \Gamma_0 \times \mathbb{R}^q)$. *Then*

$$q = M(\varphi M^{-1}_{1/2} p) \in \mathcal{A}(\mathbb{C}, S^{\mu}_{tr}(\mathbb{R}^n, \mathbb{R}^{n+q})). \tag{4.15}$$

Moreover, for every $\beta \in \mathbb{R}$,

$$q|_{\Gamma_\beta} \in S^{\mu}_{tr}(\mathbb{R}^n, \mathbb{R}^n \times \Gamma_\beta \times \mathbb{R}^q). \tag{4.16}$$

The corresponding estimates are satisfied uniformly for β *in compact intervals. The mapping* $(\varphi, p) \mapsto q$ *is separately continuous as a map from*

$$C_0^{\infty}(\mathbb{R}_+) \times S^{\mu}_{tr}(\mathbb{R}^n, \mathbb{R}^n \times \Gamma_0 \times \mathbb{R}^q)$$

to this Fréchet subspace of $\mathcal{A}(\mathbb{C}, S^{\mu}(\mathbb{R}^n, \mathbb{R}^{n+q}))$.

Proof. If it were not for the subscript "*tr*", (4.15) and (4.16) would follow from the Lemma above, because the usual symbol classes correspond to the operator-valued symbols with $E = F = \mathbb{C}$ and trivial group action.
So we have to show that the transmission property is preserved under the operation $f \mapsto f_\varphi$. This, however, is simple: a symbol $a \in S^{\mu}(\mathbb{R}^n, \mathbb{R}^n \times \Gamma_0 \times \mathbb{R}^q)$ has the transmission property iff

$$\partial^k_{x_n} a(x', 0, \xi', \langle \xi' \rangle \xi_n, z, \eta) \in S^{\mu}(\mathbb{R}^{n-1}_{x'}, \mathbb{R}^{n-1}_{\xi'} \times \Gamma_{0,z} \times \mathbb{R}^q_\eta) \hat{\otimes}_\pi H_{\xi_n}.$$

In the present situation

$$\begin{aligned} \partial^k_{x_n} q(x', 0, \xi', \langle \xi' \rangle \xi_n, z, \eta) &= M_{\rho \to z} \varphi(\rho) M^{-1}_{1/2, \zeta \to \rho} \partial^k_{x_n} p(x', 0, \xi', \langle \xi' \rangle \xi_n, \zeta, \eta) \\ &\in \mathcal{A}(\mathbb{C}, S^{\mu}(\mathbb{R}^{n-1}, \mathbb{R}^{n-1+q})) \hat{\otimes}_\pi H_{\xi_n} \end{aligned}$$

by a tensored version of the argument in 4.26. The last space coincides with $\mathcal{A}(\mathbb{C}, S^{\mu}(\mathbb{R}^{n-1}, \mathbb{R}^{n-1+q}) \hat{\otimes}_\pi H_{\xi_n})$ and (4.15) is proven. For (4.16) we can argue in the same way: Restriction to Γ_β furnishes an element in $S^{\mu}(\mathbb{R}^{n-1}, \mathbb{R}^{n-1} \times \Gamma_\beta \times \mathbb{R}^q) \hat{\otimes}_\pi H_{\xi_n}$.
Finally, the separate continuity of the mapping follows from the closed graph theorem and the continuity properties established in Lemma 4.26, since the topology of the space with the transmission property is finer than the original one. The

closed graph theorem indeed can be applied: a mapping $\Lambda : C_0^\infty(\mathbb{R}_+) \to \mathcal{Y}$, $\mathcal{Y}$ a locally convex space, is continuous if and only if its restriction to the Fréchet spaces $\mathcal{D}_K$ are continuous. □

Remark 4.28. The Mellin quantization procedure in connection with the kernel cut-off allows us to associate to an edge degenerate boundary symbol a *holomorphic* Mellin symbol.

We have now seen that, starting from an arbitrary Mellin symbol, the operation $f \mapsto f_\psi$ furnishes a holomorphic Mellin symbol. Assuming additionally that $\psi(\rho) \equiv 1$ near $\rho = 1$, the symbols f and f_ψ will differ only by a regularizing symbol along the line Γ_0, where we started. An interesting question is the following: Suppose we initially have a holomorphic Mellin symbol h. What can we say about the difference $h - h_\psi$? Will it also be small along other lines? The theorem below shows that the difference is as good as we can expect it to be.

Theorem 4.29. *Given $h \in C^\infty(\overline{\mathbb{R}}_+ \times \Omega, M_O^{\mu,d}(X;\mathbb{R}^q))$ and $\psi \in C_0^\infty(\mathbb{R}_+)$ with $\psi(\rho) \equiv 1$ near $\rho = 1$, the difference $h - h_\psi$ is an element of $C^\infty(\overline{\mathbb{R}}_+ \times \Omega, M_O^{-\infty,d}(X;\mathbb{R}^q))$.*

Proof. Choose $\beta \in \mathbb{R}$ and a nonnegative integer $M > \mu + |\beta| + 1$. Then $D_z^M h(t,\cdot,\eta)$ is integrable along Γ_β. Moreover, the analyticity of the function $z \mapsto \rho^{-z} D_z^M h(t,z,\eta)$ together with Cauchy's theorem implies that

$$\int_{-\infty}^{\infty} \rho^{-i\tau}(D_z^M h)(t,i\tau,\eta)\,d\tau = \int_{-\infty}^{\infty} \rho^{-(\beta+i\tau)}(D_z^M h)(t,\beta+i\tau,\eta)\,d\tau\,,$$

so that $M_{1/2}^{-1}(D_z^M h)(t,\rho,\eta) = \rho^{-\beta} M_{1/2,\,\zeta\to\rho}^{-1}(D_z^M h(t,\zeta+\beta,\eta))$. Hence, for $z = \beta + i\tau$,

$$\begin{aligned}
&(h - h_\psi)(t,z,\eta)\\
&= \int_0^\infty \rho^{\beta+i\tau-1}(1-\psi(\rho))(M_{1/2}^{-1}h)\,(t,\rho,\eta)\;d\rho\\
&= \int_0^\infty \rho^{\beta+i\tau-1}(1-\psi(\rho))\ln^{-M}\rho\;(M_{1/2}^{-1}(D_z^M h))(t,\rho,\eta)\;d\rho\\
&= \int_0^\infty \rho^{\beta+i\tau-1}(1-\psi(\rho))\ln^{-M}\rho\;\rho^{-\beta} M_{1/2,\,\zeta\to\rho}^{-1}(D_z^M h)(t,\zeta+\beta,\eta)\;d\rho\\
&= \int_0^\infty \rho^{i\tau-1}(1-\psi(\rho))\,M_{1/2,\,\zeta\to\rho}^{-1}h(t,\zeta+\beta,\eta)\;d\rho\\
&= \Big[M_{1/2,\rho\to z}(1-\psi(\rho))\,M_{1/2,\,\zeta\to\rho}^{-1}h(t,\zeta+\beta,\eta)\Big]\,(z-\beta).
\end{aligned}$$

On the other hand, the function $(t,z,\eta) \mapsto h(t,z+\beta,\eta)$ is an element of $C^\infty(\overline{\mathbb{R}}_+, M_O^{\mu,d}(X;\mathbb{R}^q))$; the corresponding symbol estimates hold uniformly for β in compact intervals. Applying Theorem 4.20, $h - h_\psi|_{\Gamma_\beta} \in C^\infty(\overline{\mathbb{R}}_+, \mathcal{B}^{-\infty,d}(X;\Gamma_\beta \times \mathbb{R}^q))$, uniformly for β in compact intervals. □

References

[1] AGRANOVIČ, M.S., VIŠIK, M.I.: Elliptic problems with a parameter and parabolic problems of general type; Uspekhi Mat. Nauk 19 (1964), 53–161 (Russian); Engl. transl.: Russ. Math. Surveys 19 (1964), 53–159.

[2] BEHM, S.: Pseudo-Differential Operators with Parameters on Manifolds with Edges; Dissertation, Universität Potsdam 1995.

[3] BOUTET DE MONVEL, L.: Boundary problems for pseudo-differential operators; Acta Math. 126 (1971), 11–51.

[4] BUCHHOLZ, TH., SCHULZE, B.-W.: Anisotropic edge pseudo-differential operators with discrete asymptotics; Math. Nachr. (to appear).

[5] DORSCHFELDT, CH.: An Algebra of Mellin Pseudo-Differential Operators near Corner Singularities; Dissertation, Universität Potsdam 1995.

[6] DORSCHFELDT, CH., SCHULZE, B.-W.: Pseudo-differential operators with operator-valued symbols in the Mellin-edge-approach; Ann. Global Anal. Geom. 12 (1994), 135–171.

[7] EGOROV, YU., SCHULZE, B.-W.: Pseudo-Differential Operators, Singularities, Applications; Birkhäuser, Basel (to appear).

[8] ESKIN, G.I.: Boundary Value Problems for Elliptic Pseudodifferential Equations; Moscow 1973 (Russian); Engl. transl.: Amer. Math. Soc. Transl. 52 (1981).

[9] HIRSCHMANN, T.: Functional analysis in cone and edge Sobolev spaces; Ann. Global Anal. Geom. 8 (1990), 167–192.

[10] KONDRAT'EV, V.A.: Boundary value problems in domains with conical or angular points; Trans. Moscow Math. Soc. 16 (1967), 227–313.

[11] LEWIS, J.E., PARENTI, C.: Pseudodifferential operators of Mellin type; Comm. Partial Differential Equations 8 (1983), 477–544.

[12] MELROSE, R.: The Atiyah-Patodi-Singer Index Theorem; A K Peters, Wellesley, MA 1993.

[13] REMPEL, S., SCHULZE, B.-W.: Index Theory of Elliptic Boundary Problems; Akademie-Verlag, Berlin 1982.

[14] SCHROHE, E.: Fréchet Algebras of Pseudodifferential Operators and Boundary Value Problems; Birkhäuser, Boston, Basel (to appear).

[15] SCHROHE, E., SCHULZE, B.-W.: Boundary value problems in Boutet de Monvel's algebra for manifolds with conical singularities I. Pseudo-Differential Operators and Mathematical Physics, Advances in Partial Differential Equations 1; Akademie Verlag, Berlin 1994, 97–209.

[16] SCHROHE, E., SCHULZE, B.-W.: Boundary value problems in Boutet de Monvel's algebra for manifolds with conical singularities II. Boundary Value Problems, Schrödinger Operators, Deformation Quantization, Advances in Partial Differential Equations 2; Akademie Verlag, Berlin 1995, 70–205.

[17] SCHULZE, B.-W.: Pseudo-differential operators on manifolds with edges; Symp. 'Partial Diff. Equations', Holzhau 1988, Teubner Texte zur Mathematik 112, 259 –288, Leipzig 1989.

[18] SCHULZE, B.-W.: Mellin representations of pseudo-differential operators on manifolds with corners; Ann. Global Anal. Geom. 8 (1990), 261–297.

[19] SCHULZE, B.-W.: Boundary Value Problems and Singular Pseudo-Differential Operators; John Wiley, Chichester (to appear).

[20] SCHULZE, B.-W.: Pseudo-Differential Boundary Value Problems, Conical Singularities and Asymptotics; Akademie Verlag, Berlin 1994.

[21] TRIEBEL, H.: Interpolation Theory, Function Spaces, Differential Operators; North-Holland, Amsterdam, New York, Oxford 1978.

[22] UNTERBERGER, A., UPMEIER, H.: Pseudodifferential Analysis on Symmetric Cones; Studies in Advanced Mathematics, CRC Press, Boca Raton, New York 1996.

[23] VIŠIK, M.I., ESKIN, G.I.: Normally solvable problems for elliptic systems in equations of convolution; Math. USSR Sb. 14 (116) (1967), 326–356.

Max-Planck-Arbeitsgruppe
"Partielle Differentialgleichungen und komplexe Analysis"
Universität Potsdam
14415 Potsdam
Germany
schrohe@mpg-ana.uni-potsdam.de, schulze@mpg-ana.uni-potsdam.de

1991 Mathematics Subject Classification: Primary 35S15; Secondary 58G20, 46E35, 46H35

Submitted: May 29, 1996

Operator Theory:
Advances and Applications, Vol. 102

On some global aspects of the theory of partial differential equations on manifolds with singularities

B.-W. Schulze, B. Sternin and V. Shatalov

In this paper we investigate the connection between asymptotic expansions of solutions to elliptic equations near different points of singularities of the underlying manifold. We propose the procedure of computation of the asymptotic expansion of the solution at any point of singularity via the asymptotics given at one (fixed) of these points.

1. Introduction

One of the main problems of the theory of differential equations on manifolds with singularities is the investigation of the behavior of solutions near singular points of the manifold. In fact, this problem is strongly connected with the other main problem of the theory – the solvability problem. Actually, the knowledge of the asymptotics of solutions in a neighborhood of the singularity set allows one to give the adequate function spaces in which the finiteness theorem for the corresponding operator can be proved.

It is well-known (see, for example, [1]) that a solution to a homogeneous elliptic equation near a singular (say, conical) point has a so-called *conormal asymptotic expansion*, that is, an expansion of the form

$$u(r,x) \simeq \sum_k r^{S_k} \sum_{j=0}^{m_k} a_{kj}(x) \ln^j r,$$

where (r,x) are special coordinates in a neighborhood of the conical point corresponding to the representation

$$M = ([0,1] \times X) / (\{0\} \times X)$$

of the manifold M near this point, $r \in [0,1]$, $x \in X$. Far less known is the fact that the conormal asymptotics of solutions near the points of singularity are strongly connected with one another, namely, that the asymptotics near one of these points uniquely determines the asymptotics near all of them (see [2]).

Hence, the problem arises to find out a method allowing to compute the asymptotics near any point of singularity provided that such asymptotics are known near some (fixed) point. We remark that the solution of this problem, being of interest by itself, allows one to find out the *exact function spaces* in which the considered equation is uniquely solvable (or, at least, possesses a Fredholm property).

The outline of the paper is as follows:

In the first section, we present simple examples which are aimed at the *motivation* of the general statement of the problem. Namely, in the first subsection, the connection between the asymptotic expansion of solutions at singular points and the correct statements of problems for the corresponding operators is illustrated. The second subsection illustrates (in terms of an example) the *mechanism of transport* of asymptotic expansions from one singular point to another through the complex domain.

The second section contains (as it follows from its title) the formulation of the *main problem* of investigation in this paper.

Later on, the two last sections contain the investigation of the above stated problem for the two-dimensional and multi-dimensional cases, respectively. The reason of the separate consideration of the two-dimensional case is that in this case the main ideas of the computational algorithm are not obscured by technical difficulties. Both these sections are divided, in turn, into two subsections, the first aimed at the consideration of the propagation of singularities along the regular component of the singularity set of the solution, and the second at the consideration of "jumps" of the asymptotic expansion from one such component to another.

2. Examples

2.1. Singularities of solutions to a homogeneous equation and correct statements of the problem

As it was already mentioned in the introduction, the aim of this paper is to investigate global aspects of the asymptotic theory of partial differential equations on manifolds with singularities. Namely, if the singularity set of the manifold M has two or more connected components and $\widehat{a}$ is an elliptic partial differential operator on M, then the problem is to compute the asymptotic expansion of a solution to the homogeneous equation

$$\widehat{a}u = 0 \tag{2.1}$$

on the whole singularity set of M provided that this expansion is known on one of the connected components of this set.

The stated problem has important applications to the investigation of properties of the operator $\widehat{a}$ in the weighted Sobolev spaces $H^s_\gamma(M)$ (see, e.g., [2]). Let us illustrate the connection between the asymptotic behavior of solutions to (2.1) and the investigation of the operator

$$\widehat{a} : H^s_\gamma(M) \to H^{s-m}_\gamma(M)$$

by a simple example (here m is the order of the operator $\widehat{a}$).

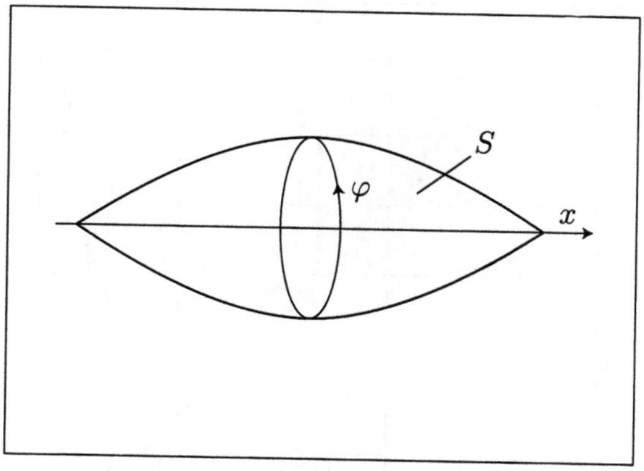

Figure 1: The spindle.

Consider the operator

$$\widehat{a} = \frac{1}{(1-x^2)^2}\left[\left((1-x^2)\frac{\partial}{\partial x}\right)^2 + \frac{\partial^2}{\partial \varphi^2}\right] \tag{2.2}$$

given on the surface of the spindle S (see Figure 1). Here x is the coordinate along the axis of the spindle, $-1 < x < 1$, and φ is the coordinate corresponding to the rotation around this axis. It is easy to check that the full system of solutions to equation (2.1) for such an operator is

$$\begin{aligned} u_k^{\pm}(x,\varphi) &= \left(\frac{1+x}{1-x}\right)^k e^{\pm ik\varphi}, \qquad k \neq 0, \\ u_0(x,\varphi) &= A + B\ln\frac{1+x}{1-x}, \end{aligned} \tag{2.3}$$

where k is an integer. So, one can see that if a solution to equation (2.1) behaves as $(1+x)^k$ at one of the vertices $x=-1$ of the spindle, then this solution necessarily behaves as $(1-x)^{-k}$ at the other vertex $x=1$. This fact allows one, in particular, to investigate the correct statements of the problem for the operator

$$\widehat{a}_1 = \left((1-x^2)\frac{\partial}{\partial x}\right)^2 + \frac{\partial^2}{\partial\varphi^2} \;:\; H^s_\gamma(S) \to H^s_\gamma(S) \tag{2.4}$$

(to be short, we had omitted the factor $(1-x^2)^{-2}$ in the expression (2.2) for the operator $\widehat{a}_1$) considered in the weighted Sobolev spaces $H^s_\gamma(S)$, $\gamma = (\gamma_0,\gamma_1)$. We recall that the latter spaces are defined with the help of the norm

$$\|u\|^2_{s,\gamma} = \int_{-1}^{1}\int_{0}^{2\pi}(1+x)^{-2\gamma_0}(1-x)^{-2\gamma_1}\left|(1-\widehat{a}_1)^{s/2}u(x,\varphi)\right|^2 d\varphi\,\frac{dx}{(1-x^2)}.$$

Here γ_0 and γ_1 are weights at the points $x=-1$ and $x=1$, respectively.

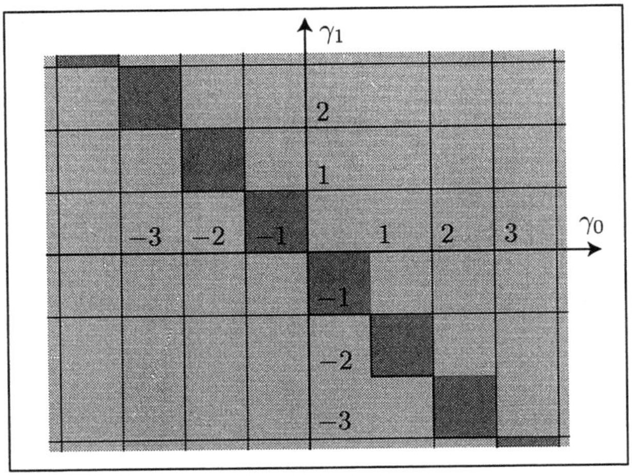

Figure 2: Kernel, cokernel, and the isomorphism region of the operator $\widehat{a}$.

One of the requirements for the operator (2.4) to be an isomorphism is that the kernel of this operator has to vanish. On the other hand, the functions $u_k^{\pm}(x,\varphi)$ given by (2.3) belong to the spaces $H_\gamma^s(S)$ for $\gamma_0 < k$, $\gamma_1 < -k$ (and arbitrary values of s). From this fact it follows that for the operator (2.4) to have trivial kernel it is necessary to require that the interval $(\gamma_0, -\gamma_1)$ does not contain any integer k. Actually, if we fix the value γ_0, then all elements $u_k^{\pm}(x,\varphi)$ of the kernel with $k > \gamma_0$ will belong to the space $H_\gamma^s(S)$ at the left vertex $x = -1$ of the spindle S. Later on, as it was already mentioned, the behavior of the solution $u_k^{\pm}(x,\varphi)$ at the left vertex prescribes the behavior of this solution at the right vertex. Namely, if the solution is of order $(1+x)^k$ at $x = -1$, then it is of order $(1-x)^{-k}$ at $x = 1$. Hence, for operator the (2.4) to have trivial kernel, it is necessary to require that all elements of its kernel $u_k^{\pm}(x,\varphi)$ which belong to the space $H_\gamma^s(S)$ at the left vertex $x = -1$ do not belong to this space at the right vertex $x = 1$ of the spindle S. This means that any value of k subject to the inequality $k > \gamma_0$ has to satisfy the condition $k \geq -\gamma_1$, because the latter inequality is equivalent to the fact that the function $u_k^{\pm}(x,\varphi)$ does not belong to the space $H_\gamma^s(S)$ at $x = 1$. So, the domain in the (γ_0, γ_1) plane where the operator $\widehat{a}$ has zero kernel is such as it is shown in Figure 2 (the upper hatched region).

To investigate the cokernel of this operator we remark that the adjoint operator[1]

$$\widehat{a}^* : H_{-\gamma}^{-s+m}(S) \to H_{-\gamma}^{-s}(S)$$

is given by the same expression. So, the domain in the (γ_0, γ_1) plane where the

[1] with respect to the pairing $\langle u, v\rangle = \int_{-1}^{1}\int_{0}^{2\pi} u(x,\varphi)\, v(x,\varphi)\, d\varphi \frac{dx}{(1-x^2)}$.

operator (2.4) has zero cokernel is of the form shown in Figure 2 (the lower hatched region).

Combining the two obtained results one can construct the *isomorphism* region for the considered operator which is crosshatched in Figure 2.

The above analysis shows that the dependence between the asymptotic behavior of one and the same solution at different points of singularity of the underlying manifold affects the correct statements of the problems of the corresponding differential operator. This paper is aimed at the investigation of this dependence.

Further, we are mostly interested in the investigation of partial differential equations on manifolds with singularities of the wedge type (see [1]). Since all these singularities can be obtained with the help of the operation of constructing a cone over a manifold and taking the direct product with a smooth manifold, it suffices to investigate the problem stated above on manifolds with conical singularities. For simplicity, we shall consider here singularities of the type of circular cones in different dimensions though the methods developed in this paper seem to be applicable to conical singularities of arbitrary type.

2.2. Propagation of singularities. Metamorphosis

Here, by simple examples, we illustrate that the investigation of the above stated problem requires the *analytic continuation* of the differential equation in question into the complex space.

1. In this example, we shall show that the singularities of solutions, being located at points of singularities of the underlying manifold in the real domain, propagate to the complex domain along the degeneration set of the analytic continuation of the considered differential operator. To do this, let us consider the Laplace equation on the surface of the two-dimensional cone C:

$$\frac{1}{r}\frac{\partial}{\partial r}\left(r\frac{\partial u}{\partial r}\right)+\frac{c^2}{r^2}\frac{\partial^2 u}{\partial \varphi^2}=0. \tag{2.5}$$

Here (r,φ) are polar coordinates on the surface of the cone, and c is a constant determined by the opening angle of the cone.

It is easy to see that the full system of solutions to equation (2.5) is

$$u_k^{\pm}(r,\varphi)=r^{\pm ck}e^{ik\varphi}, \qquad k\in\mathbb{Z}. \tag{2.6}$$

The latter formula shows that these solutions have singularities on the set

$$r^2=x^2+y^2=0$$

in the complexification of the cone C; here we use the coordinates

$$(x,y)=(r\cos\varphi, r\sin\varphi)\in\mathbb{C}^2.$$

Certainly, this observation is based on the fact that for our simple example we have written down the solutions to the homogeneous equation in explicit form. Since for general equations on manifolds with singularities this is not possible, one has to understand the reason of appearence of the singularities in terms of the considered operators. To do this, we note that, due to the relations

$$r\frac{\partial}{\partial r} = z\frac{\partial}{\partial z} + \zeta\frac{\partial}{\partial \zeta}, \; \frac{\partial}{\partial \varphi} = i\left(z\frac{\partial}{\partial z} - \zeta\frac{\partial}{\partial \zeta}\right), \tag{2.7}$$

$z = x + iy$, $\zeta = x - iy$, equation (2.5) can be rewritten in the form

$$\frac{1}{z\zeta}\left\{\left(1-c^2\right)\left(z\frac{\partial}{\partial z}\right)^2 + 2\left(1+c^2\right)z\zeta\frac{\partial^2}{\partial z \partial \zeta} + \left(1-c^2\right)\left(\zeta\frac{\partial}{\partial \zeta}\right)^2\right\}u = 0. \tag{2.8}$$

We remark that the set of singularities $x^2 + y^2 = 0$ of solutions to equation (2.5) exactly coincides with the degeneration set of this equation. Clearly, this fact is not accidental, and we shall see below that the *singularities of solutions propagate from singular points of the manifold into the complex domain along the degeneration set of the considered equation.* Hence, one can imagine that if the equation has more than one point of singularity, then the asymptotic expansion comes from one of these point to another *through the complex domain along the degeneration set of the corresponding equation.* This guess can be confirmed with the help of the following example which is a slight modification of the example considered above.

2. Consider the equation

$$\frac{1+r^2}{r^2}\left[\left(r\frac{\partial}{\partial r}\right)^2 + \frac{\partial^2}{\partial \varphi^2}\right]u = 0 \tag{2.9}$$

as an equation on the surface of the spindle. In this case one of the vertices of the spindle corresponds to the origin in the plane (x, y), and the other corresponds to infinity. We remark that the variable change $\rho = 1/r$ does not change the form of equation (2.9). In this example, the singular points $r = 0$ and $r = \infty$ are not connected by the set of singularities of the solution on the (real) spindle, but are lying on one and the same connected set $\{z = 0\} \cup \{\zeta = 0\}$ of singularities of the solution in the complexification of this spindle. Such a situation is displayed schematically in Figure 3. The form of the asymptotic expansion at one of the (real) points of singularities can be found out by that at the other one if the propagation of a singularity along the degeneration set (more exactly, along the regular part of the degeneration set) of the equation under consideration can be computed. The method of this computation will be described below.

3. However, for more general equations the situation can be quite different. Namely, in the latter example the degeneration set of solutions to the homogeneous equation was a *regular* complex manifold apart from the real singular points of M.

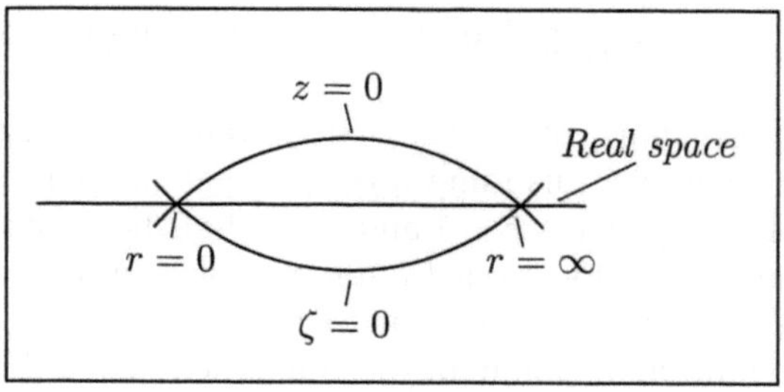

Figure 3: Degeneration set (simple configuration).

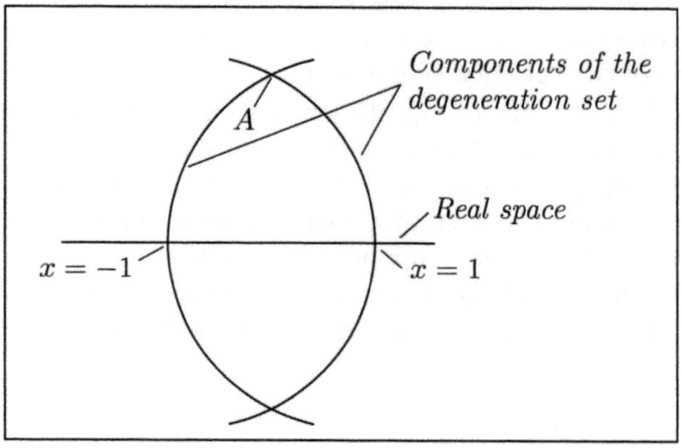

Figure 4: Degeneration set (general case).

To compute the asymptotic expansion at one of the singular points via that at the other one it was sufficient to examine the propagation of singularities along this regular part. In general, however, the regular part of the degeneration set of the equation (and, hence, the regular part of the set of singularities of solutions) can split into different connected components, and points of singularity of the manifold M can lie on different connected components. To illustrate that this situation can really take place, let us consider once again equation (2.2) on the surface of the spindle. The degeneration set for this equation can be decomposed into two irreducible components $x = 1$ and $x = -1$. Consequently, the regular part of this degeneration set splits into the (disjoint) union of two connected components (these two components intersect each other at infinity; this fact is clearly quite accidental). It is clear that one of the two points of singularity of the spindle lies on one of these components, and the other point of singularity lies on the other one; this situation is shown schematically in Figure 4.

Hence, if one knows the asymptotic expansion at one of the vertices, say, $x = 1$, then in order to compute the corresponding asymptotic expansion at the other vertex one has

1) to investigate the propagation of singularity along the component $x = 1$ of the degeneration set;

2) to examine the "jump" of the singularity from one of the connected components to the other. Clearly, this jump takes place at the point A of intersection of the two irreducible components $x = 1$ and $x = -1$ of the degeneration set (see Figure 4). So, to examine the mentioned jump one has to investigate the asymptotic expansion of the solution near the point A and then to find out the asymptotic expansion of the solution at points of the component $x = -1$ near the point A;

3) to investigate the propagation of the singularity along the component $x = -1$ from A to the vertex $x = -1$.

Thus, in general, the following two problems arise:

First, to investigate the propagation of singularities along regular parts of degeneration sets of the equation in question.

Secondly, to find out the asymptotic expansion of a singularity of the solution on one of the components of its singularity set provided that the corresponding asymptotic expansion is known on the other component. This problem can be solved by investigating the asymptotic expansion of solutions in a neighborhood of the *intersection* of the two components in question.

In the remaining part of the paper we shall consider both problems. To make the presentation more clear, we shall first consider the two-dimensional case.

3. General statement

The general statement of the problem is as follows:

Let M be an n-dimensional real-analytic manifold which is the real part of a complex-analytic manifold $M_{\mathbb{C}}$. Suppose that the manifold M has a finite number of singular points $m_1, \dots, m_k$ each being a point of conical type. This means that in a neighborhood U_j of each point m_j, $j = 1, \dots, k$, the manifold M may be identified with the cone

$$[0,1) \times S^{n-1} / \{0\} \times S^{n-1} \tag{3.1}$$

(diffeomorphically outside the vertex), where S^{n-1} is the $(n-1)$-dimensional unit sphere in the Euclidean space $\mathbb{R}^n$. So, near each point m_j there exists a "coordinate system" (r, ω), $r \in (0,1)$, $\omega \in S^{n-1}$. We suppose that this coordinate system is compatible with the complex extension $M_{\mathbb{C}}$ of the manifold M, that is, that (r, ω), $r \in D_1$, $\omega \in Q^{n-1}$ forms a coordinate system in a neighborhood of the point $m_j \in M_{\mathbb{C}}$. Here by Q^{n-1} we denote the complex quadrics

$$Q^{n-1} = \left\{ x \in \mathbb{C}^n : \sum_{j=1}^{n} \left(x^j\right)^2 = 1 \right\},$$

and by D_1 we denote the unit disk

$$D_1 = \{r \in \mathbb{C} : |r| < 1\}$$

in the complex plane $\mathbb{C}$. So, we suppose that the complex manifold $M_{\mathbb{C}}$ is biholomorphic to the complex cone

$$D_1 \times Q^{n-1} / \{0\} \times Q^{n-1}.$$

Consider an elliptic differential operator $\widehat{a}$ of order m on the manifold M. This means, in particular, that this operator has the form

$$\widehat{a} = \sum_{|\alpha| \leq m} a_\alpha (x) \left(\frac{\partial}{\partial x}\right)^\alpha \tag{3.2}$$

near each regular point of M and the form

$$\widehat{a} = \sum_{j=1}^{m} \widehat{b}_j \left(r\frac{\partial}{\partial r}\right)^j \tag{3.3}$$

near each singular point m_j, where $\widehat{b}_j$ are differential operators on the complex quadrics Q^{n-1} (we use here the above mentioned representation (3.1) of the manifold M). Suppose that the operator $\widehat{a}$ can be continued analytically on the manifold $M_{\mathbb{C}}$; the continuation is given by the same formulas (3.2) and (3.3), where $a_\alpha (x)$ are now holomorphic functions of the variable $x \in \mathbb{C}^n$, and $\widehat{b}_j$ are differential operators on the complex quadrics Q^{n-1} with holomorphic coefficients.

Now the problem is *to investigate the asymptotic expansion of analytic continuations of solutions to the homogeneous equation*

$$\widehat{a}u = 0 \tag{3.4}$$

on the complex manifold $M_{\mathbb{C}}$ near the degeneration set of the operator $\widehat{a}$.

In particular, we have to investigate the propagation of singularities along the degeneration set of the operator and the "metamorphosis" of the singularity which takes place at points of intersection of different components of the degeneration set.

4. Two-dimensional case

4.1. Propagation of singularities

It is well-known (see [1], [3]) that, under the above conditions, the asymptotics of solutions to (3.4) near each singular point of the manifold M has the form

$$u (r, \varphi) \simeq \sum_k r^{S_k} \sum_{j=0}^{m_k} a_{kj} (\varphi) \ln^j r$$

(conormal asymptotics), where r is a radial variable and φ is the angular variable along the base of the cone and the outer sum is taken over the set of values of $s = S_k$ having a finite intersection with any half-plane $\operatorname{Re} s < A$ for any real value of A.

In this subsection, we shall investigate how asymptotics of the conormal type propagate along the degeneration set of equation (3.4). Since in the two-dimensional case the sphere S^{n-1} is simply a circle, we shall use the notation φ instead of ω for the coordinate on this circle. Hence, equation (3.4) can be rewritten in the form

$$\widehat{a}u = a\left(r, \varphi, r\frac{\partial}{\partial r}, \frac{\partial}{\partial \varphi}\right) u = 0.$$

This equation can be considered as an equation on the real two-dimensional plane $\mathbb{R}^2$ with the coordinates x, y,

$$x = r\cos\varphi, \qquad y = r\sin\varphi$$

(this equation degenerates at the origin, as it ought to be). As above, we suppose that the coefficients of the obtained equation can be continued analytically on the complex domain.

To describe the degeneration set of the considered equation, it is convenient to introduce the complex variables $z = x + iy$, $\zeta = x - iy$. Using the relations (2.7) one can rewrite the equation in the form

$$a_1\left(z, \zeta, z\frac{\partial}{\partial z}, \zeta\frac{\partial}{\partial \zeta}\right) u = 0 \tag{4.1}$$

with some symbol $a_1(z, \zeta, p, q)$. The reader can notice that the latter equation is an equation of Fuchs type in the variables (z, ζ) with the degeneration set $\{z = 0\} \cup \{\zeta = 0\}$. Hence, the singularities of solutions to this equation belong to its degeneration set (see [4], [5]).

To be definite, let us consider the equation near the set $z = 0$; in this case z is a small complex variable and ζ is separated from zero. Thus, it is useful to rewrite the equation as[2]

$$a\left(z, \zeta, z\frac{\partial}{\partial z}, \frac{\partial}{\partial \zeta}\right) u = 0. \tag{4.2}$$

For example, equation (2.5) has the following form in the variables (x, y):

$$\left(x^2 + c^2y^2\right)\frac{\partial^2 u}{\partial x^2} + 2\left(1 - c^2\right)\frac{\partial^2 u}{\partial x \partial y} + \left(c^2x^2 + y^2\right)\frac{\partial^2 u}{\partial y^2} + \left(1 - c^2\right)\left(x\frac{\partial u}{\partial x} + y\frac{\partial u}{\partial y}\right) = 0.$$

In the variables (z, ζ) we obtain

$$\left\{\left(1 - c^2\right)\left(z\frac{\partial}{\partial z}\right)^2 + 2\left(1 + c^2\right)\left(z\frac{\partial}{\partial z}\right)\left(\zeta\frac{\partial}{\partial \zeta}\right) + \left(1 - c^2\right)\left(\zeta\frac{\partial}{\partial \zeta}\right)^2\right\} u = 0.$$

---

[2]In what follows we omit subscripts of symbols. So, one and the same letter a can denote different symbols.

If one examines this equation near the set $z = 0$ (under the assumption that ζ is not small), the equation can be rewritten as

$$\left\{\left(1-c^2\right)\left(z\frac{\partial}{\partial z}\right)^2 + 2\zeta\left(1+c^2\right)\left(z\frac{\partial}{\partial z}\right)\left(\frac{\partial}{\partial \zeta}\right) + \zeta^2\left(1-c^2\right)\left(\frac{\partial}{\partial \zeta}\right)^2 + \zeta\left(1-c^2\right)\left(\frac{\partial}{\partial \zeta}\right)\right\}u = 0.$$

Let us search for solutions to equation (4.2) of the form

$$u(z,\zeta) = z^S \sum_{j=0}^{k} u_j(z,\zeta)\ln^j z, \tag{4.3}$$

where $u_j(z,\zeta)$ are regular functions of z near the origin:

$$u_j(z,\zeta) = \sum_{l=0}^{\infty} u_{jl}(\zeta)\, z^l. \tag{4.4}$$

Taking into account the relations

$$z\frac{\partial}{\partial z}u(z,\zeta) = Sz^S\sum_{j=0}^{k}u_j(z,\zeta)\ln^j z + z^S\sum_{j=0}^{k}\left(z\frac{\partial}{\partial z}u_j(z,\zeta)\right)\ln^j z + \sum_{j=1}^{k} j u_j(z,\zeta)\ln^{j-1} z,$$

one can write down the result of the substitution of (4.3) into (4.2) in the form

$$\begin{aligned} a\left(z,\zeta,S+z\frac{\partial}{\partial z},\frac{\partial}{\partial \zeta}\right)u_k(z,\zeta) &= 0, \\ a\left(z,\zeta,S+z\frac{\partial}{\partial z},\frac{\partial}{\partial \zeta}\right)u_j(z,\zeta) &= \sum_{i=j+1}^{k} a_i\left(z,\zeta,z\frac{\partial}{\partial z},\frac{\partial}{\partial \zeta}\right)u_i(z,\zeta), \\ & \qquad j = k-1,\dots,0. \end{aligned} \tag{4.5}$$

Here $a_i(z,\zeta,z\partial/\partial z,\partial/\partial\zeta)$ are differential operators of order $m+j-k$ with coefficients holomorphic in (z,ζ).

Let us consider the first of the equations (4.5). Expanding the coefficients of the operator $a(z,\zeta,S+z\partial/\partial z,\partial/\partial\zeta)$ into a Taylor series in z and substituting (4.4) for $u_k(z,\zeta)$, we arrive at the following recurrent system of equations for $u_{kl}(\zeta)$:

$$\begin{aligned} a_0'\left(\zeta,S,\frac{\partial}{\partial \zeta}\right)u_{k0}(\zeta) &= 0, \\ a_0'\left(\zeta,S+l,\frac{\partial}{\partial \zeta}\right)u_{kl}(\zeta) &= -\sum_{j=0}^{l-1} a_{l-j}'\left(\zeta,S+j,\frac{\partial}{\partial \zeta}\right)u_{kj}(\zeta), \\ & \qquad l = 1,2,\dots. \end{aligned} \tag{4.6}$$

Here $a'_j(\zeta,p,q)$ are the Taylor coefficients of the symbol $a(z,\zeta,p,q)$ of the operator $a(z,\zeta,z\partial/\partial z,\zeta\partial/\partial\zeta)$:

$$a(z,\zeta,p,q)=\sum_{j=0}^{\infty} z^j a'_j(\zeta,p,q).$$

In particular,

$$a'_0(\zeta,p,q)=a(0,\zeta,p,q).$$

The recurrent system of the equations for the Taylor coefficients of the functions $u_j(z,\zeta)$ can be obtained in a quite similar way as system (4.6). Thus, one can see that in order to construct the asymptotic expansion of the form (4.2) to equation (4.3) it is sufficient to solve an ordinary differential equation for each coefficient of this expansion.

4.2. Asymptotics near the intersection

In this subsection we investigate the asymptotic expansion of solutions to equation (3.4) near a point $m_0\in M_{\mathbb{C}}$ of intersection of different components of the degeneration set. We require that the following condition is valid:

Condition 4.1. *The considered point m_0 is a (proper) point of transversal intersection of two regular components of the degeneration set of the equation in question.*

Under this condition, it is clear that the equation can be rewritten in the form (4.1) near the point m_0. However, unlike the previous subsection, one cannot neglect the factor ζ in the operator $\zeta\partial/\partial\zeta$, since at the point m_0 one has $z=\zeta=0$.

Let us show that the asymptotic expansion of a solution to equation (4.1) has a specific form in a neighborhood of the point m_0.

Lemma 4.2. *Let $u(z,\zeta)$ be a solution to equation (4.1) near $z=\zeta=0$ having the form (4.3) of conormal asymptotics apart from the origin. Then this solution has an asymptotic expansion of the form*

$$u(z,\zeta)=z^{S_1}\zeta^{S_2}\sum_{j=0}^{k_1}\sum_{l=0}^{k_2} u_{jl}(z,\zeta)\ln^j z\,\ln^l\zeta \tag{4.7}$$

with $u_{jl}(z,\zeta)$ regular near m_0.

Proof. For simplicity, we shall consider the case when both multiplicities are equal to 1. So, for the solution $u(z,\zeta)$ we have the following asymptotic expansions:

$$\begin{aligned} u(z,\zeta) &= z^{S_1}\left[u_1(z,\zeta)\ln z+u_0(z,\zeta)\right] \quad \text{for } z\to 0,\ |\zeta|>\varepsilon, \\ u(z,\zeta) &= \zeta^{S_2}\left[u'_1(z,\zeta)\ln\zeta+u'_0(z,\zeta)\right] \quad \text{for } \zeta\to 0,\ |z|>\varepsilon \end{aligned} \tag{4.8}$$

(for any positive ε) with regular functions $u_j(z,\zeta)$, $u'_j(z,\zeta)$, $j=1,2$. Denote

$$v(z,\zeta) = z^{-S_1}\zeta^{-S_2}u(z,\zeta).$$

This function has the form

$$\begin{aligned} v(z,\zeta) &= u_1(z,\zeta)\ln z + u_0(z,\zeta) \quad \text{for } z\to 0,\ |\zeta|>\varepsilon,\\ v(z,\zeta) &= u'_1(z,\zeta)\ln\zeta + u'_0(z,\zeta) \quad \text{for } \zeta\to 0,\ |z|>\varepsilon. \end{aligned}$$

Hence the function

$$v_1(z,\zeta) = v(z,\zeta) - u_1(z,\zeta)\ln z - u'_1(z,\zeta)\ln\zeta$$

possesses the following properties:

1) It is univalent both around the manifold $z=0$ and $\zeta=0$.

2) It has at most logarithmic growth as $z\to 0$ and $\zeta\to 0$.

Hence, $v_1(z,\zeta)$ is a holomorphic function in a deleted neighborhood of the point m_0. Using the theorem on removable singularity, we obtain that this function is regular at the point m_0 as well. This proves the lemma. □

Now let us search for solutions to equation (4.1) in the form (4.7). Taking into account the relations

$$\begin{aligned} z\frac{\partial u}{\partial z} &= z^{S_1}\zeta^{S_2}\left\{\sum_{j=0}^{k_1}\sum_{l=0}^{k_2}\left[\left(S_1+z\frac{\partial}{\partial z}\right)u_{jl}(z,\zeta)\right]\ln^j z\,\ln^l\zeta\right.\\ &\qquad\left.+\sum_{j=1}^{k_1}\sum_{l=0}^{k_2} j u_{jl}(z,\zeta)\ln^{j-1}z\,\ln^l\zeta\right\},\\ \zeta\frac{\partial u}{\partial\zeta} &= z^{S_1}\zeta^{S_2}\left\{\sum_{j=0}^{k_1}\sum_{l=0}^{k_2}\left[\left(S_2+\zeta\frac{\partial}{\partial\zeta}\right)u_{jl}(z,\zeta)\right]\ln^j z\,\ln^l\zeta\right.\\ &\qquad\left.+\sum_{j=0}^{k_1}\sum_{l=1}^{k_2} l u_{jl}(z,\zeta)\ln^j z\,\ln^{l-1}\zeta\right\}, \end{aligned}$$

we arrive at the following recurrent system of equations for the unknown coefficients $u_{jl}(z,\zeta)$ of the asymptotic expansion (4.7):

$$\begin{aligned} &a\left(z,\zeta,S_1+z\frac{\partial}{\partial z},S_2+\zeta\frac{\partial}{\partial\zeta}\right)u_{k_1k_2}(z,\zeta)=0,\\ &a\left(z,\zeta,S_1+z\frac{\partial}{\partial z},S_2+\zeta\frac{\partial}{\partial\zeta}\right)u_{jl}(z,\zeta)=\sum a\left(z,\zeta,z\frac{\partial}{\partial z},\zeta\frac{\partial}{\partial\zeta}\right)u_{j'l'}(z,\zeta) \end{aligned} \tag{4.9}$$

where $a_{j'l'}(z, \zeta, z\partial/\partial z, \zeta\partial/\partial\zeta)$ are differential operators of order $m+j+l-j'-l'$ and the summation is taken over all indices j', l' such that $j' \geq j$, $l' \geq l$, $(j', l') \neq (j, l)$. Let us consider the first of the equations (4.9). We shall search for a solution to this equation in the form of the Taylor series

$$u_{k_1k_2}(z, \zeta) = \sum_{\alpha,\beta=0}^{\infty} u_{k_1k_2}^{\alpha\beta} z^\alpha \zeta^\beta.$$

Substituting this relation into the first of the equations (4.9) and expanding the coefficients of the operator $a(z, \zeta, S_1 + z\partial/\partial z, S_2 + \zeta\partial/\partial\zeta)$ in powers of z and ζ, we obtain

$$\begin{aligned} &a'_{00}(S_1, S_2)\, u_{k_1k_2}^{00} = 0, \\ &a'_{00}(S_1+\alpha, S_2+\beta)\, u_{k_1k_2}^{\alpha\beta} = -\sum_{(\gamma,\delta)<(\alpha,\beta)} a'_{\alpha-\gamma,\beta-\delta}(S_1+\gamma, S_2+\delta)\, u_{k_1k_2}^{\gamma\delta}. \end{aligned} \tag{4.10}$$

Here $a'_{\gamma\delta}(p, q)$ are the Taylor coefficients of the symbol $H(z, \zeta, p, q)$ in powers of z and ζ,

$$a(z, \zeta, p, q) = \sum_{|(\gamma,\delta)|\leq m} a'_{\gamma\delta}(p, q)\, z^\gamma \zeta^\delta.$$

So, there exists a nontrivial solution to equation (4.1) of the form (4.7) only in the case when the numbers S_1 and S_2 satisfy the relation

$$a'_{00}(S_1, S_2) = 0. \tag{4.11}$$

The latter equation is a homogeneous algebraic equation with respect to (S_1, S_2) of order m. Using this equation one can find the possible values of the ratio S_1/S_2 necessary for equation (4.1) to possess a solution of the form (4.7). These values allow to compute the power of ζ in the asymptotic expansions (4.8) if the corresponding power of z is known.

The remaining equation in (4.10) can be solved in a quite similar manner if the following nondegeneracy condition is valid:

Condition 4.3. *The pair* $(S_1 + \alpha, S_2 + \beta)$ *is not a solution to equation* (4.11) *except for* $(\alpha, \beta) = 0$.

We remark that the latter condition takes place in the generic position.

Similar considerations can also be used for computing asymptotic expansions of solutions to differential equations near singular points of the manifold. We shall illustrate this by the example of the equation (2.8):

$$\frac{1}{z\zeta}\left\{(1-c^2)\left(z\frac{\partial}{\partial z}\right)^2 + 2(1+c^2)\, z\zeta\frac{\partial^2}{\partial z\partial\zeta} + (1-c^2)\left(\zeta\frac{\partial}{\partial\zeta}\right)^2\right\}u = 0.$$

Searching for solutions to this equation in the form

$$u = z^{\alpha}\zeta^{\beta}$$

we are led to the following equation for (α, β):

$$\left(1-c^2\right)\alpha^2 + 2\left(1+c^2\right)\alpha\beta + \left(1-c^2\right)\beta^2 = 0.$$

The ratio α/β can be computed from this equation:

$$\frac{\alpha}{\beta} = -\frac{1+c}{1-c} \quad \text{or} \quad \frac{\alpha}{\beta} = -\frac{1-c}{1+c}.$$

Taking into account that we are searching for solutions to (2.8) which are univalent on the real space $z = \overline{\zeta}$, we obtain the second equation for determining the numbers α and β,

$$\alpha - \beta = k \in \mathbb{Z}.$$

The last two equations give the possible values of α and β:

$$\alpha = \frac{(c+1)\,k}{2}, \qquad \beta = \frac{(c-1)\,k}{2},$$

or

$$\alpha = -\frac{(c-1)\,k}{2}, \qquad \beta = -\frac{(c+1)\,k}{2}.$$

The reader can easily verify that all the obtained values of (α, β) really determine univalent solutions to equation (2.8). Actually, the corresponding solutions to the homogeneous equation are

$$\begin{aligned} z^{\frac{(c+1)k}{2}}\zeta^{\frac{(c-1)k}{2}} &= r^{ck}e^{ik\varphi} \\ z^{-\frac{(c-1)k}{2}}\zeta^{-\frac{(c+1)k}{2}} &= r^{-ck}e^{ik\varphi}. \end{aligned}$$

The latter expressions correspond to formula (2.6) above.

5. Multi-dimensional case

In this section, we consider asymptotic expansions of solutions to the homogeneous equation (2.1) for some operator of cone type (3.3) on an n-dimensional manifold M with conical singularities. Similarly to the previous section, we divide the consideration into two parts: investigation of the propagation of singularities along regular parts of the degeneration set of the considered equation and investigation of the "metamorphosis" of the singularity on the intersection points of these regular parts.

5.1. Propagation of singularities

Considerations similar to those of Subsection 4.1 show that the complexification of the equation of corner type in the complex domain in a neighborhood of any point from the regular part of its degeneration set reads

$$a\left(z,\zeta,z\frac{\partial}{\partial z},\frac{\partial}{\partial\zeta}\right)u=0, \tag{5.1}$$

where $z\in\mathbb{C}$ is a one-dimensional complex variable transversal to the degeneration set at the considered point and $\zeta=(\zeta_1,\dots,\zeta_{n-1})\in\mathbb{C}^{n-1}$ are coordinates on the degeneration set itself.

Let us search for solutions to this equation in the form

$$u(z,\zeta)=z^S\sum_{j=0}^{k}u_j(z,\zeta)\ln^j z \tag{5.2}$$

with some regular functions $u_j(z,\zeta)$. As above, we search the functions $u_j(z,\zeta)$ in the form of the Taylor series in the variable z:

$$u_j(z,\zeta)=\sum_{l=0}^{\infty}z^l u_{jl}(\zeta). \tag{5.3}$$

Substituting the relations (5.2) and (5.3) into equation (5.1) and comparing the coefficients of powers of z, we arrive at the following recurrent system of equations for the Taylor coefficients $u_{kj}(\zeta)$ of the main term $u_k(z,\zeta)$ of expansion (5.2):

$$\begin{aligned}
a_0'\left(\zeta,S,\frac{\partial}{\partial\zeta}\right)u_{k0}(\zeta) &= 0,\\
a_0'\left(\zeta,S+l,\frac{\partial}{\partial\zeta}\right)u_{kl}(\zeta) &= -\sum_{j=0}^{l-1}a_{l-j}'\left(\zeta,S+j,\frac{\partial}{\partial\zeta}\right)u_{kj}(\zeta),\\
&\qquad l=1,2,\dots,
\end{aligned}$$

and similar recurrent systems for the Taylor coefficients of the remaining terms $u_j(z,\zeta)$, $j=0,\dots,k-1$. Here, as above, $a_j'(\zeta,p,q)$ are the Taylor coefficients of the full symbol $a(z,\zeta,p,q)$ in the variable z:

$$a(z,\zeta,p,q)=\sum_{j=0}^{\infty}z^j a_j'(\zeta,p,q).$$

We remark that in the multi-dimensional case the Taylor coefficients $u_{ij}(\zeta)$ are to be determined from a partial differential equation along the components of the degeneration set of equation (5.1).

5.2. Asymptotics near the intersection

Here we investigate the asymptotics of solutions *near the intersection* of different components of the degeneration set of the equation considered. The geometrical situation in this case is quite different from that in the two-dimensional case. The matter is that in the multi-dimensional case the intersection between different components of the degeneration set is a complex-analytic manifold of non-zero dimension whereas in the two-dimensional case this intersection is simply a discrete set of points.

So, let us consider two submanifolds X_1 and X_2 in the complexification $M_{\mathbb{C}}$ which are two irreducible components of the degeneration set of the operator $\widehat{a}$, and suppose that these two manifolds intersect each other transversally at points of their intersection $X_1 \cap X_2$. Then, similarly to the results of the previous section, one can rewrite the operator $\widehat{a}$ in the form

$$\widehat{a} = a\left(z, \zeta, \eta, z\frac{\partial}{\partial z}, \zeta\frac{\partial}{\partial \zeta}, \frac{\partial}{\partial \eta}\right),$$

where the coordinates (z, ζ, η) are chosen in such a way that $z = 0$ and $\zeta = 0$ are equations of the manifolds X_1 and X_2, respectively.

Suppose that $u(z, \zeta, \eta)$ is a solution to the homogeneous equation

$$a\left(z, \zeta, \eta, z\frac{\partial}{\partial z}, \zeta\frac{\partial}{\partial \zeta}, \frac{\partial}{\partial \eta}\right) u(z, \zeta, \eta) = 0 \tag{5.4}$$

such that $u(z, \zeta, \eta)$ has the asymptotic expansion of conormal type both near X_1 and X_2. This means that

$$u(z, \zeta, \eta) = z^{S_1} \sum_{j=0}^{k_1} u_j(z, \zeta, \eta) \ln^j z \tag{5.5}$$

with regular functions $u_j(z, \zeta, \eta)$ near X_1 apart from X_2 (that is, for $|z| < \varepsilon_1$ and $|\zeta| > \varepsilon_2$ for some positive ε_1 and ε_2) and

$$u(z, \zeta, \eta) = \zeta^{S_2} \sum_{j=0}^{k_2} v_j(z, \zeta, \eta) \ln^j \zeta \tag{5.6}$$

with regular functions $v_j(z, \zeta, \eta)$ near X_2 apart from X_1. Then, similarly to the result of Subsection 4.2, the following result holds:

Lemma 5.1. *Let $u(z, \zeta, \eta)$ be a solution to equation* (5.4) *having the asymptotic expansions* (5.5) *and* (5.6) *at points of X_1 and X_2, respectively. Then this solution has an asymptotic expansion of the form*

$$u(z, \zeta, \eta) = z^{S_1} \zeta^{S_2} \sum_{j=0}^{k_1} \sum_{l=0}^{k_2} u_{jl}(z, \zeta, \eta) \ln^j z \ln^j \zeta \tag{5.7}$$

with regular functions $u_{jl}(z,\zeta,\eta)$ near the intersection $X_1 \cap X_2$, that is, at the points with $z = \zeta = 0$.

The proof of this lemma is quite similar to that of Lemma 4.2 in Subsection 4.2.

To establish the connection between the numbers S_1 and S_2 let us search for a solution to equation (5.4) in the form (5.7). The substitution of the expansion (5.7) into equation (5.4) follows the lines of Subsection 4.2. The result is

$$a\left(z,\zeta,\eta,S_1+z\frac{\partial}{\partial z},S_2+\zeta\frac{\partial}{\partial\zeta},\frac{\partial}{\partial\eta}\right)u_{k_1k_2}(z,\zeta,\eta)=0,$$

$$a\left(z,\zeta,\eta,S_1+z\frac{\partial}{\partial z},S_2+\zeta\frac{\partial}{\partial\zeta},\frac{\partial}{\partial\eta}\right)u_{jl}(z,\zeta,\eta)$$

$$=\sum_{j'\geq j,l'\geq l,(j',l')\neq(j,l)}a_{j'l'}\left(z,\zeta,\eta,z\frac{\partial}{\partial z},\zeta\frac{\partial}{\partial\zeta},\frac{\partial}{\partial\eta}\right)u_{j'l'}(z,\zeta,\eta)$$

where $a_{j'l'}(z,\zeta,\eta,z\partial/\partial z,\zeta\partial/\partial\zeta,\partial/\partial\eta)$ are differential operators of order $m+j+l-j'-l'$. Similarly to the two-dimensional case we can construct the recurrent system of equations for the Taylor coefficients $u_{jl}^{\alpha\beta}(\eta)$ of the function $u_{jl}(z,\zeta,\eta)$ in (z,ζ):

$$u_{jl}(z,\zeta,\eta)=\sum_{\alpha,\beta=0}^{\infty}u_{jl}^{\alpha\beta}(\eta)\,z^\alpha\zeta^\beta.$$

For the main term $u_{k_1k_2}(z,\zeta,\eta)$ this system reads

$$a'_{00}\left(\eta,S_1,S_2,\frac{\partial}{\partial\eta}\right)u_{k_1k_2}^{00}(\eta)=0,$$

$$a'_{00}\left(\eta,S_1+\alpha,S_2+\beta,\frac{\partial}{\partial\eta}\right)u_{k_1k_2}^{\alpha\beta} \tag{5.8}$$

$$=-\sum_{(\gamma,\delta)<(\alpha,\beta)}a'_{\alpha-\gamma,\beta-\delta}\left(\eta,S_1+\gamma,S_2+\delta,\frac{\partial}{\partial\eta}\right)u_{k_1k_2}^{\gamma\delta}(\eta),$$

where, as above, $a'_{\gamma\delta}(\eta,p_z,p_\zeta,p_\eta)$ are the Taylor coefficients of the symbol a in powers of z and ζ:

$$a(z,\zeta,\eta,p_z,p_\zeta,p_\eta)=\sum_{|(\gamma,\delta)|\leq m}a'_{\gamma\delta}(\eta,p_z,p_\zeta,p_\eta)\,z^\gamma\zeta^\delta.$$

Clearly, similar systems can be derived for all coefficients $u_{jl}^{\alpha\beta}(\eta)$.

Thus, for the coefficients $u_{jl}^{\alpha\beta}(\eta)$ we have obtained a recurrent system of differential equations of the form (5.8). Since the first equation in this system (see (5.8)) is homogeneous (its right-hand part vanishes), for this system to admit nontrivial solutions it is necessary that the homogeneous equation

$$a'_{00}\left(\eta, S_1, S_2, \frac{\partial}{\partial \eta}\right) u = 0$$

on the manifold $X_1 \cap X_2$ has a non-zero univalent solution for the given values of S_1, S_2. This is exactly the condition for determining the connection between the values of S_1 and S_2 on the two components X_1 and X_2 of the degeneration set of the considered equation.

Remark 5.2. This condition can be formulated in more explicit terms if there exists a real-type compact submanifold X of the intersection $X_1 \cap X_2$ such that the operator expression $a'_{00}(\eta, S_1, S_2, \partial/\partial\eta)$ determines an elliptic analytic operator family on X with parameters S_1 and S_2. Denote by $\sum$ the set in the plane $\mathbb{C}_2$ with coordinates (S_1, S_2) such that the operator family $a'_{00}(\eta, S_1, S_2, \partial/\partial\eta)$ is invertible outside $\sum$. Then the connection between S_1 and S_2 is described by the inclusion $(S_1, S_2) \in \sum$.

References

[3] KONDRAT'EV, V.A.: Boundary problems for elliptic equations in domains with conical or angular points; Trans. Moscow Math. Soc. 16 (1967), 287–313.

[1] SCHULZE, B.-W.: Pseudo-Differential Operators on Manifolds with Singularities; North-Holland, Amsterdam 1991.

[2] SCHULZE, B.-W., STERNIN, B., SHATALOV, V.: Differential Equations on Manifolds with Singularities in Classes of Resurgent Functions; Max-Planck-Institut für Mathematik, Preprint MPI/95-88, Bonn 1995.

[4] STERNIN, B., SHATALOV, V.: Asymptotic solutions to Fuchsian equations in several variables; Max-Planck-Institut für Mathematik, Preprint MPI/94-124, Bonn 1994.

[5] STERNIN, B., SHATALOV, V.: Asymptotic solutions to Fuchsian equations in several variables; Proceedings of a Symposium on Singularities, Banach Center Publications, 1995, to appear.

Max-Planck-Arbeitsgruppe
"Partielle Differentialgleichungen und komplexe Analysis"
Universität Potsdam
14415 Potsdam
Germany
schulze@mpg-ana.uni-potsdam.de

Dept. of Comp. Math. and Cybern.
Moscow State University
Vorob'evy Gory
119899 Moscow
Russia

1991 Mathematics Subject Classification: Primary 30B99; Secondary 31C45, 42A16

Submitted: October 24, 1995

Operator Theory:
Advances and Applications, Vol. 102

Green's formula for elliptic operators with a shift and its applications

Z.G. Sheftel

General boundary value problems with a shift for elliptic equations were first studied in [1]. Such problems arise for instance when studying certain steady-state oscillations. In this paper we introduce the notion of normal boundary conditions with a shift and deduce the Green's formula for such boundary conditions and partial differential equations of even order. We obtain a number of applications of this formula. Namely, we introduce the notion of the adjoint problem and prove that the adjoint problem is elliptic if and only if the given problem is elliptic. We give solvability conditions for both the given and the adjoint problem in positive spaces of Sobolev type. They allow to prove isomorphism theorems (i.e., solvability theorems in complete scale of spaces) and theorems on the local increasing of smoothness. In addition we prove the existence and study the smoothness properties of the Green's function for the problem with a shift. We investigate also the approximation of functions on a manifold by solutions of the problem with a shift.

1. Elliptic problems with a shift and Green's formula

Let $G \subset \mathbb{R}^n$ be a bounded domain with boundary $\Gamma \in C^\infty$ and let $\alpha : \Gamma \to \Gamma$ be a diffeomorphism, where $\alpha(\alpha x) = x$ for any $x \in \Gamma$. Because of the smoothness of Γ the transformation α can be extended to a diffeomorphism of some neighborhood $U(\Gamma)$ in $\mathbb{R}^n$. The transformation α defines in a natural way the shift operator J in $U(\Gamma)$ which transforms any function u defined in $U(\Gamma)$ into the function Ju according to the formula $Ju(x) = u(\alpha x)$. Let

$$A(x,D) = \sum_{|\beta|\le l} a_\beta(x) D^\beta, \quad D^\beta = D_1^{\beta_1}\dots D_n^{\beta_n}, \quad D_k = i\partial/\partial x_k, \quad x \in U(\Gamma),$$

be an arbitrary linear differential expression with smooth coefficients, and let $A_0(x,\xi) = \sum_{|\beta|\le l} a_\beta(x)\xi^\beta$ be its characteristic polynomial. Then

$$JA(x,D_x)u(x) = \widehat{A}(\alpha x, D_{\alpha x})Ju(x),$$

where $\widehat{A}$ is a linear differential expression of the same order l having the characteristic polynomial

$$\widehat{A}_0(x,\xi) = A_0(\alpha x, T\xi), \tag{1.1}$$

where T is the transposed Jacobi matrix (i.e., the derivative) of the transformation α.

In G a linear differential expression $L(x,D)$ with sufficiently smooth complex-valued coefficients is assigned, ord $L = 2m$; on Γ linear differential expressions $B_{jr}(x,D)$, $j=1,\dots,2m$, $r=1,2$, with sufficiently smooth complex-valued coefficients are assigned, ord $B_{jr} \le 2m+\sigma_j$ ($\sigma_j < 0$ are given integers). We study the boundary value problem with a shift

$$Lu(x) = f(x), \qquad x \in \Gamma, \tag{1.2}$$

$$B_j u := B_{j1}u(x) + JB_{j2}u(x) = \phi_j(x), \qquad x \in \Gamma,\ j=1,\dots,2m. \tag{1.3}$$

In addition we assume that the expressions B_{jr} satisfy the following natural condition.

Matching condition. *The system of conditions* (1.3) *is invariant with respect to the substitution of x by αx. More precisely, after this substitution we obtain conditions equivalent to* (1.3).

Definition 1.1. The problem (1.2)-(1.3) is called *elliptic* [1] if the following assumptions (A) and (B) are satisfied:

(A) The expression L is properly elliptic in $\overline{G}$. It follows from here that for each $x \in \Gamma$, for any vector $\tau \neq 0$ tangential to Γ at the point x and for the unit normal ν to Γ at this point, the polynomial $L_0(x, \tau+\eta\nu)$ (with respect to η) has m roots in the upper and m roots in the lower half-plane. Hence

$$L_0(x,\tau+\eta\nu) = L^+(x,\tau+\eta\nu)L^-(x,\tau+\eta\nu),$$

where the η-roots of L^+ (L^-) have positive (negative) imaginary parts.

(B) Let us denote by Q_{jr} the principal part of the expression B_{jr} which consists only of the terms of order $2m+\sigma_j$. We assume that at each point $x \in \Gamma$ the pairs of polynomials (with respect to η)

$$\left(Q_{j1}(x,\tau+\eta\nu),\ Q_{j2}(\alpha x, T(\tau+\eta\nu))\right), \qquad j=1,\dots,2m,$$

are linearly independent modulo

$$\left(L^+(x,\tau+\eta\nu),\ L^+(\alpha x, T(\tau+\eta\nu))\right).$$

Definition 1.2. A $(2\times 4m)$-matrix $b = (b_{jr}(x,D))_{j=1,\dots,4m, r=1,2}$, $x\in\Gamma$, will be called *a Dirichlet α-matrix of order* $2m$ if it has the following properties:

a) The matrix $b(x,D)$ acts according to the equality

$$\left(b(x,D)u(x)\right)_j := b_{j1}(x,D)u(x) + Jb_{j2}(x,D)u(x), \qquad j=1,\dots,4m. \tag{1.4}$$

b) the matrix $b(x,D)$ may be decomposed (perhaps after permuting the rows) into $2m$ blocks $b_s(x,D) := (b_s^{ir}(x,D))_{i,r=1,2}$, $s=1,\dots,2m$, where ord $b_s^{ir}(x,D) \le 2m-s$ and in each point $x\in\Gamma$ and for any $s=1,\dots,2m$ the vectors

$$\left(b_{s0}^{i1}(x,\nu_x), b_{s0}^{i2}(\alpha x, \nu_{\alpha x})\right), \qquad i=1,2,$$

are linearly independent. Here ν_x is the unit normal to Γ at the point x, $b^{ir}_{s0}(x,D)$ is the principal part of $b^{ir}_s(x,D)$, which consists of the terms having order $2m-s$, $b^{ir}_{s0}(x,D)\equiv 0$ if ord $b^{ir}_s < 2m-s$.
c) The matrix $b(x,D)$ is invariant with respect to the substitution of x by αx (in the sense of its action according to the formula (1.4)).

As an example of a Dirichlet α-matrix of order $2m$ may be considered the matrix of Cauchy data:

$$\begin{pmatrix} \begin{pmatrix} D^{2m-1}_{\nu_x} & 0 \\ 0 & D^{2m-1}_{\nu_x} \end{pmatrix} \\ \cdots\cdots\cdots \\ \begin{pmatrix} 1 & 0 \\ 0 & 1 \end{pmatrix} \end{pmatrix}$$

Definition 1.3. The matrix $B(x,D) := (B_{jr}(x,D))_{j=1,\dots,2m,r=1,2}$ of the boundary conditions (1.3) will be called *α-normal* if it can be completed with new rows to a Dirichlet α-matrix of order $2m$.

Theorem 1.4. *Let the matrix $B(x,D)$ be α-normal, and let the matrix $C(x,D) := (C_{jr}(x,D))_{j=1,\dots,2m,r=1,2}$ (ord $C_{jr} \le 2m+\sigma'_j$, $\sigma'_j < 0$ are given integers) complete it to a Dirichlet α-matrix of order $2m$. Then there exist α-normal matrices $B'(x,D)$ and $C'(x,D)$ of similar form such that for any $u,v\in C^\infty(\overline{G})$ the Green's formula*

$$(Lu,v)+\sum_{1\le j\le 2m}\langle B_ju, C'_jv\rangle = (u,L^+v)+\sum_{1\le j\le 2m}\langle C_ju, B'_jv\rangle \tag{1.5}$$

holds. Here L^+ is the differential expression formally adjoint to L, $(\cdot,\cdot)$ and $\langle\cdot,\cdot\rangle$ are the inner products in $L_2(G)$ and $L_2(\Gamma)$, respectively, B_ju is defined by equality (1.3) and C_ju, B'_jv, C'_jv have similar sense. Moreover, the matrix $B'_j(x,D)$ satisfies the matching condition and

$$\text{ord } B'_{jr}\le -\sigma'_j-1,\quad \text{ord } C'_{jr}\le -\sigma_j-1,\quad j=1,\dots,2m,\ r=1,2. \tag{1.6}$$

If in addition the expression $L(x,D)$ is elliptic in $\overline{G}$, then for any j in each of the inequalities (1.6) equality takes place at least for one r.

The problem

$$L^+v(x)=g(x),\qquad x\in G, \tag{1.7}$$
$$B'_jv(x)=\psi_j(x),\qquad x\in\Gamma,\ j=1,\dots,2m, \tag{1.8}$$

is naturally called adjoint to the problem (1.2), (1.3) with respect to the Green's formula (1.5).

Theorem 1.5. *Let the problem* (1.2), (1.3) *be elliptic, and let the matrix* $B(x,D)$ *of boundary conditions be* α*-normal. Then the adjoint problem* (1.7), (1.8) *is also elliptic.*

2. Solvability conditions

Let us denote by $H^{s,p}(G)$, $s \geq 0$, $1 < p < \infty$, the space of Bessel potentials, by $H^{-s,p}(G)$ we denote the space dual to $H^{s,p'}(G)$ $(1/p + 1/p' = 1)$ with respect to the extension of the inner product in $L_2(G)$; $||\cdot||_{s,p}$ is the norm in $H^{s,p}(G)$, $s \in \mathbb{R}$. We denote by $B^{s,p}(\Gamma)$, $s \in \mathbb{R}$, the Besov space. The spaces $B^{-s,p'}(\Gamma)$ and $B^{s,p}(\Gamma)$ are dual to each other with respect to the extension of the inner product in $L_2(\Gamma)$; $\langle\langle\cdot\rangle\rangle_{s,p}$ is the norm in $B^{s,p}(\Gamma)$.

We denote by N and N^+ the kernels of the given and the adjoint problem, respectively,

$$N := \{u \in H^{2m,p}(G) : Lu = 0;\ B_j u = 0,\ j = 1,\dots,2m\},$$

$$N^+ := \{v \in H^{2m,p'}(G) : L^+ v = 0;\ B'_j v = 0,\ j = 1,\dots,2m\}.$$

In what follows we shall always assume the given problem (1.2), (1.3) to be elliptic and the matrix $B(x,D)$ to be α-normal. Then the adjoint problem (1.7), (1.8) is also elliptic. It follows from the ellipticity of these problems [1] that N and N^+ are finite-dimensional and that $N, N^+ \subset C^\infty(\overline{G})$.

Theorem 2.1. *Let the problem* (1.2), (1.3) *be elliptic and let the matrix* $B(x,D)$ *of boundary conditions be* α*-normal.*

a) *The solution* $u \in H^{2m+s,p}(G)$, $s \geq 0$, *of the problem* (1.2), (1.3) *with*

$$F := (f, \phi_1, \dots, \phi_{2m}) \in H^{s,p}(G) \times \prod_{1\leq j\leq 2m} B^{s-\sigma_j-1/p,p}(\Gamma) =: H^{s,p}(G,\Gamma) \tag{2.1}$$

exists if and only if

$$(f, v) + \sum_{1\leq j\leq 2m} \langle \phi_j, C'_j v\rangle = 0, \qquad v \in N^+. \tag{2.2}$$

b) *The solution* $v \in H^{2m+s,p}(G)$, $s \geq 0$, *of the problem* (1.7), (1.8) *with*

$$(g, \psi_1, \dots, \psi_{2m}) \in H^{s,p}(G) \times \prod_{1\leq j\leq 2m} B^{2m+s+\sigma'_j+1-1/p,p}(\Gamma)$$

exists if and only if

$$(u, g) + \sum_{1\leq j\leq 2m} \langle C_j u, \psi_j\rangle = 0, \qquad u \in N. \tag{2.3}$$

3. Theorem on isomorphism and generalized solvability in complete scales of spaces

For any $t \in \mathbb{R}$, $t \neq k + 1/p$, $k = 0, \ldots, 2m-1$, we denote by $\widetilde{H}^{t,p}(G)$ the completion of $C^\infty(\overline{G})$ with respect to the norm

$$|||u|||_{t,p} := \left(||u||_{t,p}^p + \sum_{1 \leq j \leq 2m} \langle\langle D_\nu^{j-i} u \rangle\rangle_{t-j+1-1/p,p}^p \right)^{1/p}. \tag{3.1}$$

In the case $t = k+1/p$, $k = 0, \ldots, 2m-1$, the space $\widetilde{H}^{t,p}(G)$ and the corresponding norm may be defined by means of complex interpolation between $\widetilde{H}^{[t],p}(G)$ and $\widetilde{H}^{[t]+1,p}(G)$. The spaces $\widetilde{H}^{t,p}(G)$ were introduced in [4] and studied in detail in [5]; see also [6, 2]. These spaces are very convenient for studying boundary value problems, because for any $s \in \mathbb{R}$ the closure $\Lambda_{s,p}$ of the operator

$$u \longmapsto (Lu, B_1 u, \ldots, B_{2m} u), \qquad u \in C^\infty(\overline{G}),$$

acts continuously from $\widetilde{H}^{2m+s,p}(G)$ into the space $H^{s,p}(G, \Gamma)$ defined in the relation (2.1). Now we intend to study the question of invertibility of the operator $\Lambda_{s,p}$ (cf. [4]).

Since N is finite-dimensional, any $u \in C^\infty(\overline{G})$ may be represented uniquely in the form $u = u' + u''$ where $(u', N) = 0$, $u'' \in N$. In this connection for any $s \in \mathbb{R}$ the closure P of the operator $u \mapsto u'$ acts continuously in $\widetilde{H}^{2m+s,p}(G)$. By $Q^+ H^{s,p}(G, \Gamma)$ let us denote the subspace of the elements

$$F = (f, \phi_1, \ldots, \phi_{2m}) \in H^{s,p}(G, \Gamma)$$

satisfying the condition (2.2).

Theorem 3.1. *Let the problem* (1.2), (1.3) *be elliptic and let the matrix* $B(x, D)$ *of boundary conditions be* α*-normal. Then for any* $s \in \mathbb{R}$ *and* $p \in (1, \infty)$ *the restriction* $\mathfrak{L}_{s,p}$ *of the operator* $\Lambda_{s,p}$ *onto* $P\widetilde{H}^{2m+s,p}(G)$ *realizes the isomorphism*

$$\mathfrak{L}_{s,p} : P\widetilde{H}^{2m+s,p}(G) \to Q^+ H^{s,p}(G, \Gamma).$$

Definition 3.2. The element $u \in \widetilde{H}^{2m+s,p}(G)$, $s \in \mathbb{R}$, satisfying the equality

$$\Lambda_{s,p} u = (f, \phi_1, \ldots, \phi_{2m}) \in H^{s,p}(G, \Gamma)$$

will be called *generalized (strong) solution* of the problem (1.2), (1.3).

It is not difficult to see that for $s \geq 0$ the generalized solution is also the usual solution of the problem in consideration because in this case $\widetilde{H}^{2m+s,p}(G) = H^{2m+s,p}(G)$.

The last theorem and the Green's formula easily imply the following assertion.

Theorem 3.3. *The generalized solution of the problem* (1.2), (1.3) *exists if and only if the right-hand side* $F = (f, \phi_1, \dots, \phi_{2m})$ *satisfies the condition* (2.2). *In addition any generalized solution may be found by the formula*

$$u = \mathfrak{L}_{s,p}^{-1} F + \omega, \qquad \omega \in N.$$

Assertions similar to two last theorems are valid also for the adjoint problem.

4. Some applications

4.1. Local increasing of smoothness of generalized solutions

It follows from the last formula that if $F \in H^{s_1,p_1}(G,\Gamma)$ where $s_1 \geq s, p_1 \geq p$, then the solution u is actually smoother, $u \in \widetilde{H}^{2m+s_1,p_1}(G)$. The theorem on local increasing of smoothness is also valid. This theorem states, roughly speaking, the following: Let G_0 be a subdomain of G adherent to the part $\Gamma_0 \subset \Gamma$. If F locally in G_0 up to Γ_0 pertains to H^{s_1,p_1}, $s_1 \geq s$, $p_1 \geq p$, then u is also smoother, namely, $u \in \widetilde{H}_{\text{loc}}^{2m+s_1,p_1}(G_0, \Gamma_0)$ (cf. [4], [2]).

4.2. Existence and smoothness of Green's function

Using the approach of the paper [3] and the results of the previous section (for the adjoint problem) we can establish the existence of the Green's vector-function for the problem (1.2), (1.3): There exist functions $R_0(x,y)$, $x, y \in \overline{G}$, $R_j(x,y)$, $x \in \overline{G}$, $y \in \Gamma$, $j = 1, \dots, 2m$, infinitely smooth for $x \neq y$, such that for any

$$F = (f, \phi_1, \dots, \phi_{2m}) \in Q^+ H^{s,p}(G,\Gamma)$$

with sufficiently large s the function

$$u(x) = \big(f, R_0(x, \cdot)\big) + \sum_{1 \leq j \leq 2m} \langle \phi_j, R_j(x, \cdot) \rangle$$

is a solution of the problem (1.2), (1.3). The methods of the paper [3] also allow the investigation of regularity properties of this vector-function with respect to the union of the variables and estimating its singularity for $x = y$.

4.3. Approximation by solutions of the problems with a shift

Let Λ_1 be a smooth $(n-1)$-dimensional manifold without border and $\Lambda \subset G$ be an open subset of Λ_1 with sufficiently smooth boundary. Let G_0 be a subdomain of G having arbitrarily small diameter. We put

$$\begin{aligned} M(G_0) &:= \{u \in C^\infty(\overline{G}) : \operatorname{supp} Lu \subset G_0,\ B_j u = 0,\ j = 1, \dots, 2m\}, \\ \nu_r M(G_0) &:= \{\nu_r u : u \in M(G_0)\}, \end{aligned}$$

where

$$\nu_r u := (u|_\Lambda, \ldots, D_\nu^{r-1} u|_\Lambda),$$

with $D_\nu = i\partial/\partial\nu$ and ν the unit normal to Λ.

Theorem 4.1. *Let $G \setminus \overline{\Lambda}$ be a connected set and let for the expression L^+ in G the property of uniqueness for the Cauchy problem hold: If $L^+v = 0$ in $G' \subset G$ and $v = 0$ in $G'' \subset G'$, then $v = 0$ in G'. Then the set $\nu_{2m}M(G_0)$ is dense in $\prod_{1\le j\le 2m} B^{s_j,p}(\Lambda)$ for any $s_j \ge 0$ and $1 < p < \infty$.*

Theorem 4.2. *Let*

a) *the set $\overline{\Lambda}$ be the boundary of a subdomain $G' \subset G$,*

b) *for the expression L^+ in G the property of uniqueness for the Cauchy problem hold,*

c) *the Dirichlet problem for the equation $L^+v = 0$ in G' have no more than one solution.*

Then $\nu_m M(G_0)$ is dense in $\prod_{1\le j\le 2m} B^{m-j+1-1/p,p}(\Lambda)$.

Similar questions for other types of problems, in particular for usual elliptic problems, were studied by many authors (see [7] and references therein).

Acknowledgements

The research was partially supported by the grant INTAS-94-2187 and by the International Soros Science Education Program (ISSEP), grant APU 051112.
The author would like to thank the organizers of IWOTA 95 for hospitality and financial support of his participation.

References

[1] ANTONEVICH, A.B.: On normal solvability of general boundary value problems with a shift for equations of elliptic type; Proceedings of the second conference of mathematicians of Belorussia, Minsk 1969, 253–255 (in Russian).

[2] BEREZANSKY, YU.M.: Expansions in Eigenfunctions of Selfadjoint Operators; Naukova Dumka, Kiev 1966 (in Russian); English transl.: Amer. Math. Soc. Transl. 17 (1968).

[3] BEREZANSKY, YU. M., ROITBERG, YA. A.: Theorem on homeomorphisms and Green's function for general elliptic boundary value problems; Ukrain. Mat. Zh. 19:5 (1967), 3–32 (in Russian).

[4] ROITBERG, YA.A.: Elliptic problems with non-homogeneous boundary conditions and local increasing of smoothness of generalized solutions up to the boundary; Dokl. Akad. Nauk SSSR 157:4 (1964), 798–801 (in Russian).

[5] ROITBERG, YA.A.: On the boundary values of generalized solutions of elliptic equations; Mat. Sb. 36:2 (1971), 246–267 (in Russian).

[6] ROITBERG, YA.A.: Elliptic boundary value problems in the spaces of distributions; Kluwer Akademic Publishers, Dordrecht 1996.

[7] SHEFTEL, Z.G.: On approximation of functions on manifolds by solutions of non-local elliptic problems; Dokl. Akad. Nauk Ukrain. 2 (1994), 24–28.

Pedagogical Institute
Sverdlova str.53
250038 Chernigov
Ukraine
alex@elit.chernigov.ua

1991 Mathematics Subject Classification: Primary 35R10; Secondary 35J40, 41A65

Submitted: May 31, 1996

Operator Theory:
Advances and Applications, Vol. 102

On second order linear differential equations with inverse square singularities

R. WEIKARD

We study the differential equation

$$y'' + (a/x^2 + q(x))y = Ey$$

where $a, E \in \mathbb{C}$ and where q is a complex-valued function which is locally integrable on $\mathbb{R} - \{0\}$ and analytic in a neighborhood of $x = 0$ (or nearly so). In particular, we are interested in the behavior of solutions of initial value problems as the initial point varies and in the asymptotic behavior of solutions as the parameter E tends to infinity.

1. Introduction

The spectrum of Hill's differential operator, i.e., of the self-adjoint operator associated with the equation $y'' + qy = Ey$ where q is a real-valued, periodic, continuous function consists of a countable number of intervals, called (spectral) bands. Those potentials for which the number of bands is finite are called finite-band potentials. They have lately received a great deal of attention due to their connection with nonlinear integrable systems and the Korteweg-de Vries hierarchy (see, e.g., Belokolos et al. [2] and the literature cited there). The earliest example for a finite-band potential was given by Ince [8]: when $\wp$ denotes Weierstrass' elliptic function with real and purely imaginary half-periods ω and ω' and when g is a nonnegative integer then the Lamé potential

$$q(x) = -g(g+1)\wp(x + \omega')$$

is a real-valued, continuous, and periodic function of the real variable x and the self-adjoint $L^2(\mathbb{R})$-operator associated with the differential expression $d^2/dx^2 + q$ gives rise to a spectrum consisting of g compact bands and one closed ray. These and other elliptic finite-band potentials have been the subject of investigation in the papers [4] – [7]. From the point of view taken in these papers the requirement that q should be real-valued and continuous is very unnatural. In fact, we prove in [7] that so called Picard potentials which are in general complex-valued and which, for instance, include the Lamé potentials even when ω' is not purely imaginary, have finite-band structure. According to Rofe-Beketov [9] the bands will then be analytic arcs which are not confined to the real axis anymore. The results of [7], however, should even extend to $q(x) = -g(g+1)\wp(x)$ where the potential is not even continuous but has singularities of the form $-g(g+1)/(x - 2n\omega)^2$ when x is close to a period $2n\omega$ of $\wp$. This fact is the motivation for the present investigation in which we treat perturbations of a potential with an inverse square singularity. Periodic, complex-valued potentials with such singularities and their relationships with integrable systems will be treated in a subsequent investigation [10].

The differential equation $y'' + qy = Ey$ where the potential q has singularities of the form a/x^2 has also been treated by Gesztesy and Kirsch [3] when a is real and not larger than $1/4$. However, they were interested in solutions which are square-integrable even near the singularity. This forces Dirichlet boundary conditions when $a \leq -3/4$ while arbitrary boundary conditions may be chosen when $-3/4 < a \leq 1/4$ (in which case the differential expression is in the limit circle case at $x = 0$). In the present treatment, however, we assume that q is analytic in a neighborhood of $x = 0$ with the nonpositive imaginary axis removed. This allows to continue solutions in a unique way from one side of the singular point to the other. In particular, no boundary conditions can be applied at $x = 0$.

In Section 2 we treat the case $q = a/x^2$ by reducing the equation $y'' + qy = Ey$ to Bessel's equation. In Section 3 we treat certain perturbations of a/x^2 by constructing and studying solutions of an appropriate integral equation.

2. The unperturbed case

The substitutions $z = kx$, $y(x) = x^{1/2}w(z)$, and $4\nu^2 = 1 - 4a$ transform the differential equation

$$Ly = y'' + \frac{a}{x^2}y = -k^2 y$$

into Bessel's equation

$$z^2 w'' + zw' + (z^2 - \nu^2)w = 0.$$

A fundamental system of solutions of Bessel's equation is given by the Bessel functions $J_\nu(z)$ and $Y_\nu(z)$. Hence a fundamental system of solutions of $Ly = Ey$ is given by $c_0(E, x_0, \cdot)$ and $s_0(E, x_0, \cdot)$ where x_0 is any nonzero complex number and

$$\begin{aligned}
c_0(-k^2, x_0, x) &= \frac{\pi}{4}(x/x_0)^{1/2}\left(J_\nu(kx)Y_\nu(kx_0) - J_\nu(kx_0)Y_\nu(kx)\right) \\
&\quad + k\frac{\pi}{2}(xx_0)^{1/2}\left(J_\nu(kx)Y_\nu'(kx_0) - J_\nu'(kx_0)Y_\nu(kx)\right), \\
s_0(-k^2, x_0, x) &= \frac{\pi}{2}(xx_0)^{1/2}\left(J_\nu(kx_0)Y_\nu(kx) - J_\nu(kx)Y_\nu(kx_0)\right).
\end{aligned}$$

Note that $c_0(E, x_0, \cdot)$ and $s_0(E, x_0, \cdot)$ are those solutions of $Ly = Ey$ which satisfy initial conditions $y(x_0) = 1$, $y'(x_0) = 0$ and $y(x_0) = 0$, $y'(x_0) = 1$, respectively. In $\Omega_0 = \{x \in \mathbb{C} : \Re(x) = 0 \Rightarrow \Im(x) > 0\}$ these functions are single-valued and analytic. In particular, we thus have a unique solution of the initial value problem $Ly = Ey$, $y(x_0) = y_0$, $y'(x_0) = y_0'$ in $\mathbb{R} - \{0\}$.

For any choice of x and x_0 in Ω_0 the functions $c_0(\cdot, x_0, x)$ and $s_0(\cdot, x_0, x)$ as well as their derivatives with respect to x, i.e., $c_0'(\cdot, x_0, x)$ and $s_0'(\cdot, x_0, x)$ are entire functions of order $1/2$.

Upon interchanging the roles played by x and x_0 one obtains the following relationships:

$$\begin{aligned} c_0(E, x_0, x) &= s_0'(E, x, x_0), \\ s_0(E, x_0, x) &= -s_0(E, x, x_0), \\ c_0'(E, x_0, x) &= -c_0'(E, x, x_0). \end{aligned}$$

Note that for $a = 0$ we have $c_0(-k^2, x_0, x) = \cos(k(x - x_0))$ and $s_0(-k^2, x_0, x) = \sin(k(x - x_0))/k$.

3. Perturbations

Next we consider perturbations of the potential a/x^2 which are locally integrable and analytic near zero. More precisely, for some real number $b > 0$ let $U = \{x \in \mathbb{C} : |x| < b, \Re(x) = 0 \Rightarrow \Im(x) > 0\}$ and $\Omega = U \cup (\mathbb{R} - \{0\})$. Then suppose that $q : \Omega \to \mathbb{C}$ satisfies the following three conditions:

1. $q \in L^1_{\text{loc}}(\mathbb{R} - \{0\})$,
2. q is analytic in U,
3. there exists an $\varepsilon \in (0, 1]$ and a $Q > 0$ such that $|q(x)| \leq Q|x|^{\varepsilon - 1}$ for all $x \in U$.

For $E, y_0, y_0' \in \mathbb{C}$ and $x_0, x \in \Omega$ define

$$\phi_0(E, y_0, y_0', x_0, x) = y_0 c_0(E, x_0, x) + y_0' s_0(E, x_0, x).$$

Theorem 3.1. *If q satisfies the above hypotheses then there exists a function $\phi : \mathbb{C}^3 \times \Omega^2 \to \mathbb{C}$ with the following properties:*
(a) *$\phi(E, y_0, y_0', x_0, \cdot)$ is the unique solution of the integral equation*

$$y(x) = \phi_0(E, y_0, y_0', x_0, x) - \int_{\gamma_s} s_0(E, x', x) q(x') y(x') dx' \tag{3.1}$$

where γ_s is a piecewise continuously differentiable simple path in Ω connecting x_0 and x. In particular, $\phi(E, y_0, y_0', x_0, x_0) = y_0$.
(b) *$\phi(E, y_0, y_0', x_0, \cdot)$ is analytic in U and continuously differentiable on $\mathbb{R} - \{0\}$. In fact, ϕ' (where the prime denotes differentiation with respect to the last argument) is given by*

$$\phi'(E, y_0, y_0', x_0, x) = \phi_0'(E, y_0, y_0', x_0, x) - \int_{\gamma_s} s_0'(E, x', x) q(x') \phi(E, y_0, y_0', x_0, x') dx'$$

and therefore $\phi'(E, y_0, y_0', x_0, x_0) = y_0'$. The function $\phi'(E, y_0, y_0', x_0, \cdot)$ is locally absolutely continuous in $\mathbb{R} - \{0\}$.

(c) *The function* ϕ'' *is given by*

$$\phi''(E, y_0, y_0', x_0, x) = (E - \frac{a}{x^2} - q(x))\phi(E, y_0, y_0', x_0, x)$$

(for $x \in \mathbb{R} - \{0\}$ *this holds almost everywhere). Hence* $\phi(E, y_0, y_0', x_0, \cdot)$ *is the unique solution of the initial value problem*

$$y'' + (a/x^2 + q(x))y = Ey,\ y(x_0) = y_0,\ y'(x_0) = y_0'.$$

(d) $\phi(E, y_0, y_0', \cdot, x)$ *and* $\phi'(E, y_0, y_0', \cdot, x)$ *are locally absolutely continuous in* $\mathbb{R} - \{0\}$. *Moreover,*

$$\frac{\partial \phi}{\partial x_0}(E, y_0, y_0', x_0, x) = \phi(E, -y_0', (a/x_0^2 + q(x_0) - E)y_0, x_0, x), \tag{3.2}$$

$$\frac{\partial \phi'}{\partial x_0}(E, y_0, y_0', x_0, x) = \phi'(E, -y_0', (a/x_0^2 + q(x_0) - E)y_0, x_0, x). \tag{3.3}$$

(e) *There exist positive constants* C *and* Λ *depending on* x_0 *and* x *but not on* E, y_0 *and* y_0' *such that*

$$|\phi(E, y_0, y_0', x_0, x) - \phi_0(E, y_0, y_0', x_0, x)| \leq Ce^{|\Re(\sqrt{E})(x-x_0)|}\frac{|y_0||\sqrt{E}| + |y_0'|}{|\sqrt{E}|^{1+\varepsilon}},$$

$$|\phi'(E, y_0, y_0', x_0, x) - \phi_0'(E, y_0, y_0', x_0, x)| \leq Ce^{|\Re(\sqrt{E})(x-x_0)|}\frac{|y_0||\sqrt{E}| + |y_0'|}{|\sqrt{E}|^{\varepsilon}}.$$

when $x_0, x \in \mathbb{R} - \{0\}$ *and when* $|E| \geq \Lambda$. *In particular, the functions* $\phi(\cdot, y_0, y_0', x_0, x)$ *and* $\phi'(\cdot, y_0, y_0', x_0, x)$ *are entire and have order* $1/2$.

Proof. Let K be a compact subset of Ω and B a compact subset of $\mathbb{C}$. Since ϕ_0 is continuous in all its arguments there exists a positive constant M_1 such that $|\phi_0|$, $|s_0|$, and $|s_0'|$ are bounded by M_1 on $B^3 \times K^2$ and $B \times K^2$, respectively.
Let $\gamma : [0, 1] \to K$ be a piecewise continuously differentiable simple path with initial point $\gamma(0) = x_0$ and let Γ be a positive constant such that $|\gamma'(t)| \leq \Gamma$. For $s \in [0, 1]$, let $\gamma_s : [0, 1] \to \Omega$ be defined by $\gamma_s(t) = \gamma(st)$.

Now, for $x = \gamma(s)$ and for $n \in \mathbb{N}$ define recursively

$$\phi_n(E, y_0, y_0', x_0, x) = -\int_{\gamma_s} s_0(E, x', x)q(x')\phi_{n-1}(E, y_0, y_0', x_0, x')dx'.$$

Since $s_0(E, \cdot, x)q(\cdot)\phi_0(E, y_0, y_0', x_0, \cdot)$ is analytic in U and since U is simply connected the induction principle shows that $\phi_n(E, y_0, y_0', x_0, x)$ is independent of the path chosen to connect x_0 and x and that it is analytic in U when it is regarded as a function of x.
Defining $R(t) = \Gamma M_1|q(\gamma(t))|$ one obtains the estimate

$$|\phi_n(E, y_0, y_0', x_0, x)| \leq \int_0^s R(s_1) \int_0^{s_1} R(s_2) \ldots \int_0^{s_{n-1}} R(s_n)M_1 ds_n \ldots ds_1.$$

Defining $\hat{R}(t) = \int_0^t R(t')dt'$ gives then

$$|\phi_n(E, y_0, y_0', x_0, x)| \leq M_1 \frac{\hat{R}(s)^n}{n!}.$$

Hence, by the Weierstrass M-test $\sum_{n=0}^{\infty} \phi_n$ converges absolutely and uniformly on $B^3 \times K^2$ and hence on any compact subset of $\mathbb{C}^3 \times \Omega^2$.
Because of uniform convergence it follows next that the function

$$\phi(E, y_0, y_0', x_0, \cdot) = \sum_{n=0}^{\infty} \phi_n(E, y_0, y_0', x_0, \cdot)$$

is a solution of the integral equation (3.1). In order to show uniqueness of solutions of (3.1) assume that there are two solutions y and $\tilde{y}$. Defining

$$L = \sup\{|y(x) - \tilde{y}(x)| : x \in \gamma([0, 1])\}$$

we obtain firstly $|y(x) - \tilde{y}(x)| \leq L\hat{R}(s)$. A repeated approximation shows now that

$$|y(x) - \tilde{y}(x)| \leq L \frac{\hat{R}(s)^n}{n!}$$

for all $n \in \mathbb{N}$. Since the right hand side in this inequality becomes arbitrarily small this shows that $y = \tilde{y}$ and hence that solutions of (3.1) are unique. This completes the proof of part (a) of the theorem.
$\phi(E, y_0, y_0', x_0, \cdot)$ is analytic in U as the uniform limit of analytic functions. Equation (3.1) shows that $\phi(E, y_0, y_0', x_0, \cdot)$ is differentiable in $\mathbb{R} - \{0\}$ and that its derivative is given as stated in (b).
Since $\phi_0(E, y_0, y_0', x_0, \cdot)$ is infinitely often differentiable in $\mathbb{R} - \{0\}$ we have to show the locally absolute continuity of

$$g(x) = \int_{\gamma_s} s_0'(E, x', x) q(x') \phi(E, y_0, y_0', x_0, x') dx'$$

in order to finish the proof of part (b). Hence assume that $[\alpha, \beta]$ is a subset of $\mathbb{R} - \{0\}$ and that

$$\alpha \leq t_0 < t_1 \leq t_2 < \cdots < t_{2N-1} \leq t_{2N} < t_{2N+1} \leq \beta.$$

When γ_1 is a piecewise continuously differentiable simple path connecting x_0 and α define γ_2 by $\gamma_2(t) = \alpha + (\beta - \alpha)t$ and let γ be the product path $\gamma_1\gamma_2$. Then

$$\text{(3.4)} \quad \sum_{k=0}^{N} |g(t_{2k+1}) - g(t_{2k})| \leq \sum_{k=0}^{N} \int_{s_{2k}}^{s_{2k+1}} M_1 e^{\hat{R}(1)} R(s') ds'$$
$$+ \int_0^1 \sum_{k=0}^{N} |s_0'(E, \gamma(s'), t_{2k+1}) - s_0'(E, \gamma(s'), t_{2k})| \, |q(\gamma(s'))| \, M_1 e^{\hat{R}(1)} \Gamma ds'$$

where $s_k = \gamma^{-1}(t_k)$. Because of the integrability of R the first term on the right hand side can be made arbitrarily small provided $\sum_{k=0}^N (t_{2k+1} - t_{2k})$ is suitably small. To treat the second term note that $\Re s_0'(E,x',\cdot)$ and $\Im s_0'(E,x',\cdot)$ are continuously differentiable on $[\alpha,\beta]$. Hence

$$|s_0'(E,x',t_{2k+1}) - s_0'(E,x',t_{2k})| \le M_2(t_{2k+1} - t_{2k})$$

for all $x' \in \gamma([0,1])$ when M_2 is a suitably chosen constant. Hence the second term on the right hand side of (3.4) can be bounded by a constant times $\sum_{k=0}^N (t_{2k+1} - t_{2k})$. This finishes the proof of part (b) of the theorem.
Differentiating the equation

$$\phi'(E,y_0,y_0',x_0,x) = \phi_0'(E,y_0,y_0',x_0,x) - \int_{\gamma_s} s_0'(E,x',x)q(x')\phi(E,y_0,y_0',x_0,x')dx'$$

with respect to x shows now part (c) of the theorem.
Again, let $[\alpha,\beta]$ be a subset of $\mathbb{R} - \{0\}$. Since

$$\frac{\partial \phi_0(E,y_0,y_0',x_0,x)}{\partial x_0} = \phi_0(E,-y_0',y_0(a/x_0^2 - E),x_0,x) \tag{3.5}$$

$\phi_0(E,y_0,y_0',\cdot,x)$ is continuously differentiable and hence absolutely continuous in $[\alpha,\beta]$. In fact, for any $\varepsilon > 0$, there exists $\delta > 0$ such that

$$\sum_{k=0}^N |\phi_0(E,y_0,y_0',t_{2k+1},x) - \phi_0(E,y_0,y_0',t_{2k},x)| < \varepsilon$$

whenever $x \in K$ and $t_0,\ldots,t_{2N+1}$ are points satisfying

$$\alpha \le t_0 < t_1 \le t_2 < \cdots < t_{2N-1} \le t_{2N} < t_{2N+1} \le \beta$$

and $\sum_{k=0}^N (t_{2k+1} - t_{2k}) < \delta$. Since

$$\begin{aligned} &|\phi_n(E,y_0,y_0',t_{2k+1},x) - \phi_n(E,y_0,y_0',t_{2k},x)| \\ &\le \int_{s_{2k}}^{s_{2k+1}} R(s')|\phi_{n-1}(E,y_0,y_0',t_{2k},\gamma(s'))|ds' \\ &\quad + \int_0^s R(s')|\phi_{n-1}(E,y_0,y_0',t_{2k+1},\gamma(s')) - \phi_{n-1}(E,y_0,y_0',t_{2k},\gamma(s'))|ds' \end{aligned}$$

and because of the integrability of R one obtains for suitably small δ

$$\sum_{k=0}^N |\phi_1(E,y_0,y_0',t_{2k+1},x) - \phi_1(E,y_0,y_0',t_{2k},x)| \le \varepsilon(M_1 + \hat{R}(s)).$$

An induction shows now that

$$\sum_{k=0}^{N} |\phi_n(E, y_0, y_0', t_{2k+1}, x) - \phi_n(E, y_0, y_0', t_{2k}, x)|$$
$$\leq \varepsilon \left(M_1 \frac{(\hat{R}(1) + \hat{R}(s))^{n-1}}{(n-1)!} + \frac{\hat{R}(s)^n}{n!} \right) \leq \varepsilon M_3 \frac{M_3^{n-1}}{(n-1)!}$$

for a suitable positive constant M_3. In particular, for every $n \in \mathbb{N}_0$ the function $\phi_n(E, y_0, y_0', \cdot, x)$ is locally absolutely continuous in $\mathbb{R} - \{0\}$.
Now consider $\phi(E, y_0, y_0', \cdot, x)$. We have

$$\sum_{k=0}^{N} |\phi(E, y_0, y_0', t_{2k+1}, x) - \phi(E, y_0, y_0', t_{2k}, x)|$$
$$= \sum_{k=0}^{N} |\lim_{M \to \infty} \sum_{n=0}^{M} (\phi_n(E, y_0, y_0', t_{2k+1}, x) - \phi_n(E, y_0, y_0', t_{2k}, x))|$$
$$\leq \lim_{M \to \infty} \sum_{n=0}^{M} \sum_{k=0}^{N} |\phi_n(E, y_0, y_0', t_{2k+1}, x) - \phi_n(E, y_0, y_0', t_{2k}, x)|$$
$$\leq \lim_{M \to \infty} \left(\varepsilon + \sum_{n=1}^{M} \varepsilon M_3 \frac{M_3^{n-1}}{(n-1)!} \right) = \varepsilon(1 + M_3 \exp(M_3))$$

which proves that $\phi(E, y_0, y_0', \cdot, x)$ is locally absolutely continuous in $\mathbb{R} - \{0\}$.
Differentiating the integral equation (3.1) with respect to x_0 and using (3.5) gives now:

$$\begin{aligned} \frac{\partial \phi(E, y_0, y_0', x_0, x)}{\partial x_0} &= \phi_0(E, -y_0', (a/x_0^2 + q(x_0) - E)y_0, x_0, x) \\ &\quad - \int_{\gamma_s} s_0(E, x', x) q(x') \frac{\partial \phi(E, y_0, y_0', x_0, x')}{\partial x_0} dx'. \end{aligned}$$

Because solutions of the integral equation are unique this proves the validity of (3.2). Equation (3.3) follows now straight from part (b). This concludes the proof of part (d) of the theorem.
The Bessel functions and their derivatives may be expressed by

$$\begin{aligned} J_\nu(z) &= \sqrt{\frac{2}{\pi z}} \left(P_\nu(z) \cos \chi - Q_\nu(z) \sin \chi \right), \\ Y_\nu(z) &= \sqrt{\frac{2}{\pi z}} \left(P_\nu(z) \sin \chi + Q_\nu(z) \cos \chi \right), \\ J_\nu'(z) &= \sqrt{\frac{2}{\pi z}} \left(-R_\nu(z) \sin \chi - S_\nu(z) \cos \chi \right), \\ Y_\nu'(z) &= \sqrt{\frac{2}{\pi z}} \left(R_\nu(z) \cos \chi - S_\nu(z) \sin \chi \right) \end{aligned}$$

where $\chi = z - \nu\pi/2 - \pi/4$. With these expressions one obtains

$$\begin{aligned} c_0(-k^2, x_0, x) &= \cos(k(x-x_0))(f_2(kx, kx_0) + g_1(kx, kx_0)/(2kx_0)) \\ &\quad + \sin(k(x-x_0))(g_2(kx, kx_0) - f_1(kx, kx_0)/(2kx_0)), \\ s_0(-k^2, x_0, x) &= \frac{\sin(k(x-x_0))}{k} f_1(kx, kx_0) - \frac{\cos(k(x-x_0))}{k} g_1(kx, kx_0) \end{aligned}$$

where

$$\begin{aligned} f_1(z, z_0) &= P_\nu(z)P_\nu(z_0) + Q_\nu(z)Q_\nu(z_0), \\ f_2(z, z_0) &= P_\nu(z)R_\nu(z_0) + Q_\nu(z)S_\nu(z_0), \\ g_1(z, z_0) &= P_\nu(z)Q_\nu(z_0) - Q_\nu(z)P_\nu(z_0), \\ g_2(z, z_0) &= P_\nu(z)S_\nu(z_0) - Q_\nu(z)R_\nu(z_0). \end{aligned}$$

The functions P_ν, R_ν, Q_ν, and S_ν have the following asymptotic behavior:

$$\begin{aligned} P_\nu(z), R_\nu(z) &= 1 + O(z^{-2}), \\ Q_\nu(z) &= \frac{4\nu^2 - 1}{8z} + O(z^{-3}), \\ S_\nu(z) &= \frac{4\nu^2 + 3}{8z} + O(z^{-3}) \end{aligned}$$

as z tends to infinity if $|\arg z| < \pi$ (see, e.g., Abramowitz and Stegun [1]). If $z = we^{i\pi}$ one finds

$$\begin{aligned} P_\nu(z) &= P_\nu(w) + i\cos(\nu\pi)\mathrm{e}^{-2iw}(P_\nu(w) - iQ_\nu(w)), \\ Q_\nu(z) &= -Q_\nu(w) - \cos(\nu\pi)\mathrm{e}^{-2iw}(P_\nu(w) - iQ_\nu(w)), \\ R_\nu(z) &= R_\nu(w) - i\cos(\nu\pi)\mathrm{e}^{-2iw}(R_\nu(w) - iS_\nu(w)), \\ S_\nu(z) &= -S_\nu(w) + \cos(\nu\pi)\mathrm{e}^{-2iw}(R_\nu(w) - iS_\nu(w)). \end{aligned}$$

This implies that $f_k(z, z_0)$ and $g_k(z, z_0)$, $k = 1, 2$, are bounded by a constant C_1 as long as z and z_0 are bounded away from zero and their arguments are in $[-\pi, \pi]$. Let $k = \sqrt{-E}$ have its argument in $(-\pi, 0]$, i.e., $k = \kappa - i\eta$ where $\kappa \in \mathbb{R}$ and $\eta \geq 0$ and let $r = 1/|k|$. For $x_0 < 0$ and $x_1 > 0$ let $\gamma : [0, 1] \to \Omega_0$ be defined by

$$\gamma(t) = \begin{cases} x_0 - 3t(r + x_0) & \text{if } \ 0 \leq t \leq 1/3, \\ -r\exp(-i\pi(3t-1)) & \text{if } \ 1/3 \leq t \leq 2/3, \\ r + (3t-2)(x_1 - r) & \text{if } \ 2/3 \leq t \leq 1. \end{cases}$$

Then $\arg(k\gamma(t)) \in (-\pi, \pi]$ and $|k\gamma(t)| \geq 1$ for all $t \in [0, 1]$. Therefore, if $x, x' \in \gamma([0, 1])$

$$\begin{aligned} |c_0(-k^2, x', x)| &\leq 4C_1 \exp(1 + \eta|\Re(x) - \Re(x')|), \\ |s_0(-k^2, x', x)| &\leq \frac{2C_1}{|k|} \exp(1 + \eta|\Re(x) - \Re(x')|). \end{aligned}$$

Now define

$$f(k,x) = \exp(-\eta(x-x_0))\phi(-k^2, y_0, y_0', x_0, x).$$

From (3.1) we get

$$\begin{aligned} f(k,x) &= \mathrm{e}^{-\eta(x-x_0)}\phi_0(-k^2, y_0, y_0', x_0, x) \\ &\quad - \int_{\gamma_s} s_0(-k^2, x', x) q(x') \mathrm{e}^{-\eta(x-x')} f(k,x') dx'. \end{aligned}$$

Therefore

$$|kf(k,x)| \leq 4C_1\mathrm{e}\left(|y_0||k| + |y_0'|\right) + 2C_1\mathrm{e}\int_0^1 |q(\gamma(s'))| F(k) |\gamma'(s')| ds'$$

where

$$F(k) = \max\{|f(k,x')| : x' \in \gamma([0,1])\}.$$

Now consider $\int_0^1 |q(\gamma(s'))||\gamma'(s')|ds'$. Since the integral associated with the part of γ which lies outside of U depends only on x_0 and x_1, since $|\gamma'(t)|$ is bounded by a constant depending only on x_0 and x_1, and since $|q(x')|$ is bounded by $Q|x'|^{\varepsilon-1} \leq Q|k|^{1-\varepsilon}$ we obtain the existence of a constant $\tilde{Q}$ such that

$$\int_0^1 |q(\gamma(s'))||\gamma'(s')|ds' \leq \tilde{Q}|k|^{1-\varepsilon}$$

for suitably large $|k|$. Hence

$$|f(k,x)| \leq 4C_1\mathrm{e}\left(|y_0| + \frac{|y_0'|}{|k|}\right) + \frac{2C_1\mathrm{e}\tilde{Q}}{|k|^\varepsilon} F(k).$$

Since the right hand side of this inequality is independent of x it is, in fact, also a bound for $F(k)$. Hence

$$F(k) \leq 4C_1\mathrm{e}\left(|y_0| + \frac{|y_0'|}{|k|}\right)\left(1 - 2C_1\mathrm{e}\tilde{Q}|k|^{-\varepsilon}\right)^{-1} \leq 8C_1\mathrm{e}\left(|y_0| + \frac{|y_0'|}{|k|}\right)$$

provided $|k|$ is so large that $2C_1\mathrm{e}\tilde{Q}|k|^{-\varepsilon} \leq 1/2$. Inserting this estimate into (3.1) and the derivative of this equation and taking into account that $s_0'(E, x', x) = c_0(E, x, x')$ we obtain the desired estimates. □

Acknowledgements

This paper is based upon work supported by the US National Science Foundation under Grant No. DMS-9401816.

References

[1] M. Abramowitz and I. A. Stegun: Handbook of Mathematical Functions; Dover, New York 1972.

[2] E. D. Belokolos, A. I. Bobenko, V. Z. Enol'skii, A. R. Its, and V. B. Matveev: Algebro-Geometric Approach to Nonlinear Integrable Equations; Springer, Berlin 1994.

[3] F. Gesztesy and W. Kirsch: One-dimensional Schrödinger operators with interactions singular on a discrete set; J. Reine Angew. Math. 362 (1985), 28–50.

[4] F. Gesztesy and R. Weikard: Lamé potentials and the stationary (m)KdV hierarchy; Math. Nachr. 176 (1995), 73–91.

[5] F. Gesztesy and R. Weikard: Treibich-Verdier potentials and the stationary (m)KdV hierarchy; Math. Z. 219 (1995), 451–476.

[6] F. Gesztesy and R. Weikard: On Picard potentials; Diff. Int. Eq. 8 (1995), 1453–1476.

[7] F. Gesztesy and R. Weikard: Picard potentials and Hill's equation on a torus; Acta Math. 176 (1996), 73–107.

[8] E. L. Ince: Further investigations into the periodic Lamé functions; Proc. Roy. Soc. Edinburgh 60 (1940), 83–99.

[9] F. S. Rofe-Beketov: The spectrum of non-self-adjoint differential operators with periodic coefficients; Sov. Math. Dokl. 4 (1963), 1563–1566.

[10] R. Weikard: Hill's equation with a singular, complex-valued potential; Proc. London Math. Soc. (3), to appear.

Department of Mathematics
University of Alabama at Birmingham
Birmingham, AL 35294-1170
USA
rudi@math.uab.edu

1991 Mathematics Subject Classification: Primary 34A30; Secondary 34A12, 34E10

Submitted: April 11, 1996

Titles previously published in the series

OPERATOR THEORY: ADVANCES AND APPLICATIONS

BIRKHÄUSER VERLAG

Edited by

I. Gohberg,

School of Mathematical Sciences, Tel-Aviv University, Ramat Aviv, Israel

This series is devoted to the publication of current research in operator theory, with particular emphasis on applications to classical analysis and the theory of integral equations, as well as to numerical analysis, mathematical physics and mathematical methods in electrical engineering.

75. **C.B. Huijsmans, M.A. Kaashoek, B. de Pagter**: Operator Theory in Function Spaces and Banach Lattices. The A.C. Zaanen Anniversary Volume, 1994 (ISBN 3-7643-5146-2)
76. **A.M. Krasnosellskii**: Asymptotics of Nonlinearities and Operator Equations, 1995, (ISBN 3-7643-5175-6)
77. **J. Lindenstrauss, V.D. Milman** (Eds): Geometric Aspects of Functional Analysis Israel Seminar GAFA 1992–94, 1995, (ISBN 3-7643-5207-8)
78. **M. Demuth, B.-W. Schulze** (Eds): Partial Differential Operators and Mathematical Physics, 1995, (ISBN 3-7643-5208-6)
79. **I. Gohberg, M.A. Kaashoek, F. van Schagen**: Partially Specified Matrices and Operators: Classification, Completion, Applications, 1995, (ISBN 3-7643-5259-0)
80. **I. Gohberg, H. Langer** (Eds): Operator Theory and Boundary Eigenvalue Problems. International Workshop in Vienna, July 27–30, 1993, 1995, (ISBN 3-7643-5275-2)
81. **H. Upmeier**: Toeplitz Operators and Index Theory in Several Complex Variables, 1996, (ISBN 3-7643-5282-5)
82. **T. Constantinescu**: Schur Parameters, Factorization and Dilation Problems, 1996, (ISBN 3-7643-5285-X)
83. **A.B. Antonevich**: Linear Functional Equations. Operator Approach, 1995, (ISBN 3-7643-2931-9)
84. **L.A. Sakhnovich**: Integral Equations with Difference Kernels on Finite Intervals, 1996, (ISBN 3-7643-5267-1)

85/ **Y.M. Berezansky, G.F. Us, Z.G. Sheftel**: Functional Analysis, Vol. I + Vol. II, 1996,
86. Vol. I (ISBN 3-7643-5344-9), Vol. II (3-7643-5345-7)

87. **I. Gohberg, P. Lancaster, P.N. Shivakumar** (Eds): Recent Developments in Operator Theory and Its Applications. International Conference in Winnipeg, October 2–6, 1994, 1996, (ISBN 3-7643-5414-5)
88. **J. van Neerven** (Ed.): The Asymptotic Behaviour of Semigroups of Linear Operators, 1996, (ISBN 3-7643-5455-0)
89. **Y. Egorov, V. Kondratiev**: On Spectral Theory of Elliptic Operators, 1996, (ISBN 3-7643-5390-2)
90. **A. Böttcher, I. Gohberg** (Eds): Singular Integral Operators and Related Topics. Joint German-Israeli Workshop, Tel Aviv, March 1–10, 1995, 1996, (ISBN 3-7643-5466-6)
91. **A.L. Skubachevskii**: Elliptic Functional Differential Equations and Applications, 1997, (ISBN 3-7643-5404-6)
92. **A.Ya. Shklyar**: Complete Second Order Linear Differential Equations in Hilbert Spaces, 1997, (ISBN 3-7643-5377-5)
93. **Y. Egorov, B.-W. Schulze**: Pseudo-Differential Operators, Singularities, Applications, 1997, (ISBN 3-7643-5484-4)
94. **M.I. Kadets, V.M. Kadets**: Series in Banach Spaces. Conditional and Unconditional Convergence, 1997, (ISBN 3-7643-5401-1)
95. **H. Dym, V. Katsnelson, B. Fritzsche, B. Kirstein** (Eds): Topics in Interpolation Theory, 1997, (ISBN 3-7643-5723-1)
96. **D. Alpay, A. Dijksma, J. Rovnyak, H. de Snoo**: Schur Functions, Operator Colligations, and Reproducing Kernel Pontryagin Spaces, 1997, (ISBN 3-7643-5763-0)
97. **M.L. Gorbachuk / V.I. Gorbachuk**: M.G. Krein's Lectures on Entire Operators, 1997, (ISBN 3-7643-5704-5)
98. **I. Gohberg / Yu. Lyubich** (Eds): New Results in Operator Theory and Its Applications The Israel M. Glazman Memorial Volume, 1997, (ISBN 3-7643-5775-4)
99 **T. Ayerbe Toledano / T. Dominguez Benavides / G. López Acedo**: Measures of Noncompactness in Metric Fixed Point Theory, 1997, (ISBN 3-7643-5794-0)
100 **C. Foias / A.E. Frazho / I. Gohberg / M.A. Kaashoek**: Metric Constrained Interpolation, Commutant Lifting and System, 1998, (ISBN 3-7643-5889-0)
101 **S.D. Eidelman / N.V. Zhitarashu**: Parabolic Boundary Value Problems, 1998, (ISBN 3-7643-2972-6)

MIX
Papier aus verantwortungsvollen Quellen
Paper from responsible sources
FSC® C105338

If you have any concerns about our products,
you can contact us on
ProductSafety@springernature.com

In case Publisher is established outside the EU,
the EU authorized representative is:
Springer Nature Customer Service Center GmbH
Europaplatz 3, 69115 Heidelberg, Germany

Printed by Libri Plureos GmbH
in Hamburg, Germany